AF371630

Brown Rice

Annamalai Manickavasagan
Chandini Santhakumar • N. Venkatachalapathy
Editors

Brown Rice

 Springer

Editors
Annamalai Manickavasagan
School of Engineering
University of Guelph
Guelph, Ontario, Canada

N. Venkatachalapathy
Indian Institute of Food
    Processing Technology
Thanjavur, Tamil Nadu, India

Chandini Santhakumar
College of Agricultural and Marine
    Sciences
Sultan Qaboos University
Muscat, Oman

ISBN 978-3-319-59010-3   ISBN 978-3-319-59011-0   (eBook)
DOI 10.1007/978-3-319-59011-0

Library of Congress Control Number: 2017947830

© Springer International Publishing AG 2017
This work is subject to copyright. All rights are reserved by the Publisher, whether the whole or part of
the material is concerned, specifically the rights of translation, reprinting, reuse of illustrations, recitation,
broadcasting, reproduction on microfilms or in any other physical way, and transmission or information
storage and retrieval, electronic adaptation, computer software, or by similar or dissimilar methodology
now known or hereafter developed.
The use of general descriptive names, registered names, trademarks, service marks, etc. in this publication
does not imply, even in the absence of a specific statement, that such names are exempt from the relevant
protective laws and regulations and therefore free for general use.
The publisher, the authors and the editors are safe to assume that the advice and information in this book
are believed to be true and accurate at the date of publication. Neither the publisher nor the authors or the
editors give a warranty, express or implied, with respect to the material contained herein or for any errors
or omissions that may have been made. The publisher remains neutral with regard to jurisdictional claims
in published maps and institutional affiliations.

Printed on acid-free paper

This Springer imprint is published by Springer Nature
The registered company is Springer International Publishing AG
The registered company address is: Gewerbestrasse 11, 6330 Cham, Switzerland

Preface

All over the world diet related non-communicable diseases are increasing at an alarming rate. The World Health Organization and other health authorities recommend following healthy eating behavior on a daily basis. In many countries, rice is a staple food, and brown rice is becoming its wholegrain counterpart. In this book, we have sincerely attempted to provide comprehensive coverage of brown rice in 16 chapters authored by experts in the field. This book is divided into six parts and covers various aspects of brown rice.

The first part explains in detail the milling process used to produce brown rice from paddy. The physical, chemical and engineering properties of brown rice and rice bran are elaborated on in the second part. The nutritional, medicinal and health benefits of brown rice and rice bran oil are covered in the third part. The glycemic value of brown rice is also explained through a comprehensive chapter in this part. The fourth part focuses on value addition to brown rice and describes products such as germinated brown rice. In the fifth part authors describe a novel storage method to extend the shelf life of brown rice and storage entomology of brown rice. Finally, the hurdles in brown rice consumption and opportunities and challenges in the marketing of brown rice are covered in the sixth part.

We believe this comprehensive collection will benefit students, scientists, nutritionists, dieticians, professionals in the rice industry, and many others. We are grateful to all the contributors for promptly submitting their chapters. We also thank the staff of the editorial and production departments of Springer for their unstinted support and efforts to bring about this book in its present form.

<table>
<tr><td>Guelph, Canada</td><td>Annamalai Manickavasagan</td></tr>
<tr><td>Muscat, Oman</td><td>Chandini Santhakumar</td></tr>
<tr><td>Thanjavur, Tamil Nadu, India</td><td>N. Venkatachalapathy</td></tr>
</table>

Contents

Contributors

V. Alex Albert Agricultural Engineering College and Research Institute, Tamil Nadu Agricultural University, Tiruchirappalli, Tamil Nadu, India

Amanat Ali Department of Food Science and Nutrition, College of Agricultural and Marine Sciences, Sultan Qaboos University, Muscat, Sultanate of Oman

A. Anandan ICAR-National Rice Research Institute, Cuttack, Odisha, India

R.M. Anjana Department of Diabetology, Madras Diabetes Research Foundation, Chennai, Tamil Nadu, India

Sila Bhattacharya Grain Science & Technology Department, CSIR-Central Food Technological Research Institute, Mysore, India

Asish K. Binodh Agricultural College and Research Institute, Killikulam, Tuticorin, Tamil Nadu, India

Sowriappan John Don Bosco Department of Food Science and Technology, Pondicherry University, Puducherry, India

Sankar Devarajan Nutrition & Dietetics Program, Department of Human Sciences, University of Arkansas at Pine Bluff, Pine Bluff, Arkansas, USA

V. Eyarkai Nambi Central Institute of Post Harvest Engineering and Technology, Ludhiana, India

R. Gayathri Department of Foods, Nutrition & Dietetics Research, Madras Diabetes Research Foundation, Chennai, Tamil Nadu, India

Shumaila Jan Department of Food Engineering & Technology, Sant Longowal Institute of Engineering & Technology, Longowal, Punjab, India

A. Jayadeep Department of Grain Science and Technology, CSIR-Central Food Technological Research Institute (CFTRI), Mysore, Karnataka, India

M. Jayanthan Department of Foods, Nutrition & Dietetics Research, Madras Diabetes Research Foundation, Chennai, Tamil Nadu, India

R.P. Kingsly Ambrose Department of Agricultural and Biological Engineering, Purdue University, West Lafayette, IN, USA

Upendra Kumar ICAR- National Rice Research Institute, Cuttack, Odisha, India

M.N. Lavanya Indian Institute of Food Processing Technology, Thanjavur, Tamil Nadu, India

K.R. Lijini Department of Grain Science and Technology, CSIR-Central Food Technological Research Institute (CFTRI), Mysore, Karnataka, India

M. Loganathan Indian Institute of Food Processing Technology, Thanjavur, Tamil Nadu, India

Annamalai Manickavasagan School of Engineering, University of Guelph, Guelph, ON, Canada

P. Masilamani Agricultural Engineering College and Research Institute, Tamil Nadu Agricultural University, Tiruchirappalli, Tamil Nadu, India

R. Meenatchi Indian Institute of Food Processing Technology, Thanjavur, Tamil Nadu, India

Shabir Ahmad Mir Department of Food Technology, Islamic University of Science and Technology, Awantipora, Jammu and Kashmir, India

V. Mohan Department of Diabetology, Madras Diabetes Research Foundation, Chennai, Tamil Nadu, India

Zhongli Pan Department of Biological and Agricultural Engineering, University of California, Davis, CA, USA

Healthy Processed Foods Research Unit, USDA-ARS-WRRC, Albany, CA, USA

Shruti Pandey Department of Grain Science and Technology, CSIR-Central Food Technological Research Institute (CFTRI), Mysore, Karnataka, India

P. Panneerselvam ICAR- National Rice Research Institute, Cuttack, Odisha, India

H.A. Pushpadass ICAR-National Dairy Research Institute, Bengaluru, India

M. Ramya Bai Department of Foods, Nutrition & Dietetics Research, Madras Diabetes Research Foundation, Chennai, Tamil Nadu, India

V. Ruchi Department of Foods, Nutrition & Dietetics Research, Madras Diabetes Research Foundation, Chennai, Tamil Nadu, India

D.C. Saxena Department of Food Engineering & Technology, Sant Longowal Institute of Engineering & Technology, Longowal, Punjab, India

S. Selvam Anbil Dharmalingam Agricultural College and Research Institute, Tamil Nadu Agricultural University, Tiruchirappalli, Tamil Nadu, India

Manzoor Ahmad Shah Department of Food Science and Technology, Pondicherry University, Puducherry, India

S. Shahir Department of Food Technology, Bannari Amman Institute of Technology, Sathyamangalam, India

S. Shobana Department of Foods, Nutrition & Dietetics Research, Madras Diabetes Research Foundation, Chennai, Tamil Nadu, India

M.S. Shwetha Indian Institute of Food Processing Technology, Thanjavur, Tamil Nadu, India

V.R. Sinija Indian Institute of Food Processing Technology, Thanjavur, Tamil Nadu, India

V. Sudha Department of Foods, Nutrition & Dietetics Research, Madras Diabetes Research Foundation, Chennai, Tamil Nadu, India

T. Sugitha Agricultural College and Research Institute, Killikulam, Tuticorin, Tamil Nadu, India

J. Alice R.P. Sujeetha National Institute of Plant Health Management (NIPHM), Hyderabad, Telungana, India

S. Sulochana Indian Institute of Food Processing Technology, Thanjavur, Tamil Nadu, India

P.T. Umashankar Consultant, Asian Development Bank, Adayar, Chennai, Tamil Nadu, India

R. Unnikrishnan Department of Diabetology, Madras Diabetes Research Foundation, Chennai, Tamil Nadu, India

N. Venkatachalapathy Indian Institute of Food Processing Technology, Thanjavur, Tamil nadu, India

Chandrasekar Venkitasamy Department of Biological and Agricultural Engineering, University of California, Davis, Davis, CA, USA

Idrees Ahmed Wani Department of Food Science and Technology, University of Kashmir, Srinagar, Jammu and Kashmir, India

About the Editors

 **Annamalai Manickavasagan,** Ph.D., P.Eng. (Canada) obtained his Ph.D. in Biosystems Engineering from the University of Manitoba, Canada. At present, he is an Associate Professor at the University of Guelph, Canada. He has published more than 50 scientific papers in peer-reviewed journals and edited 4 books. He has supervised more than 10 postgraduate students (Ph.D. and M.Sc.).

He is the Founder and President of the Whole Grains Research Foundation in India. Through this foundation, he is promoting whole grains research in Asia region. He has organized several conferences, symposiums and workshops in whole grains.

 **Chandini Santhakumar,** Ph.D., obtained her Ph.D. (Food Science) from the Central Food Technological Research Institute (CSIR), Mysore, India. She has 11 years of research experience in several organizations at various capacities. She has several publications in internationally reputed journals and a book chapter. Besides her research experience, she has also worked as a nutrition consultant.

N. Venkatachalapathy, Ph.D., received his Ph.D. from the Gandhigram Rural University, Gandhigram, India. At present, he is working as an Associate Professor and Head, Department of Food Engineering at the Indian Institute of Crop Processing Technology, Thanjavur, an autonomous institution under the Ministry of Food Processing Industries, Government of India. He is a life member of the Indian Society of Agricultural Engineers and Association of Food Scientists and Technologists (India).

The primary focus of his recent research work has been directed towards addressing the improvement of farmers and primary processors at farm level. This includes development of gadgets for increasing the income level of farmers and processors at farm level. Also, his research interest is on development of novel technologies for production of quality rice and energy efficient drying of farm produce.

Part I
Milling

Chapter 1
Rice Milling Technology to Produce Brown Rice

V. Eyarkai Nambi, Annamalai Manickavasagan, and S. Shahir

Introduction

Rice is one of the cereals that belongs to the grass family and considered as an important plant for the supply of staple food over half of the world's population (Babu et al. 2009). In general, rice is considered as the source of carbohydrates and supplies energy to body through diet. The rice is obtained from paddy grains after many processing operations. The paddy crop is harvested after desired maturity and threshed for separation of paddy from panicles. Threshing only detaches the spikelets from the plant along with the glumes. Once the threshing is over, the paddy has to undergo many unit operations to become edible rice. This chapter discusses the various unit operations involved in conversion of paddy into brown rice and its effect on the quality of brown rice.

Technical Terms Definitions

Before going into detail about milling aspects, the technical terms related to rice milling need to be understood, and the definition of each term is given below as per the codex standards 198-1995:

V. Eyarkai Nambi (✉)
Central Institute of Post Harvest Engineering and Technology, Ludhiana, India
e-mail: eyarkainambi@gmail.com

A. Manickavasagan
School of Engineering, University of Guelph, Guelph, ON, Canada
e-mail: manicks33@hotmail.com

S. Shahir
Department of Food Technology, Bannari Amman Institute of Technology, Sathyamangalam, India

© Springer International Publishing AG 2017
A. Manickavasagan et al. (eds.), *Brown Rice*,
DOI 10.1007/978-3-319-59011-0_1

Paddy rice is rice which has retained its husk after threshing.

Husked rice, **brown rice** or **cargo rice** is paddy rice from which the husk only has been removed.

Milled rice or **white rice** is husked rice from which all or part of the bran and germ has been removed by milling. Milled rice (white rice) may be further classified into the following degrees of milling:

- *Well-milled rice* is obtained by milling husked rice in such a way that some of the germ and all the external layers and most of the internal layers of the bran have been removed.
- *Extra-well-milled rice* is obtained by milling husked rice in such a way that almost all of the germ, all of the external layers and the largest part of the internal layers of the bran and some of the endosperm have been removed.
- *Under-milled rice* is obtained by milling husked rice but not to the degree necessary to meet the requirements of well-milled rice.

Parboiled rice is paddy or husked rice that has been soaked in water and subjected to a heat treatment so that the starch is fully gelatinized, followed by a drying process.

Glutinous rice or waxy rice is kernels of special varieties of rice which have a white and opaque appearance. The starch of glutinous rice consists almost entirely of amylopectin. It has a tendency to stick together after cooking.

Structure of Paddy Grain

It is essential to understand the physiological structure of paddy grain for effective processing and milling to obtain brown rice/white rice. In paddy, caryopsis is enclosed in a tough husk/hull which is woody and siliceous in nature and is not edible. The hull protects the rice caryopsis and grain from insect infestation and fungal damage.

The hull is composed of two modified leaves: the palea and larger lemma. Structurally, the hull can be divided into four structural layers, viz. outer epidermis of highly silicified cells; sclerenchyma or hypoderm fibres two or three cell layers thick; crushed, spongy parenchyma cells; and inner epidermis of generally isodiametric cells.

The main edible portion of the paddy is endosperm. Botanically and compositionally, different layers are present as outer covering to endosperm. The first layer is called as pericarp which is composed of epicarp, cross cells and cuticle. Next to pericarp is testa also called as seed coat, followed by nucellus and aleurone layers. The aleurone layer is strongly attached with endosperm. In addition to these layers, embryo is attached to one end of the endosperm which is also similarly covered (Fig. 1.1).

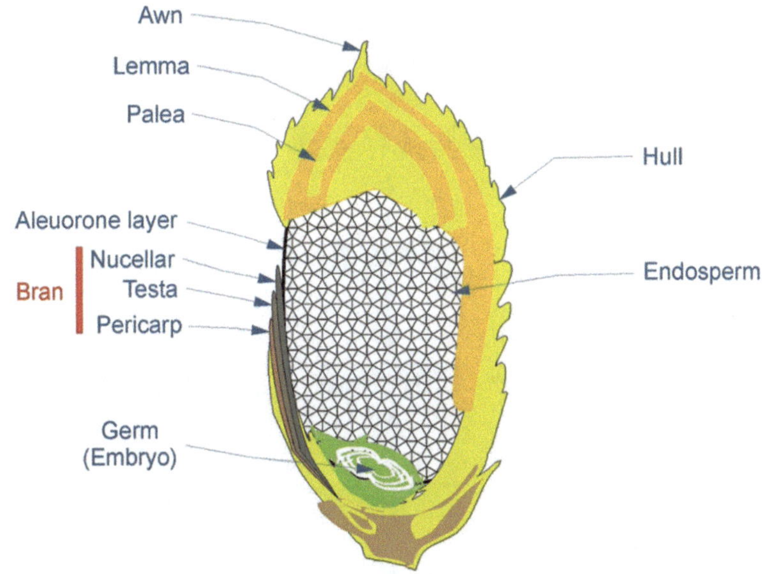

Fig. 1.1 Sectional view of paddy spikelet

Processing of Paddy for Brown Rice

In paddy processing, there are two main operations, viz. shelling or dehulling and milling or polishing. In case of parboiled rice, the paddy undergoes parboiling and drying before shelling and milling. The overall process flow and detailed step by step unit operations in paddy processing is given in flowchart (Fig. 1.2) and briefed below.

Cleaning Principle of cleaning paddy is based on different physical properties such as weight, size, density and properties of impurities, and paddy impurities lighter than paddy are removed by aspirators.

Parboiling Parboiling is a hydrothermal process that includes soaking, steeping and steaming, usually done to get parboiled white or brown rice. In soaking, the paddy is soaked either in normal or hot water for predefined duration. The soaking is the result of molecular absorption, capillary absorption and hydration. Then the soaked paddy is steamed/partially cooked in hot water or using wet steam. During steaming, the enzymes are inactivated; the endosperm becomes compact, translucent and pasty due to gelatinization; and crack in the caryopsis is also sealed.

Drying The parboiled paddy is required to be dried to moisture of 14–16% to obtain the desirable milling and storing properties. Traditionally, drying is carried out on open yards (sun drying), and the process usually goes up to 5 days depending upon the ambient condition. In modern rice mills, continuous drying systems using hot air are used to make the process short and hygienic.

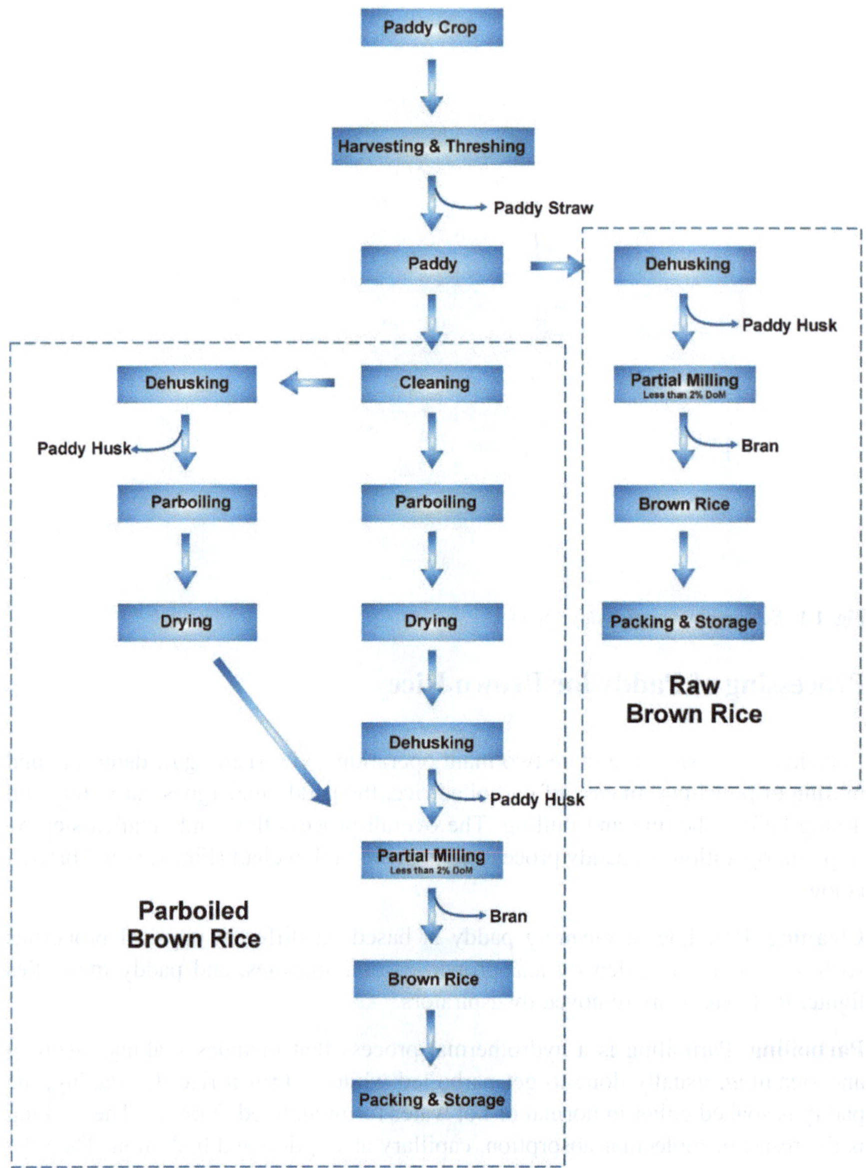

Fig. 1.2 Flow chart for brown rice production

Dehusking The process of removing the husk is referred as dehusking or dehulling or more commonly as shelling. Separation of hull or husk from the paddy grains is carried out mechanically. Many machines have been invented with different working principles in order to make the dehusking process more effective. Some of the machines used for dehusking of paddy are discussed later in this chapter.

Separation The output material obtained from the dehusking machine would be the mixture of dehusked rice (brown rice), husks and unhusked paddy. This mixture is subjected to sieving-cum aspiration to separate the desirables from the undesirables.

Milling or Polishing Removal of bran layers is called as milling or polishing. There are two types of polishers, viz. abrasive and friction polishers, which are commonly used at commercial level in rice milling.

Grading After polishing operations, the milled rice contains head rice (unbroken rice), and brokens of different size. Separation of brokens from head rice is done with stage sieving machines.

Bagging and Storage Predetermined quantities (by weight) of rice are packed using automatic weighers and baggers. Continuous and enclosed flow through conveyors and pipes ensures dust-free bagging and finished products.

Dehusking

As discussed earlier, paddy has hull or husk as outer cover. The process of removing the outer cover (hull) from paddy is called as dehusking. Dehusking is a mechanical process and many machines are involved. Depending upon the working principle, the machines are classified and explained below.

Dehusking Machine

Centrifugal Shellers

Using centrifugal force, dehusking is carried out in these machines. Through rotating impeller, the paddy grains are subjected to centrifugal force and thrown towards the outer casing. In the outer casing, shelling takes place due to hitting with great force. Significant characteristics of these centrifugal shellers are high capacity and simple constitutional features as there is only one moving part, i.e. impeller. The inner surface of the casing is lined with rubber where the paddy grains hit. The paddy is fed to the centre of the rotor. This machine can be modified easily to obtain a compact multistage mill by combining several units. Heat generation does not occur in this type of machine which is another added advantage (Fig. 1.3).

Disc Sheller

Disc shellers mainly consist of two abrasive discs either with one stationary or both moving in opposite directions; the dehusking takes place between the discs. The best commercially available machine of this type is under-runner disc sheller. It

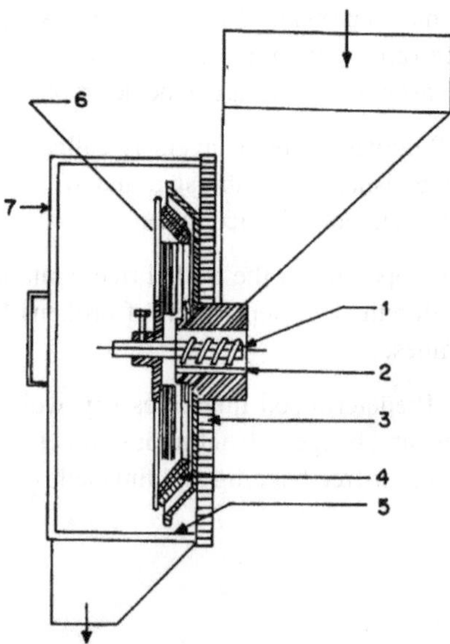

1.Feed screw, 2.Coller, 3.Crank gear, 4.Rubber ring, 5.Lining 6.Impeller 7.Lid

Fig. 1.3 Line diagram of sectional view of centrifugal sheller (Reproduced from Sahay and Singh (2004)). *1* Feed screw, *2* collar, *3* crank gear, *4* rubber ring, *5* lining, *6* impeller, *7* lid

consists of two discs with their inner faces lined with emery roller; bottom roller is rotating and the other is stationary. The paddy passes in between the two discs and gets dehusked. There are provisions in disc shellers to adjust the clearance between the discs, which make the machine suitable to many varieties of paddy. It gives a marked increase (2–4%) in rice output over conventional huller in almost all milling conditions (Fig. 1.4).

Rubber Roll Sheller

Paddy grains are fed between two rubber rollers, and the rollers rotate in opposite direction at differential speed to increase the shearing force on the paddy grains which loosens and removes the husk form the paddy grains. During process of dehulling, the grains are handled gently, due to compressible and flexible nature of rubber rolls. This is the reason to get more head rice recovery in rubber roll shellers than other type of shellers. The functional coefficient between paddy grain surface and rubber is lower than that of paddy and steel. This facilitates easy dehulling. The rubber roller mill gives better-quality brown rice compared to other types of shellers which fetch more price in the market (Fig. 1.5).

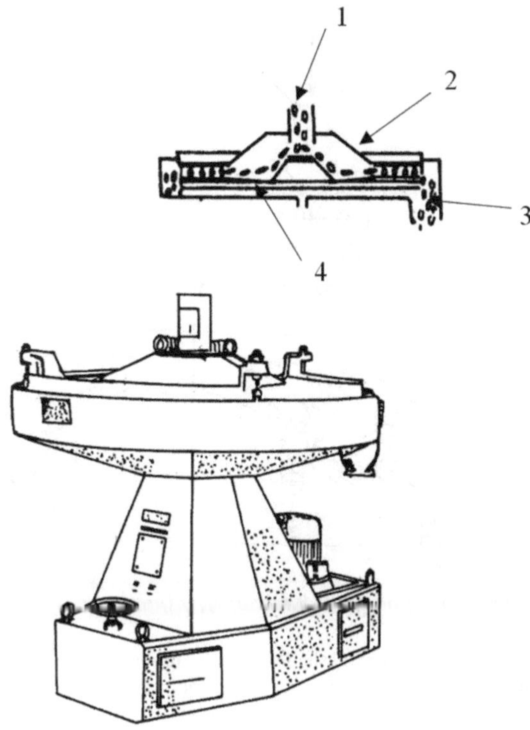

1.Feeding chute, 2.Stationary upper disc, 3. dehusked grain, 4.rotational lower disc

Fig. 1.4 Line diagram of under-runner disc huller with internal discs with rice flow (Reproduced from Sahay and Singh (2004)). *1* feeding chute, *2* stationary upper disc, *3* dehusked grain, *4* rotational lower disc

Comparison of Shellers

The comparison between abovementioned shelling machines with many parameters especially in terms of brown rice recovery is given below (Table 1.1) in the form of weightage points. More number of hashtags represents higher range/value.

From Table 1.1, the appropriate sheller for brown rice production can be finalized. Though comparable dehusking efficiency could be obtained from disc type shellers with rubber roll sheller, partial polishing takes place in the disc type. In brown rice production, partial polishing is an added advantage; at the same time, broken percentage has to be taken into account. Centrifugal sheller may yield lower dehusking efficiency and higher broken percentage, and this may be the reason for it not being used commercially for rice milling. These are the reasons to prompt the brown rice producers to go for rubber roll shellers.

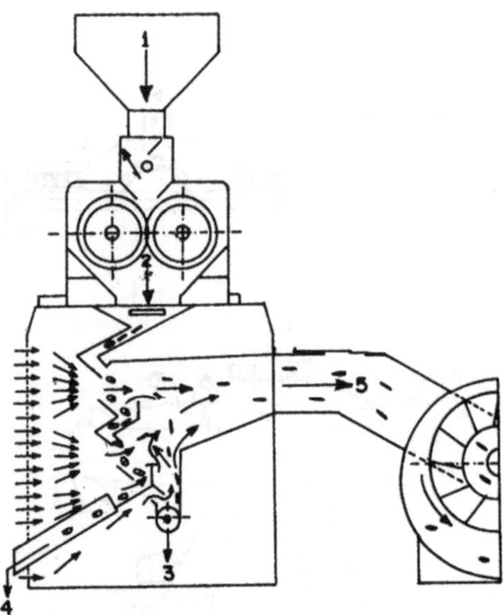

1.Paddy, 2.Brown rice paddy husk mixture, 3. Immature grains,4 Brown rice,5 Husk

Fig. 1.5 Line diagram of rubber roll sheller with separation unit along with grain flow (Reproduced from Sahay and Singh (2004)). *1* paddy, *2* brown rice paddy husk mixture, *3* immature grains, *4* brown rice, *5* husk

Table 1.1 Comparison of three types of paddy dehusking equipment

S. no	Parameter	Centrifugal shellers	Disc shellers	Rubber roll sheller
1	Brown rice recovery	##	#	###
2	Broken percentage	##	###	#
3	Dehusking efficiency	#	##	###
4	Man power requirement	#	#	##
5	Energy requirements	##	##	##
6	Occurrence rice polishing	#	###	#

#Indicating the weightage. More number of # represents higher value

Milling

Many of us get confused whether milling is necessary for brown rice production or not. This section provides the detailed information about milling of rice and its relation with brown rice production.

As discussed in the previous section, paddy grain after dehusking, has many layers like pericarp, testa nucellus and aleurone layers. These bran layers consist mainly of fibre, hemicelluloses, cellulose, proteins, minerals and phytate, but very little or no starch. The rice milling process involves removal of bran layers from husked rice to produce whole white rice kernels. The distribution of major constituents in the bran

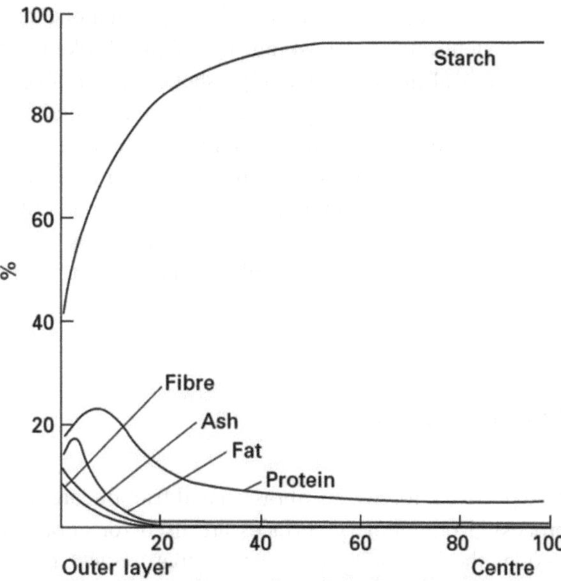

Fig. 1.6 Distribution of major constituents within the brown rice grain (Reproduced from Barber (1972))

layers is shown in Fig. 1.6. So, sufficient care is required to prevent excessive breakage of the kernel (broken rice) and improve recovery (head rice) during milling of rice.

Is Milling Needed for Brown Rice?

Many controversial discussions have been made regarding the question that whether brown rice needs to be polished/milled to any level. Nutritionists say that important micronutrients which are essential for human metabolism are present in the outer layer/bran and germ of rice. Extensive milling of husked rice leads to reduction of these micronutrients.

On the other hand, though bran layers are edible, they resist to heat penetration and protect the grain from cooking which leads to more cooking time and energy loss while cooking. Prolonged cooking of brown rice as a whole with all bran layers leads to bursting and opening up of grains which is not acceptable in general (Bhattacharya 2011). Desikachar et al. (1965) reported that a certain amount of scratching to the intact bran was essential to enable the brown rice to hydrate adequately in boiling water during cooking. Hence partial removal of bran layers is inevitable to make the rice edible without compromising the appearance, texture and consumers' acceptability.

Nutritionists have therefore long held, often passionately, that rice should be minimally milled, only 2–4% by weight of the brown rice at the most (Bhattacharya 2011). So it can be termed as "under-milled rice" instead of brown rice. Ajimilah and Rosniyana (1994) have reported that under-milled rice can also be termed as brown rice.

As discussed earlier, using disc-type shellers, removal of bran takes place to some extent during dehusking. But higher broken percentage and no control over bran removal make the disc-type shellers unsuitable for production of brown rice. Hence employing milling machine to some extend can be recommended for brown rice. Some of the milling/polishing machines related to rice milling are briefly explained below. Interested readers can refer to Sahay and Singh (2004), especially for additional information about peripheral speed of rollers, capacities, clearances and size and power requirement. The hardness of the bran layers increases from outer to inner bran layers (Lamberts et al. 2006), so appropriate selection of milling machine has to be made accordingly.

Milling Machines

Two principle forces, i.e. abrasion and friction, are involved in bran removal/milling process. Rough/emery surfaces are used for abrasion to peel off the bran layers, besides the friction between grains that induces the removal of bran layers during milling.

Vertical Milling Machine

A conical-shaped cast iron cylinder with abrasive surface is fixed in a vertical shaft, and a sieve is fixed around the conical roller with the adjustable clearances. The average clearance between the cone and the sieve is about 10 mm. Adjustable rubber brake divides the screen at regular intervals to reduce the breakage of rice. The husked rice is fed into top centre of cone through a hopper. The centrifugal force generated by rotation of cone moves the brown rice between the cone and wire mesh/sieve. Rubber brakes restrict the movement of rice, thus, applies pressure. As a result of pressure, brown rice is pressed against the abrasive coating of the cone, and this friction removes bran layer. Partly or fully whitened rice leaves the machine through a self-unloading discharge spout. The residence time of grains inside the mill and the clearance between cone and sieve are inversely proportional. Heat generation will be high at lower clearance and higher residence time, but effective bran removal would be obtained. For obtaining brown rice with minimum removal of bran layers, the clearance and residence time of rice has to be optimized as per the requirement (Fig. 1.7).

Horizontal Milling Machine

A stationary horizontal sieve and rotating emery roller are both mounted concentrically in a milling chamber. Uniform clearance is maintained between roller and sieve provided. The husked rice is fed between roller and sieve; polishing is obtained

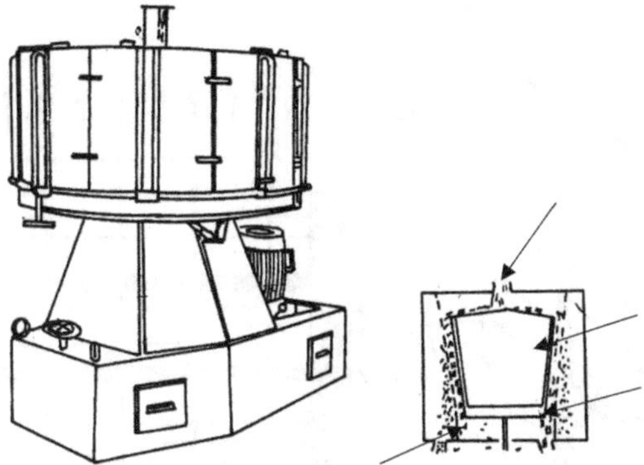

1.Feeding chute, 2.internal rotating cone, 3. Outer stationary cylindrical sieve, 4.polished grains

Fig. 1.7 Line diagram of vertical rice milling/polishing machine along with internal cone arrangement and grain flow (Reproduced from Sahay and Singh (2004)). *1* feeding chute, *2* internal rotating cone, *3* outer stationary cylindrical sieve, *4* polished grains

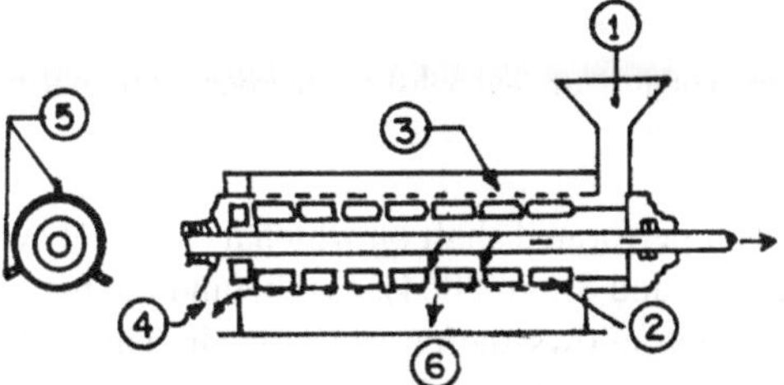

1.Hopper, 2.Abrasive roller, 3.Screened steel cylinder4.white rice,5.Brakes 6.Bran

Fig. 1.8 Sectional diagram of horizontal rice milling/polishing machine (Reproduced from Sahay and Singh (2004)). *1* hopper, *2* abrasive roller, *3* screened steel cylinder *4* white rice, *5* brakes, *6* bran

by abrasion and friction with emery roll, screen and rice grain. Degree of milling (DoM) is controlled by adjustable weight at the end of outlet and adjustable brakes to screen cylinder. The removed bran layers come out through holes provided on the sieve. A blower is provided to cool rice grains, emery roll and screen cage which collects and carries the bran in the outlet duct (Fig. 1.8).

Air-Jet-Based Polisher

In this type, air jet streams are used to remove the bran layer from the husked rice. Since no mechanical rotation or friction is involved in this process, the rice will not heat up to higher temperature which happens with other two machines. The performance of this machine is good with short grains; for medium and long grain varieties, the machine considerably increases in brokens. In this, it consists mainly of a

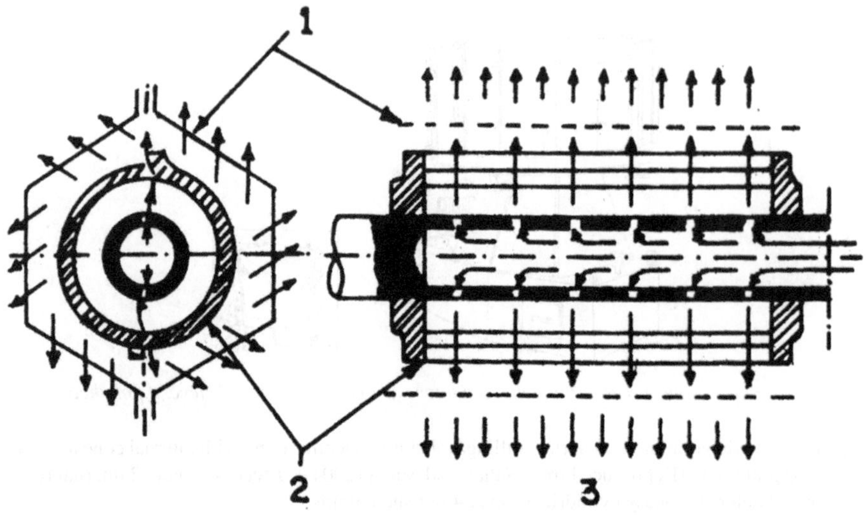

1.Screen 2. Milling Roller, 3.Bran

Fig. 1.9 Sectional diagram of air-jet-based polisher (Reproduced from Sahay and Singh (2004)). *1* screen, *2* milling roller, *3* bran

horizontal partly hollow perforated shaft on which a cast steel cylinder with friction ridges is clamped (Sahay and Singh 2004). Just behind the two ridges, the cylinder has a long opening which allows the passage for air. This cylinder runs inside a hexagonal chamber consisting of two halves hexagonal screens with slotted perforations. A feeding screw with horizontal shaft feeds the rice into the press chamber of the machine. The clearance between the hexagonal screen and the cast steel cylinder is adjustable by a screw controlling the distance between the two halves of the screen (Fig. 1.9).

Degree of Milling (DoM) of Rice

The bran layer can be removed to varying extents including the aleurone layer or even a part of the subaleurone. The extend of bran layer removal is referred as degree of milling (DoM) of rice.

Many researchers worked and reported various methods for estimating DoM (Barber and Benedito de Barber 1978; Desikachar 1955a, b, 1956; Kik 1951; Wadsworth 1994). The fundamental principle for estimating DoM is to determine how much of the grain or any of its constituents has been lost (i.e. removed) during milling. Theoretically, the primary method is based on the extent of weight loss. DoM is defined as the extent of degree of removing the different bran layers (Webb 1985).

$$DoM = \frac{w_1 - w_2}{w_1} \times 100$$

where

w_1 = weight of grains taken for milling (dehusked rice) and w_2 = weight of milled grains (polished rice).

For example, 100 g of dehusked rice has been milled and the final weight of milled rice is 95 g; then the DoM is 5%, which means 5.0% of grain weight has been removed during milling as bran.

Effect of DoM on Brown Rice Quality

Rice bran contains many nutritional components like vitamins, mineral, proteins, dietary fibre, essential unsaturated fatty acids and phenolic compounds (Kahlon 2009; Yılmaz 2016). In addition to these components, the bran contains considerable amount of many antioxidants like tocopherols and c-oryzanol. Gamma-oryzanol is a unique mixture of the ferulic acid esters of triterpene alcohols and plant sterols; it has a cholesterol-reducing effect (Yılmaz 2016). Hence removal of bran obviously alters the quality characteristics of final brown rice/polished white rice.

Many researchers have reported about the effect of milling at different degrees on various quality parameters of brown rice or milled rice (Bhattacharya 2011; Chen and Siebenmorgen 1997; Ellis et al. 1986; Lamberts et al. 2007; Liu et al. 2009; Lyon et al. 1999; Pan et al. 2005; Rosniyana et al. 2006). Some of the relevant and important changes are discussed below.

Liu et al. (2009) investigated the relationships between physical properties of brown rice and degree of milling (DoM). The properties like length, width, thickness, aspect ratio, equivalent diameter, sphericity, surface area, volume, bulk density, true density, porosity and thousand-seed weight were observed at different degrees of milling with ten rice cultivars. The study concluded that degree of milling had significant influence on physical properties except true density and porosity. Rosniyana et al. (2006) studied the effect of degree of milling on brown rice of Q34 variety at different degrees of milling (DoM) from 0–8%. The chemical composition of brown rice was significantly affected by the milling process. The vitamin and mineral contents were found high in brown rice followed by partially milled rice at 4, 6 and 8% DoM. Brown rice milled at 8% DoM had the lowest protein, fat and crude fibre content. Brown rice took longer time to cook (28 min) which significantly differed from the cooking time of partially milled and milled rice. The detailed results reported in that study are reproduced below (Tables 1.2, 1.3, 1.4). It was concluded that brown rice and partially milled rice offer healthier benefits than "well-milled rice".

Champagne et al. (1990) studied the effects of milling (0–9.1%) on lipid content and starch gelation temperature (Onset and Peak). Milling reduced both the

Table 1.2 Proximate composition of Q34 variety rice at different DoM

Composition (% wet basis)	Degree of milling (DoM)			
	0 (Brown rice)	4%	6%	8%
Moisture	10.50 ± 0.80	10.76 ± 0.05	10.91 ± 0.05	10.25 ± 0.35
Protein	8.80 ± 0.10	8.76 ± 0.01	8.58 ± 0.05	8.25 ± 0.05
Fat	1.80 ± 0.0	1.12 ± 0.01	0.67 ± 0.25	0.39 ± 0.01
Ash	0.86 ± 0.03	0.69 ± 0.03	0.45 ± 0.03	0.40 ± 0.03
Crude fibre	0.68 ± 0.02	0.50 ± 0.02	0.48 ± 0.25	0.40 ± 0.02
Carbohydrate	78.25 ± 0.05	78.77 ± 0.25	79.97 ± 0.05	81.31 ± 0.30

Reproduced from Rosniyana et al. (2006)

Table 1.3 Mineral and vitamin compositions of Q34 variety rice at different DoM

Properties (mg/100 g)	Degree of milling (DoM)			
	0 (Brown rice)	4%	6%	8%
Phosphorus	201.00 ± 3.00	191.41 ± 0.25	160.8 ± 0.50	108.50 ± 4.50
Potassium	58.50 ± 0.50	52.65 ± 0.7	46.00 ± 0.25	35.00 ± 1.00
Sodium	53.00 ± 0.00	50.86 ± 0.68	40.69 ± 0.01	20.50 ± 0.50
Calcium	49.00 ± 1.00	47.00 ± 0.20	32.9 ± 0.10	10.9 ± 0.90
Iron	6.05 ± 0.25	4.81 ± 0.72	2.95 ± 0.10	1.60 ± 0.10
Thiamine	0.35 ± 0.01	0.30 ± 0.01	0.22 ± 0.25	0.16 ± 0.02
Niacin	5.50 ± 0.10	4.80 ± 0.25	3.82 ± 0.01	3.45 ± 0.05
Riboflavin	0.16 ± 0.10	0.14 ± 0.01	0.09 ± 0.05	0.07 ± 0.10

Reproduced from Rosniyana et al. (2006)

Table 1.4 Physico-chemical and cooking properties of Q34 variety rice at DoM

Properties	Degree of milling (DoM)			
	0 (Brown rice)	4%	6%	8%
Amylose content (%)	21.80 ± 0.50	22.25 ± 0.25	23.00 ± 0.25	24.4 ± 0.50
Gelatinization temperature (°C)	High	High	High	Intermediate
Gel consistency (mm)	28.55 ± 0.35	33.25 ± 0.50	33.75 ± 0.35	43.80 ± 0.50
Elongation ratio	1.24 ± 0.25	1.63 ± 0.75	1.73 ± 0.50	1.84 ± 4.50
Cooking time (min.)	28.18 ± 0.50	23.30 ± 0.25	18.80 ± 0.25	15.00 ± 1.00
Volume of expansion	3.51 ± 0.75	4.30 ± 0.50	4.32 ± 0.01	4.80 ± 0.50
Water uptake ratio	2.61 ± 0.35	3.14 ± 0.20	3.80 ± 0.10	3.90 ± 0.50
Solid loss	0.48 ± 0.25	0.45 ± 0.20	0.44 ± 0.10	0.56 ± 0.10

Reproduced from Rosniyana et al. (2006)

lipid content and the gelatinization temperature (onset and peak). Higher reduction was found with high DoM rice samples. Effect of degrees of milling (DoM of 0–25%) on brown rice (variety *Puntal*) was studied on colour and nutritional properties by Lamberts et al. (2007). The yellow and red pigments present in the brown rice were found decreasing till DoM of 15%, and after that no significant colour change was observed which indicated that 15% of DoM is sufficient to remove bran completely and outer endosperm to some extent. Proteins and minerals found in the brown rice started decreasing after DoM of 9%. Proteins were mostly concentrated (84%) in the outer endosperm and bran contained most of the minerals (61.0%). The changes in the quality parameter over various degrees of milling are given in Fig. 1.10.

Nutritional and sensory profiles of two Indian rice varieties (Bapatla and Uma varieties in both raw and parboiled forms) with different degrees of milling (0–8%) were studied and reported by Shobana et al. (2011). The protein, fat, dietary fibre, g-oryzanol, polyphenols, vitamin E, total antioxidant activity and free radical scavenging abilities had decreased during milling, while the available carbohydrates had increased. Flávia et al. (2014) studied about changes in proximate composition, colour, total flavonoids, anthocyanins and proanthocyanidins contents, total phenolics and antioxidant activity of IAC-600 and MPB-10 rice cultivars as a function of degree of milling (DoM of 0–15%). Milling at 4% DoM reduced 47% of the fat content in IAC-600; similar fat reduction was obtained in MPB-10 rice at 7% DoM (Fig. 1.11).

Effect of milling on phenolic profiles and cellular antioxidant activity of brown rice was studied by Liu et al. (2015), and they reported that the total phenolic content decreased around 55.6%, and the total CAA value decreased around 92.8%, while milling the brown rice up to 9.6% of DoM. In addition to that, bound forms to total phenolics and flavonoids decreased with increasing in DoM.

Sensory evaluation of whole brown rice/husked rice is not favourable to the consumers, since they are accustomed to the polished or milled rice. Shobana et al. (2011) evaluated the sensory of brown rice at different degrees of milling with the panel of experts and reported that the whole husked rice was not acceptable due to its branny taste and chewiness. Milling had decreased the branny taste and chewiness which is favourable in consumer point of view. On the whole, sensory score of brown rice with minimum milling was found accepted. These findings indicate that milling should be carefully controlled for acceptable sensory quality and retention of phytochemicals during brown rice milling.

Conclusion

The conversion of paddy into brown rice involves many processes, and each has to be optimized according to different cultivars. Among the processes, milling is an important process to control the whole quality and nutrient loss of brown rice and at the same time to increase the consumers' acceptability. More research and development are required to investigate the effect of various degrees of milling on quality parameters and shelf life of brown rice.

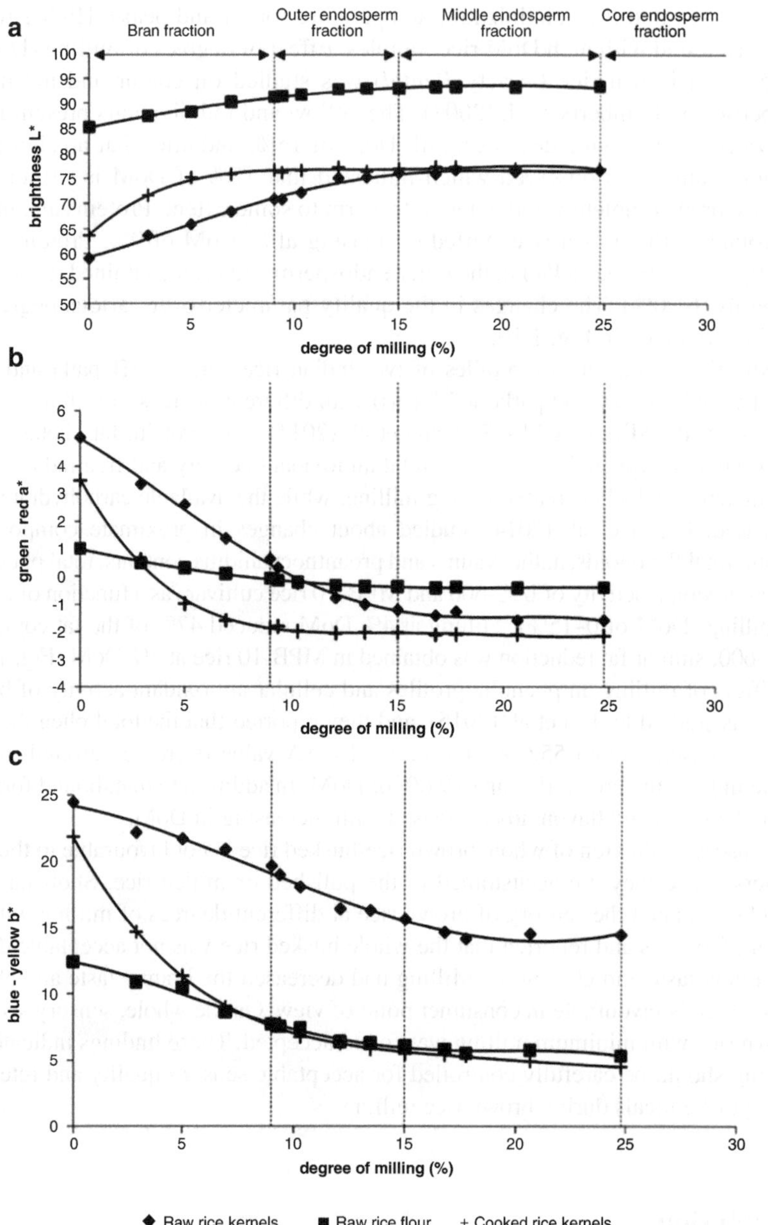

Fig. 1.10 L* (brightness), a* (redness) and b* (yellowness) colour values of brown rice as a function of degree of milling (Reproduced with permission from Lamberts et al. (2007))

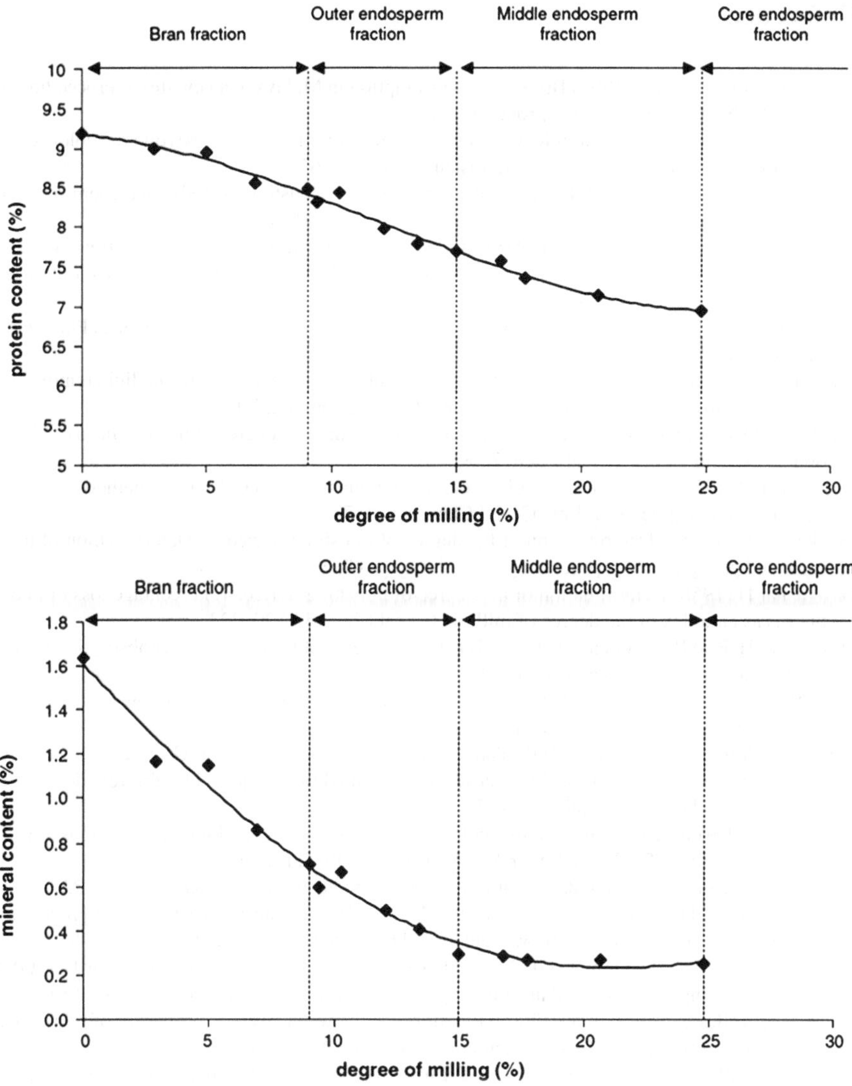

Fig. 1.11 Protein and mineral content of rice, as a function of degree of milling (Reproduced with permission from Lamberts et al. (2007))

References

Ajimilah N, Rosniyana A (1994) Brown rice consumption in Malaysia: a new trend. Proceeding of 5th ASEAN FOOD Conference, Kuala Lumpur

Babu PD, Subhasree R, Bhakyaraj R, Vidhyalakshmi R (2009) Brown rice-beyond the color reviving a lost health food-a review. Am Eurasian J Agro 2(2):67–72

Barber S (1972) Milled rice and changes during aging. In: Houston DF (ed) Rice chemistry and technology. American Association of Cereal Chemists, St. Paul, p 215

Barber S, Benedito de Barber C (1978) Outlook for rice milling quality evaluation systems. Proceedings of the Workshop on Chemical Aspects of Rice Grain Quality. International Rice Research Institute, Los Banos, 1979

Bhattacharya KR (2011) Rice quality: a guide to rice properties and analysis. Woodhead Publishing Limited, Cambridge

Champagne E, Marshall W, Goynes W (1990) Effects of the degree of milling and lipid removal on starch gelatinization in the brown rice kernel. Cereal Chem 67:570–574

Chen H, Siebenmorgen T (1997) Effect of rice kernel thickness on degree of milling and associated optical measurements. Cereal Chem 74:821–825

Desikachar H (1955a) Determination of the degree of polishing in rice. 1. Some methods for comparison of the degree of milling. Cereal Chem 32:71–77

Desikachar H (1955b) Determination of the degree of polishing in rice. 2. Determination of thiamine and phosphorus for processing control. Cereal Chem 32:78–80

Desikachar H (1956) Determination of the degree of polishing in rice. 4. Percentage loss of phosphorus as an index of the degree of milling. Cereal Chem 33:320–323

Desikachar H, Rao R, Ananthachar T (1965) Effect of degree of milling on water absorption of rice during cooking. J Food Sci Technol 2:110

Ellis JR, Villareal CP, Juliano BO (1986) Protein content, distribution and retention during milling of brown rice. Plant Foods Hum Nutr 36:17–26

Flávia PF, Vanier NL, Berrios JDJ, Pan J, de Almeida VF, Takeoka G, Elias MC (2014) Physicochemical and nutritional properties of pigmented rice subjected to different degrees of milling. J Food Compos Anal 35:10–17

Kahlon TS (2009) Rice bran: production, composition, functionality and food applications, physiological benefits. CRC Press/Taylor & Francis Group, Boca Raton

Kik M (1951) Determining the degree of milling by photoelectric means. Rice J 54:18–22

Lamberts L, Bie ED, Derycke V, Veraverbeke W, De Man W, Delcour J (2006) Effect of processing conditions on color change of brown and milled parboiled rice. Cereal Chem 83:80–85

Lamberts L, De Bie E, Vandeputte GE, Veraverbeke WS, Derycke V, De Man W, Delcour JA (2007) Effect of milling on colour and nutritional properties of rice. Food Chem 100:1496–1503

Liu K, Cao X, Bai Q, Wen H, Gu Z (2009) Relationships between physical properties of brown rice and degree of milling and loss of selenium. J Food Eng 94:69–74

Liu L, Guo J, Zhang R, Wei Z, Deng Y, Guo J, Zhang M (2015) Effect of degree of milling on phenolic profiles and cellular antioxidant activity of whole brown rice. Food Chem 185:318–325

Lyon BG, Champagne ET, Vinyard BT, Windham WR, Barton FE, Webb BD, McClung AM, Moldenhauer KA, Linscombe S, McKenzie KS (1999) Effects of degree of milling, drying condition, and final moisture content on sensory texture of cooked rice. Cereal Chem 76:56–62

Pan Z, Thompson J, Amaratunga K, Anderson T, Zheng X (2005) Effect of cooling methods and milling procedures on the appraisal of rice milling quality. Trans of the ASAE 48:1865–1871

Rosniyana A, Rukunudin I, Norin SS (2006) Effects of milling degree on the chemical composition, physico-chemical properties and cooking characteristics of brown rice. J Trop Agric Food Sci 34:37

Sahay K, Singh K (2004) Unit operations of agricultural processing. Vikas Publishing House Private Limited, New Delhi

Shobana S, Malleshi N, Sudha V, Spiegelman D, Hong B, Hu F, Willett W, Krishnaswamy K, Mohan V (2011) Nutritional and sensory profile of two Indian rice varieties with different degrees of polishing. Int J Food Sci Nutr 62:800–810
Wadsworth JI (1994) Degree of milling. CRC Press, New York
Webb BD (1985) Criteira of rice quality in the United States. American Association of Cereal Chemists, St. Paul
Yılmaz N (2016) Middle infrared stabilization of individual rice bran milling fractions. Food Chem 190:179–185

Part II
Physical and Chemical Properties

Chapter 2
Variations in Brown Rice Quality Among Cultivars

Shabir Ahmad Mir, Manzoor Ahmad Shah, and Sowriappan John Don Bosco

Introduction

In Asia, more than 2000 million people obtain their 60–70% calories from a wide variety of rice cultivars and its products (Lin et al. 2011). The chemical composition of rice grain varies widely, depending on variety, soil and environment. Brown rice is the name given to dehusked paddy which retains the embryo and bran layers of the grain. Brown rice is rich in fibres, minerals, vitamins and phytochemicals mostly present in the bran layer (Min et al. 2012).

There are about more than 40,000 varieties of rice, but only a few varieties are familiar to most of us. Diversity in quality parameters of rice varieties depends mostly on the genetic background and climatic and soil conditions during the rice grain development, which affect the physicochemical, functional and nutritional properties of a particular variety and have a great impact on consumer preferences (Dendy 2005; Singh et al. 2011). The physicochemical properties of grains have been widely investigated with respect to the variety difference. Large pools of rice genotypes differ in their amylose content which was investigated for the variability of their flour by examining their functional properties (Lin et al. 2011). Moreover the development of specific rice-based products depends on physicochemical properties of rice varieties. Hence, the knowledge about the characterization of rice varieties is important for the industry to develop the desired rice products with

S.A. Mir (✉)
Department of Food Technology, Islamic University of Science and Technology,
Awantipora, Jammu and Kashmir, India
e-mail: shabirahmir@gmail.com

M.A. Shah • S.J.D. Bosco
Department of Food Science and Technology, Pondicherry University, Puducherry, India
e-mail: mashahft@gmail.com; sjdbosco@yahoo.com

© Springer International Publishing AG 2017
A. Manickavasagan et al. (eds.), *Brown Rice*,
DOI 10.1007/978-3-319-59011-0_2

high consumer acceptability. Recently, new rice varieties have been developed to improve the eating quality of cooked rice, to develop food products with enhanced nutritional properties and to improve processing characteristics of rice-based food products.

Origin and Botany of Rice

The cultivation of rice extends over a wide range of climatic, soil and hydrological conditions – from tropical to semi-arid and warm temperate regions, from heavy clay to poor sandy soils and from dry land to swampy land in fresh or brackish water. Geographically, rice is grown between the latitudes of 53°N (in Northeast China) and 35°S (in New South Wales, Australia). However, it grows best between the latitudes of 45°N and 40°S (Dendy 2005). Rice cultivation extends from below sea level to an altitude of more than 2500 m in the Himalayas. Topographical conditions vary from uplands in hilly or plateau regions, with problems of deficit soil moisture, to medium lands with efficient water management and lowlands with excess water up to a depth of 5 m (Kent and Evers 1994). In South and Southeast Asian countries, two to three crops are harvested in a year by extending cultivation to the dry season with the help of irrigation (Laborte et al. 2012).

Rice is harvested as paddy with the hull comprising of about 20%, the bran and embryo 8–12% and the endosperm part 70–72% (Gujral et al. 2012). Paddy after removal of husk gives brown rice that is further polished to remove the bran and germ resulting in white rice. The rice produced in different parts of geographical conditions varies significantly in physical and functional properties, composition and cooking quality. Genetic diversity and environmental factors are mainly responsible for variation in the quality characteristics of rice (Singh et al. 2005; Kesarwani et al. 2013).

Rice is divided into two types – *japonica* and *indica* – and these differ in their characteristics and appearance. The grains of *japonica* rice are round and do not break easily. The cooked rice is sticky and moist and is produced mostly in Japan. The grains of *indica* rice are long and tend to break easily. When cooked, the rice is fluffy and does not stick together. Most of the rice produced in Southern Asia, including India, Southern China, Thailand and Vietnam is *indica* rice. Both the *japonica* and *indica* types of rice include non-glutinous and glutinous forms. Non-glutinous rice is popularly used generally for cooking purpose which is somewhat transparent and when cooked is less sticky than glutinous rice. Glutinous rice tends to be white, opaque and very sticky when cooked. It is commonly used to make rice products like cakes and various kinds of desserts, or processed to make rice snacks (Kent and Evers 1994).

Factors Affecting Variety Difference

Many types of rice which have evolved through the centuries of extensive rice culture vary widely in their range of adaptability. In tropical humid regions, tall rank-growing long-season varieties are usually cultivated. In tropical Asian countries, varieties requiring a growing season of over 180 days are common. These tropical varieties grow tall, tiller profusely and produce rank vegetative growth on soils that are annually cropped to rice. In temperate regions, short-season varieties maturing in 85–146 days are grown. They are usually hardy, somewhat short stature types that can withstand wide variations in daily temperatures and produce satisfactory yield even when grown in cold irrigation water (Abrol and Gadgil 1999; Nanda and Agrawal 2006).

The rapid rise in rice production in the recent years stemmed from expanded irrigation areas, increased use of fertilizers, effective control of pests, double rice cropping, use of innovative technologies and the adoption of improved genetic materials. The combined use of nitrogen fertilizers and high-yielding varieties has made possible the cultivation of high yield in irrigated areas. The new genetic information and genetic diversity provide the foundation for more efficient breeding programs for rice production (Luh 1980).

Varieties

The first concern of the farmer in selecting seed must always be the reliability of the cultivar; will the variety sprout, survive to maturity and yield well under the conditions it will encounter in his fields? The actual yield under the conditions of fertilizer application and cultural practices which will be applied is the next consideration. Hybridization programs conducted at the rice experiment stations have brought about the development of new varieties by intercrossing commercially grown strains and foreign ones. The main objective of breeding experiments is to increase yield and stabilizing productions, as well as providing a range of grain types.

The variety of rice cultivar is very important which significantly affects the quality of brown rice. The variety of rice cultivars is selected according to the need of consumers and the industry. The selection of particular variety also depends on the geographic region and climatic conditions. Characteristics that influence rice quality include those under genetic control. The modern technologies of rice breeding programs continuously refine and improve genetic characteristics influencing quality traits. Breeding and selection for desirable physicochemical, cooking and processing qualities in hybrid selections, breeding lines and new cultivars are essential components of varietal improvement programs. New varieties developed in these programs should meet the quality parameters before being released for commercial production (Luh 1980).

Climatic Conditions

A major factor influencing the quality of rice is the environment in which the crop is grown. Primarily, rice production is controlled by climatic variables that ideally should provide adequate water during the growing season, relatively high air and soil temperatures, adequate solar radiation, a moderately long growing season and relatively rain-free conditions during the ripening period. Once a new variety is released for commercial production, it spreads to wherever it can be produced advantageously compared to already grown varieties. Therefore, before release, new varieties are tested agronomically and for quality traits over their likely production area. These trials permit the-+evaluation of the quality parameters of new varieties over wide ranges of environmental conditions such as soil, climate and cultural practices (Abrol and Gadgil 1999).

Soil Types

Rice is grown on a wide range of soil types. The lack of adequate irrigation water is more apt to be a limiting factor than is soil type. Soil pH can vary from 4.0 to 8.0 without serious damage to the plant's growth. The rice plant has a high tolerance for alkaline soils and is sometimes grown for reclaiming such lands. Soils with impervious subsoils capable of holding floodwater or level prairies that can be readily flooded make ideal rice lands. Deep soils tend to give higher yields than soils that have an impenetrable layer at a shallow depth. Medium or heavy clays, clay loams, silt loams, or fine sandy loams with slowly permeable subsoils are preferred for rice cultivation.

Fertilizers

Fertilizers are widely used throughout the rice regions and are essential for crop growth. The fertilizer type and dose significantly affect the quality of rice. The proper mixture and percentage of fertilizer at appropriate period are necessary for rice plant which affects the quality of rice variety. If the nutrients are not provided in appropriate quantity to the rice crop, the variations were observed in landmark properties of rice and ultimately affect the consumer and industrial acceptability (Nanda and Agrawal 2006).

In addition to genetic differences, fertilizer, soil type and climatic conditions affect the bran quality such as bran oil content, biochemical properties of bran and overall brown rice quality. However, the effect of other climatic factors on brown rice quality is not available in public domain. More research is required in this area to encourage the agricultural scientists to include these factors in the crop improvement programs which could increase the productivity without losing grain characteristics.

Brown Rice Qualities in Different Varieties

Rice produced in different parts of the world differs in physical properties, composition and functional and cooking properties. The differences in these properties have been attributed to the differences in their genetic make-up and climatic conditions. The modern technologies of rice breeding programs continuously refine and improve genetic characteristics influencing quality traits.

Physical Properties of Brown Rice

Rice varieties with different properties are available in the market. Rice is marketed and preferred by the industry and consumers according to grain size and shape, i.e. long, medium and short. Grain size and shape are among the first quality properties considered in developing new varietie0073 (Table 2.1). The kernel dimensions are primary quality factors in most operations of processing, breeding and grading. Intensive genetic selection is practiced to eliminate heritable abnormalities in rice grain such as deep creases, which tend to leave bran streaks on milling; irregularly shaped kernels; sharp-pointed extremities, which break easily in milling; and oversized germs, which detract from milling quality and grain appearance.

Significant variation in physical and mechanical properties has been shown among rice varieties produced in different parts of the world with the influence of diverse genetic and environmental factors (Izawa 2008; Mir et al. 2013). Physical and mechanical properties play an important role in deciding the rice processing operations, which directly affect the quality of rice at industrial scale and hence determine its consumer acceptability. Several authors have reported about the physical and mechanical characteristics of rice, which influence the milling, cooking quality and consumer preference (Mohapatra and Bal 2006; Correa et al. 2007; Varnamkhasti et al. 2008).

The differences in the physical properties of rice grain are of practical concern for processing, handling and storage. The knowledge of physical properties of grain is helpful for designing appropriate machineries for process operations like sorting, drying, heating and milling and finds the solutions to problems associated with these processes (Sahay and Singh 1994; Mir et al. 2013). These properties are also important in the construction of storage structures and the calculation of the dimensions of storage bins of a particular capacity (Thompson and Ross 1983). The axial dimensions of rice grains are useful in selecting sieve separators and calculating the

Table 2.1 Physical properties of brown rice from various varieties

Variety	Length (mm)	Width (mm)	Thickness (mm)	Bulk density (kg/m^3)	Surface area (mm^2)	Thousand kernel weight (g)	Reference
Pusa-3	8.32	2.02	1.79	736	35.84	21.84	Mir et al. (2013)
Jehlum	6.29	2.55	1.84	794	25.85	21.84	
K-332	4.74	2.80	2.04	790	23.80	18.81	
SKAU-345	6.62	2.62	2.00	805	28.77	22.92	
Koshar	5.00	3.03	2.10	817	26.43	21.65	
ITA 150	7.76	2.80	2.05	860	35.56	33.85	Shittu et al. (2012)
WAB 189	6.87	2.94	2.00	870	31.76	28.39	
WAB 450	7.11	2.56	1.83	840	28.48	25.42	
Pusa Basmati	8.23	1.86	1.64	716	25.26	18.47	Mohapatra and Bal (2012)
Swarna	5.75	2.27	1.71	749	21.41	15.72	
ADT37	5.21	2.75	1.96	807	24.45	19.62	
Y9	4.83	2.90	2.09	765	25.11	17.97	Liu et al. (2009)
N42	4.96	3.05	2.05	766	26.04	19.54	
T559	5.41	3.18	2.40	757	31.46	22.34	
K818	6.33	2.50	1.89	728	26.07	19.19	
S5	6.73	1.98	1.78	708	23.03	15.06	
PR 116	7.01	2.17	1.73	857	11.83	19.87	Thakur and Gupta (2006)

energy requirement during the milling process. They can also be used to calculate surface area and volume of kernels which are important during modelling of grain drying, heating and cooling operations (Shittu et al. 2012).

Seven cultivars of brown rice, namely, Jehlum, K-332, Koshar, Pusa-3, SKAU-345, SKAU-382 and SR-1, grown in the Indian temperate climates, were studied for the variety difference in their physical properties. Results reported that the significant difference in the physical properties including length (4.74–8.32 mm), width (2.02–3.03 mm), thickness (1.79–2.10 mm), equivalent diameter (3.03–3.28 mm), surface area (23.80–35.84 mm^2), sphericity (37.40–63.44%), aspect ratio (0.24–0.61), volume (14.54–18.51 mm^3), bulk density (736.49–817.85 kg/m^3), true density (1270.81–1484.42 kg/m^3), porosity (41.06–46.70%), thousand kernel weight (18.81–22.92 g) and angle of repose (29.93–33.04°) was observed among the different brown rice varieties. The hardness value significantly varied from 131.48 to 73.99 N, with the highest hardness found in the brown rice of Jehlum variety (Mir et al. 2013). Shittu et al. (2012) observed the significant difference in physical properties of brown rice of some improved rice varieties. The length of brown rice varied from 6.87 to 7.76 mm, while their width varied from 2.46 to 2.94 mm, respectively. The thickness ranged from 1.82 to 2.05 mm. The thousand grain weight ranged from 25.42 to 33.85 g, while the bulk and true density ranged from 0.83 to 0.87 g/cm^3 and 1.41 to 1.57 g/cm^3, respectively. The wide variations were observed in the physical properties of brown rice in different geographical areas of the world, which affects their industrial and consumer acceptability.

Joshi et al. (2014) studied the physical properties of 12 different varieties of *indica* rice. The length of the brown rice grains varied from 5.34 to 6.67 mm, width from 1.76 to 2.24 mm and thickness from 1.46 to 1.8 mm, classifying them into long- and medium-grain varieties. The true density and bulk density of the brown rice varied from 1373 to 1743 kg/m^3 and 730 to 813 kg/m^3, respectively. The compressive hardness as measured by textural analyser varies between 86 and 160 N. The rice hardness is important as harder grains tend to prevent breakage during milling.

Razavi and Farahmandfar (2008) studied the physical properties of three rice varieties, namely, Fajr, Neda and Tarom Mahali, at three levels of processing, viz. paddy, brown rice and milled rice. The thousand kernel weight, porosity and volume decreased significantly with the level of rice processing; however, the bulk density value increased. The paddy of each variety showed the lowest value of true density. The results showed that static coefficient of friction was affected by cultivars, levels of processing and frictional materials. The angle of repose for all rice varieties decreased with different levels of processing.

Colour Properties of Brown Rice

Brown rice exists in different colours, which varies among the varieties (Table 2.2). There are some varieties with a distinctly darken, even almost black, pericarp. Such coloured-pericarp brown rice is not entirely uncommon, being more often found among upland rice in hilly and mountainous regions, for instance, in the northeastern mountainous states in India. It is generally believed that some of these

Table 2.2 Colour properties of brown rice from various varieties

Variety	L*	a*	b*	Reference
Chindeul	56.89	5.69	25.24	Lee et al. (2016)
Jehlum	66.98	4.19	23.42	Mir et al. (2016a)
SKAU-345	63.95	4.69	23.59	
SKAU-382	62.42	5.04	22.28	
Khao Dawk Mali 105	59.00	3.1	18.00	Sirisoontaralak et al. (2015)
SR-1	58.10	7.73	23.95	Mir et al. (2013)
Pusa-3	55.99	7.44	25.33	
Koshar	63.02	6.25	26.29	
ITA 150	81.31	1.81	21.80	Shittu et al. (2012)
ITA 301	82.90	2.29	21.21	
WAB 189	72.35	2.58	19.79	
WAB 450	80.34	2.09	20.53	

coloured pericarp rice varieties have medicinal properties. Most consumers are already aware that conventional brown rice is nutritionally superior to white rice in terms of fibre and beneficial functional components because of its outer layer and mostly select the rice by the colour.

The colour of rice kernels is one of the important physical properties from the industry and consumer point of view. Coloured rice varieties have been found to be potent source of functional components (Sompong et al. 2011). The colour varieties have better antioxidant properties than colourless varieties. Thus it can be concluded that colour varieties could be used as a natural antioxidant source (Moko et al. 2014). The colour analysis of brown rice showed that L*, a* and b* values ranged from 55.99 to 67.19, 4.23 to 7.73 and 22.41 to 26.29, respectively (Mir et al. 2013). Itagi and Singh (2015) also reported a significant difference in colour values of brown rice. The L* value for brown rice varied from 23.5 to 59.3, a* from 4.4 to 14.2 and b* from 4.8 to 22.6. The variation in colour of the rice kernels may be due to the difference in genetic make-up, composition and coloured pigments.

Cooking Properties of Brown Rice

Rice is mostly consumed in cooked form. However, its cooking properties differ due to the diversified cultivation of rice which affects its physicochemical properties and results in varied cooking properties. The demand of brown rice for cooking purpose is increasing nowadays due to consumer consciousness about their health, and they prefer the foods which have functional properties in addition to the nutritional value. As cooked rice is the staple food for most of the population, it is the easiest way to enhance the consumption of functional components as in the form of cooked brown rice.

Cooking quality of rice is one of the important factors influencing the acceptability of consumers (Soponronnarit et al. 2008). Cooking is the most important processing step to provide desirable texture to the rice grain. The rice grains are boiled in limited or excess amount of water during cooking. The chief constituent of rice is starch, which is made up of two major components, amylose and amylopectin. The starch of grain absorbs moisture and swells during cooking due to its gelatinization (Yadav and Jindal 2007). During cooking amylose leaches out from the starch granule and retrogrades when cooled, whereas amylopectin remains in the gelatinized granule. Amylose content is one of the key determinants of cooking and eating quality of rice (Juliano 1985). The wide varietal difference in cooking rice is mainly due to the bran layer which varies among the cultivars and provides the significant effect on the cooking properties.

Cooking quality preferences vary within the country and geographical regions, within ethnic groups and from one culture to another (Soponronnarit et al. 2008; Mir et al. 2016b). Desired quality of rice may also vary from one geographical region to another and consumer demand of certain varieties and favours specific quality traits of rice for cooking. The cooking methods and textural preferences of rice vary from place to place (Suwannaporn and Linnemann 2008; Son et al. 2013).

Cooking time varies for rice varieties and is the time when 90% of the starch in the grain no longer shows opaque centre when pressed between two glass plates. Cooking time is an important parameter which determines tenderness of cooked rice as well as its stickiness (Shinde et al. 2014). Brown rice is also used for cooking purpose in some parts of the world, but it takes longer time to cook as compared to polished rice. Cooking time of rice depends on coarseness of the grain and its gelatinization temperature. Rice with low gelatinization temperature require less than 20 min for cooking, while rice with intermediate gelatinization temperature require more than 20 min for cooking (Singh et al. 2005).

Variety differences were observed in cooking properties of brown rice. Cooking time, water uptake ratio, gruel solid loss and elongation ratio varied from 33.63 to 46.71 min, 1.74–2.85, 1.97–3.32% and 1.14–1.81, respectively. Textural attributes of cooked brown rice showed significant variation among the varieties with the hardness varied from 2.27 to 6.00 N (Mir et al. 2016b). The water absorption by rice during cooking is considered as an economic quality parameter, because it gives the estimate of the volume increase during cooking. During cooking, rice grains absorb sufficient water and increase in volume through increase in length and breadth. Lengthwise increase without increase in girth is desirable characteristic in high-quality rice (Shinde et al. 2014).

Gujral and Kumar (2003) reported that the cooking time for brown rice varies from 35.0 to 51.0 min, elongation ratio from 32.3 to 35.4%, water uptake from 84.69 to 136.75%, solid loss from 1.1 to 6.8% and hardness from 6.995 to 10.939 N. Deepa et al. (2008) reported that the cooking time of dehusked Jyothi and IR 64 rice varieties was found to be 30 min, while Njavara needed longer time to cook (38 min).

Chemical Composition of Brown Rice

The chemical composition of brown rice varies widely, depending on variety, soil and environment. Brown rice consists of bran layers (6–7%), embryo (2–3%) and endosperm (about 90%) (Chen et al. 1998). Starch is the major constituent of brown rice, whereas other components are also present in significant proportion. Non-starch constituents like protein, fat, ash, fibre and lignin are higher in brown rice than in milled rice. In addition to that, free sugars, free amino acids and aroma compounds are also present which are more concentrated in the bran fraction of the rice kernel (Itani et al. 2002; Ohtsubo et al. 2005; Lamberts et al. 2007). Brown rice is also a rich source of vitamins, minerals and rare amino acids (Liu et al. 2009; Mir et al. 2016b). During the milling process, from brown rice to white rice, the losses of proteins and total minerals reached 28.6% and 84.7%, respectively (Lamberts et al. 2007).

Brown rice retains the bran layer (containing many vitamins and minerals as well as fibres), as this has not been polished off to produce white rice. Red rice is known to be rich in iron and zinc, while black and purple rice are especially high in protein, fat and crude fibre. Red, black and purple rice get their colour from anthocyanin pigments, which are known to have free radical scavenging and antioxidant capacities, as well as other health benefits.

Protein is the second highest component after starch in rice kernel. Protein is available in varying amounts in brown rice, mostly ranging from 6.5% to 8.7% with some exceptions, where it varies from the main range (Cao et al. 2004; Mir et al. 2016b). Protein content varies among the rice varieties and decreased linearly with the increase in the degree of polishing as it is mainly concentrated in the peripheral layers of the grain (Zhou et al. 2002). Protein is the most abundant in the subaleurone layers of rice grain. In addition to that, small quantities are also present in aleurone cells (Azhakanandam et al. 2000). Rice protein is more nutritious because of its relatively well-balanced amino acid profile and is superior in lysine content than other cereal crops (Mohan et al. 2010).

The lipid or fat content of rice is concentrated in the bran layers where it can contribute up to 20% by mass, particularly as lipid bodies or spherosomes. The lipid content of brown rice varies and is available in the varieties from 0.5% to 3.5% (Dendy 2005). Singh et al. (1998) reported the lipid contents of six varieties of brown rice ranged from 2.1% to 3.2% and 0.61% to 0.95%. Charoenthaikij et al. (2012) reported the lipid content ranged from 2.65% to 3.24% in brown rice flour, whereas Mir et al. (2016b) reported in the range 2.38–2.84%. In brown rice, 51% of crude oil is found in germ, 32% in the bran layer and only 17% in the endosperm. In the endosperm part of the rice grain, lipids are unevenly distributed having highest percentage in the outer layers and decreasing progressively towards the centre of the grain (Dendy 2005).

Brown rice contains notable amount of vitamins which are essential for the human health. Vitamins are mostly present in higher levels in brown rice than in milled rice. Rice is a good source of vitamin B_1, vitamin B_2, vitamin B_3 and vitamin B_6, and a wide variation is observed among the different varieties. Deepa et al. (2008) investigated the level of vitamins in brown rice of three varieties, namely, Njavara, Jyothi and IR 64 Njavara. They reported that the vitamin B_1 content varied from 0.04 to 0.05 mg/100g, vitamin B_2 from 0.053 to 0.071 mg/100g, vitamin B_3 from 4.68 to 7.32 mg/100g and folic acid from 0.04 to 0.05 mg/100g. However, Njavara rice contained higher amounts of vitamin B_1 (27–32%), vitamin B_2 (4–25%) and vitamin B_3 (2–36%) as compared to the other two rice varieties.

Mir et al. (2016b) observed the wide variation in vitamin content of brown rice varieties. The vitamin B_1 content ranged from 0.09 to 0.16 mg/100 g, B_2 from 0.10 to 0.27 mg/100 g, B_3 from 4.02 to 5.41 mg/100 g and B_6 from 0.08 to 0.19 mg/100 g. Kyritsi et al. (2011) reported that brown rice contains 0.403, 0.065, 5.433 and 0.563 mg/100 g of vitamin B_1, B_2, B_3 and B_6, respectively. The difference in vitamin content of rice may be due to the variation in genetic background among the rice varieties.

Minerals Composition of Brown Rice

Minerals are essential nutrients for human being, and they play a vital role in the effective functioning of the body activities. The mineral composition of the rice grain depends considerably on the availability of soil nutrients during crop growth and variety (Heinemann et al. 2005; Wang et al. 2011). It is noteworthy that the analysed brown rice contains, on an average, significantly greater concentrations of copper, potassium, magnesium, manganese, sodium, phosphorus and zinc in comparison to polished rice samples. Wide genetic diversity in the mineral elements of brown rice accessions was observed (Table 2.3). These mineral elements were affected by genotype as well as environment (Huang et al. 2016). Antoine et al. (2012) analysed 25 rice brands for 36 essential and non-essential elements using four different instrumental techniques. The mean values of minerals as reported by them are as follows: for calcium (127 mg/kg; 104 mg/kg), copper (1.65 mg/kg; 2.96 mg/kg), iron (22.3 mg/kg; 20.1 mg/kg), magnesium (371 mg/kg; 1205 mg/kg), manganese (10.5 mg/kg; 26.5 mg/kg), molybdenum (0.790 mg/ kg; 0.770 mg/kg), phosphorus (1203 mg/kg; 3361 mg/kg), potassium (913 mg/kg; 2157 mg/kg), selenium (0.108 mg/kg; 0.131 mg/kg), sodium (6.00 mg/kg; 15.1 mg/kg), sulphur (1131 mg/kg; 1291 mg/kg) and zinc (15.6 mg/kg; 20.2 mg/kg) for polished and brown rice, respectively.

Wavelength dispersive X-ray fluorescence spectrometer analysis showed that brown rice is rich source of minerals including calcium, iron, potassium, magnesium, manganese, phosphorus, sulphur and zinc. Significant difference was observed in the mineral composition of brown rice from different varieties. The most abundant mineral was potassium (93.15–110.35 mg/100 g) followed by phosphorus (76.30–89.80 mg/100 g), sulphur (20.75–26.90 mg/100 g) and magnesium (17.15–20.90 mg/100 g), whereas the lowest was zinc (2.10–2.45 mg/100 g). The results showed that amounts of minerals were significantly different among the varieties which depend on the genetic make-up of the variety. The brown rice rich in minerals can be considered as cost-effective and promising approach to alleviate malnutrition and other health-related problems (Mir et al. 2016b).

Starch Properties of Rice

Even though the brown rice contains significant amounts of functional and nutritional components which are concentrated in the bran layer, it also contains the starch in high proportion which affects its physicochemical properties. Physicochemical properties of rice starch depend largely on the variety of the rice and their genetic background (Bao et al. 2004; Wani et al. 2012; Mir and Bosco 2014). Literature has reported the starch characterization from rice varieties grown in different areas of India (Sodhi and Singh 2003; Singh et al. 2007), Thailand (Noosuk et al. 2005), Africa (Lawal et al. 2011) and China (Wang et al. 2010). Starch properties also

Table 2.3 Mineral composition of brown rice from various varieties

Variety	Calcium (mg/100 g)	Iron (mg/100 g)	Potassium (mg/100 g)	Magnesium (mg/100 g)	Manganese (mg/100 g)	Zinc (mg/100 g)	Sodium (mg/100 g)	Copper (mg/100 g)	Reference
Azucena	10.6	0.92	312.10	156.66	–	3.15	4.45	0.28	Huang et al. (2016)
Dular	9.09	0.92	295.30	144.10	–	2.50	1.14	0.13	
Pokkali	12.00	0.40	346.30	167.60	–	2.09	1.70	0.18	
Swarna	8.18	0.49	281.10	162.80	–	1.90	0.99	0.24	
K-332	5.05	2.75	104.10	19.70	2.30	2.10	–	–	Mir et al. (2016b)
Kohsar	5.35	2.95	110.35	19.45	3.20	2.45	–	–	
Pusa-3	6.65	3.05	104.25	20.40	2.70	2.15	–	–	
SKAU-382	5.70	3.25	93.15	20.90	1.80	2.30	–	–	
SR-1	6.40	2.76	100.95	17.15	2.55	2.40	–	–	
ZN7	2.25	2.28	–	171.7	3.05	2.73	–	–	Wang et al. (2011)
ZN60	1.67	2.43	–	181.4	4.24	2.91	–	–	
ZN34	1.42	1.55	–	150.8	2.81	2.40	–	–	
Pusa basmati	2.65	2.72	–	15.82	–	0.69	–	–	Das et al. (2008)
Njavara	11.6	1.93	304	216	–	–	30.9	–	Deepa et al. (2008)
Jyothi	9.70	3.95	268	150	–	–	22.6	–	
IR 64	9.20	2.73	248	163	–	–	27.8	–	
Brazilian brown rice	6.85	0.57	181.71	16.88	0.36	1.98	0.54	0.16	Heinemann et al. (2005)
Australian brown rice	3–11	0.5–5.7	210–300	100–130	2.5–6	1.3–2.1	0–19	0.14–1.3	Marr et al. (1995)

depend upon the soil and climatic conditions during rice grain development (Wani et al. 2012). The rice starch contains amylose and amylopectin components with α 1–4 and α 1–4 and α 1–6 linkages, respectively, whose percentage varies among the varieties (Vandeputte et al. 2003; Fitzgerald et al. 2009).

Starch granules of rice are the smallest known to exist in cereal grains, with the size observed in the range of 2–7 μm (Vandeputte and Delcour 2004). Starch granules of rice varieties are mainly polyhedral in shape. These granules may also be oval, angular, irregular or smooth in shape. The size of granules varies between non-waxy and waxy rice starches, and it also differs from variety to variety. Starch properties depend mostly on the factors including amylose/amylopectin ratio, granule shape and size and the presence of other constituents present in the starch (Wani et al. 2012). The variability in amylose and amylopectin molecules is due to the complexity of starch biosynthesis which reflected the diversity of granule morphology. The variation, particularly in granular shape and size, is associated with various techno-functional properties in different food products (Lu et al. 2009; Lawal et al. 2011).

Amylose and amylopectin are the major constituents affecting the physicochemical properties of rice starch, and their role in the properties of rice starch has been widely investigated (Li et al. 2008; Wang et al. 2010). Large pools of rice varieties differing in their amylose content are found in different geographical regions of the world. Amylose is controlling almost all the properties of rice starch due to its influence on thermal properties, pasting properties, syneresis, solubility, swelling and other techno-functional properties (Lu et al. 2009). Waxy rice starches have high solubility and swelling properties and larger crystallinity degree than non-waxy starches. However, non-waxy rice starches have observed a higher gelatinization temperature as compared to waxy starches (Zavareze et al. 2010). From the industrial standpoint, it is a practical approach to simplify the rice variety categorization in order to control the rice quality. Hence, it is reliable to classify the rice starches categorized on the basis of their similarity after selecting on the basis of food applicable properties for appropriate control of rice starch quality.

Antioxidant Properties of Brown Rice

Many epidemiological studies have shown that consumption of brown rice is highly correlated to reduce incidences of chronic diseases (Anderson 2003). Epidemiological investigations reported that the low incidence of certain chronic diseases in brown rice-consuming areas of the world might be associated with the antioxidant properties of rice. Rice has been recognized as an excellent source of unique complex naturally occurring antioxidant compounds (Chotimarkorn et al. 2008). The scavenging effect of bioactive components, such as phenolic compounds, flavonoids, anthocyanins, proanthocyanidins, tocopherols and oryzanol, which are rich in the rice grains, may be a mechanism whereby whole grains have their protective effects (Slavin 2003). Most of the phytochemicals in the brown rice grain are present in the bran fraction consisting of bran layers and the germ. Rice, as one of the main foods in the diet of most populations, has an important role in the concentration of antioxidants ingested daily. In addition to phytochemicals, brown rice also contains dietary fibre that adds bulk to the gastrointestinal tract in humans which indirectly promotes cardiovascular health. These functional components exist mainly in the bran and germ layers in the brown rice and are lost when it is polished to white rice (Champagne et al. 2004).

The health benefit properties of brown rice are attributed mainly due to the phenolic compounds, which has received the increasing attention because of its potent antioxidant properties (Liu 2004; Kim et al. 2012). Brown rice was reported to have higher phenolic content and stronger antioxidant capacity than polished rice (Shen et al. 2009; Zhang et al. 2010). Major anthocyanin components of brown rice were identified as cyanidin-3-gluoside and peonidin-3-glucoside, and these compounds possessed prominent antioxidant activities (Hu et al. 2003; Zhu et al. 2010). Therefore, brown rice is more preferred by the consumers because of its high functional properties.

The markable difference in antioxidant properties of brown rice was observed in different varieties (Table 2.4). Brown rice was proved to have potent antioxidant activity with a significant variation in total phenolic (0.81–1.64 mg gallic acid equivalent/g), flavonoid content (50.67–79.41 µg catechin eq/g), 2,2-diphenyl-1-picrylhydrazyl radical scavenging activity (46.18–70.51%) and total reducing power (7.34–17.14 µmol ascorbic acid equivalent/g) among varieties (Mir et al. 2016b).

Table 2.4 Antioxidant properties of brown rice from various varieties

| Variety | Phenolics (mg gallic acid equivalent/100 g) | | | Flavonoids (mg catechin equivalent/100 g) | | | Reference |
	Free	Bound	Total	Free	Bound	Total	
Longjing 25	59.85	60.28	120.13	65.51	44.76	110.27	Gong et al. (2017)
Sonjing 16	63.77	51.52	115.29	80.70	31.33	112.03	
Tianyouhuazhan	42.09	30.36	72.45	52.03	23.87	75.90	
Wuyou 308	67.86	46.43	114.29	76.82	33.56	110.38	
Fenghuazhan	58.09	44.70	102.78	62.15	41.89	104.04	
Japonica rice	65.60	34.80	100.40	42.60	34.30	76.90	Liu et al. (2015)
Indica rice	62.00	37.30	99.30	56.30	55.70	112.10	
Tianyou 998	100.30	73.70	174.00	61.10	63.90	124.90	Ti et al. (2014)
DV 123	67.00	41.00	108.00	28.00	13.00	41.00	Min et al. (2014)
HB1	220.00	60.00	280.00	99.00	17.00	116.00	
IAC 600	490.00	75.00	565.00	180.00	18.00	198.00	
Kechengnuo4	44.00	61.00	105.00	20.00	24.00	44.00	Min et al. (2012)
Cocodrie	62.00	63.00	125.00	23.00	26.00	49.00	
Bengal	58.00	46.00	104.00	22.00	24.00	46.00	
Heugjinjubyeo	1640.00	176.00	1820.00	317.00	22.00	339.00	Kong and Lee (2010)
Heugkwangbyeo	1180	153	1330	197.00	16.00	213.00	

The brown rice had phenolic contents ranging from 108.1 to 1244.9 mg gallic acid equivalent/100 g (Choi et al. 2007). Goffman and Bergman (2004) reported that the phenolic contents in the red, purple and white rice ranged from 34 to 424, 69 to 535 and 25 to 246 mg gallic acid equivalent/100 g, respectively.

In brown rice, phenolic acids are mainly present in three forms, soluble free, soluble conjugated and insoluble bound (Butsat et al. 2009; Huang and Ng 2012). Most of the phenolic acids in the grains are in the bound form (Adom and Liu 2002; Irakli et al. 2012). The most abundant bound phenolic acid in brown rice was ferulic acid, which accounts for almost 50–65% of total bound phenolic acids (Qiu et al. 2010). Ferulic acid has a wide range of therapeutic effects against many chronic conditions such as inflammation, cancer, apoptosis, cardiovascular, diabetes and neurodegenerative diseases, so the consumption of brown rice is helpful to reduce the incidence of chronic diseases in human beings (Srinivasan et al. 2007).

The antioxidant properties of eight whole rice grain varieties varying in colour, total phenolics, flavonoids and 2,2-diphenyl-1-picrylhydrazyl radical scavenging activity of solvent-extractable free and bound fractions were studied by Min et al. (2012). The free and soluble phenolic fraction content varied from 0.44 to 6.97 and 0.46 to 2.28 mg gallic acid equivalent/g, respectively. The total free flavonoid content showed the variation from 0.16 to 2.28 and 0.24 to 0.43 mg catechin equivalents/g and radical scavenging content from 1.19 to 41.95 and 0.99 to 10.55 μmol Trolox equivalents/g, for free and soluble fraction, respectively. Red and purple rice grains showed the higher phenolic content, flavonoid content and antioxidant capacities than light-coloured rice varieties.

Ti et al. (2014) quantified free and bound phytochemicals in the endosperm and bran/embryo of different *indica* rice varieties. Phytochemicals mainly existed as free form in the bran/embryo and as both free and bound forms in the endosperm. The average values of total phenolic content and flavonoid content in the bran/embryo were 3.1 and 10.4 times higher than those in the endosperm, respectively. In whole brown rice, the bran contributed 59.2% and 53.7% of total phenolics and flavonoids, respectively. Seven individual phenolics including gallic, protocatechuic, chlorogenic, caffeic, syringic, coumaric and ferulic acids were detected with most coumaric and ferulic acids in the bran.

Niu et al. (2013) investigated the antioxidant properties of 22 red rice samples grown in Zhejiang. The total phenolic contents ranged from 433 to 2213 mg ferulic acid equivalents/g, whereas the cyanidin-3-O-glucoside concentration was 11.6–16.5 mg/g in the red rice samples. The distribution of phenolic acids and anthocyanins in endosperm, embryo and bran of white, red and black rice grains was investigated by Shao et al. (2014). Total phenolic content was highest in the bran, averaging 7.35 mg gallic acid equivalent/g and contributing 60%, 86% and 84% of phenolics in white, red and black rice, respectively. The average total phenolic content of the embryo and endosperm was 2.79 and 0.11 mg gallic acid equivalent/g accounting for 17% and 23%, 4% and 10% and 7% and 9% in white, red and black rice, respectively. The antioxidant capacity determined using 2,2-diphenyl-1-picrylhydrazyl radical scavenging activity and oxygen radical scavenging capacity shows a similar trend to total phenolic content. Free/conjugated phenolic acids in white, red and black rice bran accounted for 41%, 65% and 85% of total acids. Bound phenolic acids in bran of brown rice accounted for 90% of total acids in rice grain.

Conclusion

The wide differences in rice varieties were observed in physical and cooking properties, composition, minerals, starch and antioxidant properties. The diversity in properties of brown rice may be due to the difference in genetic make-up and climatic and soil conditions of grain during development. The quality characteristics desired vary considerably, being ultimately related to final consumer acceptance of each rice product. The brown rice from different varieties provides opportunities to promote the nutritional benefit of food products from rice varieties and could have implications for commercial practice in the food industries. Further research is required on the effect of agronomic and climatic conditions on different rice varieties.

References

Abrol YP, Gadgil S (1999) Rice in a variable climate. APC Publishing House, New Delhi, pp 243–252

Adom KK, Liu RH (2002) Antioxidant activity of grains. J Agric Food Chem 50(21):6182–6187

Anderson JW (2003) Whole grains protect against atherosclerotic cardiovascular disease. Proc Nutr Soc 62(1):135–142

Antoine JMR, Fung LAH, Grant CN, Dennis HT, Lalor GC (2012) Dietary intake of minerals and trace elements in rice on the Jamaican market. J Food Comp Anal 26(1):111–121

Azhakanandam K, Power JB, Lowe KC, Cocking EC, Tongdang T, Jumel K, Davey MR (2000) Qualitative assessment of aromatic Indica rice (*Oryza sativa* L.): proteins, lipids and starch in grain from somatic embryo-and seed-derived plants. J Plant Physiol 156(5):783–789

Bao J, Sun M, Zhu L, Corke H (2004) Analysis of quantitative trait loci for some starch properties of rice (*Oryza sativa* L.): thermal properties, gel texture and swelling volume. J Cereal Sci 39(3):379–385

Butsat S, Weerapreeyakul N, Siriamornpun S (2009) Changes in phenolic acids and antioxidant activity in Thai rice husk at five growth stages during grain development. J Agric Food Chem 57(11):4566–4571

Cao W, Nishiyama Y, Koide S (2004) Physicochemical, mechanical and thermal properties of brown rice grain with various moisture contents. Int J Food Sci Tech 39(9):899–906

Champagne ET, Wood DF, Juliano BO, Bechtel DB (2004) The rice grain and its gross composition. In: Champagne ET (ed) *Rice: chemistry and technology*, 3rd edn. Inc. St. Paul, American Association of Cereal Chemists, pp 77–107

Charoenthaikij P, Jangchud K, Jangchud A, Prinyawiwatkul W, No HK (2012) Composite wheat–germinated brown rice flours: selected physicochemical properties and bread application. Int J Food Sci Tech 47(1):75–82

Chen H, Siebenmorgen TJ, Griffin K (1998) Quality characteristics of long-grain rice milled in two commercial systems. Cereal Chem 75(4):560–565

Choi Y, Jeong HS, Lee J (2007) Antioxidant activity of methanolic extracts from some grains consumed in Korea. Food Chem 103(1):130–138

Chotimarkorn C, Benjakul S, Silalai N (2008) Antioxidant components and properties of five long-grained rice bran extracts from commercial available cultivars in Thailand. Food Chem 111(3):636–641

Correa PC, Silva FS, Jaren C, Junior PCA, Arana I (2007) Physical and mechanical properties in rice processing. J Food Eng 79(1):137–142

Das M, Banerjee R, Bal S (2008) Evaluation of physicochemical properties of enzyme treated brown rice (part B). LWT-Food Sci Tech 41(10):2092–2096

Deepa G, Singh V, Naidu KA (2008) Nutrient composition and physicochemical properties of Indian medicinal rice-Njavara. Food Chem 106(1):165–171

Dendy DAV (2005) Rice. In: Dendy DAV, Dobraszczyk BJ (eds) Cereal and cereal products-chemistry and technology. Springer, New Delhi, pp 266–314

Fitzgerald MA, McCouch SR, Hall RD (2009) Not just a grain of rice: the quest for quality. Trends Plant Sci 14(3):133–139

Goffman FD, Bergman CJ (2004) Rice kernel phenolic content and its relationship with antiradical efficiency. J Sci Food Agric 84(10):1235–1240

Gong ES, Luo SJ, Li T, Liu CM, Zhang GW, Chen J et al (2017) Phytochemical profiles and antioxidant activity of brown rice varieties. Food Chem 227:432–443

Gujral HS, Kumar V (2003) Effect of accelerated aging on the physicochemical and textural properties of brown and milled rice. J Food Eng 59(2):117–121

Gujral HS, Sharma P, Kumar A, Singh B (2012) Total phenolic content and antioxidant activity of extruded brown rice. Int J Food Prop 15(2):301–311

Heinemann RJB, Fagundes PL, Pinto EA, Penteado MVC, Lanfer-Marquez UM (2005) Comparative study of nutrient composition of commercial brown, parboiled and milled rice from Brazil. J Food Comp Anal 18(4):287–296

Hu C, Zawistowski J, Ling W, Kitts DD (2003) Black (*Oryza sativa L. indica*) pigmented fraction suppresses both reactive oxygen species and nitric oxide in chemical and biological model system. J AgriFood Chem 51(18):5271–5277

Huang SH, Ng LT (2012) Quantification of polyphenolic content and bioactive constituents of some commercial rice varieties in Taiwan. J Food Comp Anal 26(1):122–127

Huang Y, Tong C, Xu F, Chen Y, Zhang C, Bao J (2016) Variation in mineral elements in grains of 20 brown rice accessions in two environments. Food Chem 192:873–878

Irakli MN, Samanidou VF, Biliaderis CG, Papadoyannis IN (2012) Simultaneous determination of phenolic acids and flavonoids in rice using solid-phase extraction and RP-HPLC with photodiode array detection. J Sep Sci 35(13):1603–1611

Itagi HN, Singh V (2015) Status in physical properties of coloured rice varieties before and after inducing retro-gradation. J food sci tech 52(12):7747–7758

Itani T, Tamaki M, Arai E, Horino T (2002) Distribution of amylose, nitrogen, and minerals in rice kernels with various characters. J Agri Food Chem 50(19):5326–5332

Izawa T (2008) The process of rice domestication: a model based on recent data. Rice 1(2):127–134

Joshi ND, Mohapatra D, Joshi DC (2014) Varietal selection of some indica rice for production of puffed rice. Food Bioprocess Tech 7(1):299–305

Juliano BO (1985) *Rice: Chemistry and Technology*, 2nd edn. American Association of Cereal Chemists, St. Paul

Kent NL, Evers AD (1994) Technology of Cereals: an introduction for students of food science and agriculture, 4th edn. Woodhead publishing Limited, Abington Hall, Abington, Cambridge, England

Kesarwani A, Chiang PY, Chen SS, Su PC (2013) Antioxidant activity and total phenolic content of organically and conventionally grown rice cultivars under varying seasons. J Food Biochem 37(6):661–668

Kim HY, Hwang IG, Kim TM, Woo KS, Park DS, Kim JH et al (2012) Chemical and functional components in different parts of rough rice (*Oryza sativa* L.) before and after germination. Food Chem 134(1):288–293

Kong S, Lee J (2010) Antioxidants in milling fractions of black rice cultivars. Food Chem 120(1): 278–281

Kyritsi A, Tzia C, Karathanos VT (2011) Vitamin fortified rice grain using spraying and soaking methods. LWT Food Sci Technol 44(1):312–320

Laborte AG, De Bie KC, Smaling E, Moya PF, Boling AA, Van Ittersum MK (2012) Rice yields and yield gaps in Southeast Asia: past trends and future outlook. Eur Jf Agron 36(1):9–20

Lamberts L, De Bie E, Vandeputte GE, Veraverbeke WS, Derycke V, De Man W, Delcour JA (2007) Effect of milling on colour and nutritional properties of rice. Food Chem 100(4):1496–1503

Lawal OS, Lapasin R, Bellich B, Olayiwola TO, Cesàro A, Yoshimura M, Nishinari K (2011) Rheology and functional properties of starches isolated from five improved rice varieties from West Africa. Food Hydrocoll 25(7):1785–1792

Lee KH, Kim HJ, Woo KS, Jo C, Kim JK, Kim SH et al (2016) Evaluation of cold plasma treatments for improved microbial and physicochemical qualities of brown rice. LWT-Food Sci Technol 73:442–447

Li Y, Shoemaker CF, Ma J, Moon KJ, Zhong F (2008) Structure-viscosity relationships for starches from different rice varieties during heating. Food Chem 106(3):1105–1112

Lin JH, Singh H, Chang YT, Chang YO (2011) Factor analysis of the functional properties of rice flours from mutant genotypes. Food Chem 126(3):1108–1114

Liu RH (2004) Whole grain phytochemicals and health. J Cereal Sci 46(3):207–219

Liu K, Cao X, Bai Q, Wen H, Gu Z (2009) Relationships between physical properties of brown rice and degree of milling and loss of selenium. J Food Eng 94(1):69–74

Liu L, Guo J, Zhang R, Wei Z, Deng Y, Guo J, Zhang M (2015) Effect of degree of milling on phenolic profiles and cellular antioxidant activity of whole brown rice. Food Chem 185:318–325

Lu ZH, Sasaki T, Li YY, Yoshihashi T, Li LT, Kohyama K (2009) Effect of amylose content and rice type on dynamic viscoelasticity of a composite rice starch gel. Food Hydrocoll 23(7): 1712–1719

Luh BS (1980) Rice: production and utilization. AVI Publishing Co, Inc, Westport

Marr KM, Batten GD, Blakeney AB (1995) Relationships between minerals in Australian brown rice. J Sci Food Agric 68(3):285–291

Min B, Gu L, McClung AM, Bergman CJ, Chen MH (2012) Free and bound total phenolic concentrations, antioxidant capacities, and profiles of proanthocyanidins and anthocyanins in whole grain rice (*Oryza sativa* L.) of different bran colours. Food Chem 133(3):715–722

Min B, McClung A, Chen MH (2014) Effects of hydrothermal processes on antioxidants in brown, purple and red bran whole grain rice (*Oryza sativa* L.) Food Chem 159:106–115

Mir SA, Bosco SJD (2014) Cultivar difference in physicochemical properties of starches and flours from temperate rice of Indian Himalayas. Food Chem 157:448–456

Mir SA, Bosco SJD, Sunooj KV (2013) Evaluation of physical properties of rice cultivars grown in the temperate region of India. Int Food Res J 20(4):152–1527

Mir SA, Bosco SJD, Shah MA, Mir MM (2016a) Effect of puffing on physical and antioxidant properties of brown rice. Food Chem 191:139–146

Mir SA, Bosco SJD, Shah MA, Mir MM, Sunooj KV (2016b) Variety difference in quality characteristics, antioxidant properties and mineral composition of brown rice. J Food Measur Charact 10(1):177–184

Mohan BH, Malleshi NG, Koseki T (2010) Physico-chemical characteristics and non-starch polysaccharide contents of *Indica* and *Japonica* brown rice and their malts. LWT-Food Sci Technol 43(5):784–791

Mohapatra D, Bal S (2006) Cooking quality and instrumental textural attributes of cooked rice for different milling fractions. J Food Eng 73(3):253–259

Mohapatra D, Bal S (2012) Physical properties of Indica rice in relation to some novel mechanical properties indicating grain characteristics. Food Bioprocess Technol 5(6):2111–2119

Moko EM, Purnomo H, Kusnadi J, Ijong FG (2014) Phytochemical content and antioxidant properties of colored and non colored varieties of rice bran from Minahasa, North Sulawesi, Indonesia. Int Food Res J 21(3):1053–1059

Nanda JS, Agrawal PK (2006) Nutrient management in rice, In: rice. Kalyani Publishers, New Delhi, pp 221–243

Niu Y, Gao B, Slavin M, Zhang X, Yang F, Bao J, Yu LL (2013) Phytochemical compositions, and antioxidant and anti-inflammatory properties of twenty-two red rice samples grown in Zhejiang. LWT-Food Sci Technol 54(2):521–527

Noosuk P, Hill SE, Farhat IA, Mitchell JR, Pradipasena P (2005) Relationship between viscoelastic properties and starch structure in rice from Thailand. Starch-Starke 57(12):587–598

Ohtsubo K, Suzuki K, Yasui Y, Kasumi T (2005) Bio-functional components in the processed pregerminated brown rice by a twin-screw extruder. J Food Comp Anal 18(4):303–316

Qiu Y, Liu Q, Beta T (2010) Antioxidant properties of commercial wild rice and analysis of soluble and insoluble phenolic acids. Food Chem 121(1):140–147

Razavi SMA, Farahmandfar R (2008) Effect of hulling and milling on the physical properties of rice grains. Int Agrophys 22(4):353–359

Sahay KM, Singh KK (1994) Unit operations of agricultural processing, 1st edn. Vikas Publishing House Pvt. Ltd., New Delhi

Shao Y, Xu F, Sun X, Bao J, Beta T (2014) Identification and quantification of phenolic acids and anthocyanins as antioxidants in bran, embryo and endosperm of white, red and black rice kernels (*Oryza sativa* L.) J Cereal Sci 59(2):211–218

Shen Y, Jin L, Xiao P, Lu Y, Bao J (2009) Total phenolic, flavonoids, antioxidant capacity in rice grain and their relations to grain color, size and weight. J Cereal Sci 49(1):106–111

Shinde YH, Vijayadwhaja A, Pandit AB, Joshi JB (2014) Kinetics of cooking of rice: a review. J Food Eng 123:113–129

Shittu TA, Olaniyi MB, Oyekanmi AA, Okeleye KA (2012) Physical and water absorption characteristics of some improved rice varieties. Food Bioprocess Tech 5(1):298–309

Singh S, Dhaliwal YS, Nagi HPS, Kalia M (1998) Quality characteristics of six rice varieties of Himachal Pradesh. J Food Sci Technol 35(1):74–78

Singh N, Kaur L, Sodhi NS, Sekhon KS (2005) Physicochemical, cooking and textural properties of milled rice from different Indian rice cultivars. Food Chem 89(2):253–259

Singh N, Nakaura Y, Inouchi N, Nishinari K (2007) Fine structure, thermal and viscoelastic properties of starches separated from *Indica* rice cultivars. Starch-Starke 59(1):10–20

Singh N, Pal N, Mahajan G, Singh S, Shevkani K (2011) Rice grain and starch properties: effects of nitrogen fertilizer application. Carbohyd Polym 86(1):219–225

Sirisoontaralak P, Nakornpanom NN, Koakietdumrongkul K, Panumaswiwath C (2015) Development of quick cooking germinated brown rice with convenient preparation and containing health benefits. LWT-Food Sci Technol 61(1):138–144

Slavin J (2003) Why whole grains are protective: biological mechanisms. Proc Nutr Soc 62(1): 129–134

Sodhi NS, Singh N (2003) Morphological, thermal and rheological properties of starches separated from rice cultivars grown in India. Food Chem 80(1):99–108

Sompong R, Siebenhandl-Ehn S, Linsberger-Martin G, Berghofer E (2011) Physicochemical and antioxidative properties of red and black rice varieties from Thailand, China and Sri Lanka. Food Chem 124(1):132–140

Son JS, Do VB, Kim KO, Cho MS, Suwonsichon T, Valentin D (2013) Consumers' attitude towards rice cooking processes in Korea, Japan, Thailand and France. Food Qual Prefer 29(1):65–75

Soponronnarit S, Chiawwet M, Prachayawarakorn S, Tungtrakul P, Taechapairoj C (2008) Comparative study of physicochemical properties of accelerated and naturally aged rice. J Food Eng 85(2):268–276

Srinivasan M, Sudheer AR, Menon VP (2007) Ferulic acid: therapeutic potential through its antioxidant property. J Clin Biochem Nutr 40(2):92–100

Suwannaporn P, Linnemann A (2008) Rice-eating quality among consumers in different rice grain preference countries. J Sens Stud 23(1):1–13

Thakur AK, Gupta AK (2006) Water absorption characteristics of paddy, brown rice and husk during soaking. J Food Eng 75(2):252–257

Thompson SA, Ross IJ (1983) Compressibility and frictional coefficient of wheat. Trans Am Soc Agric Eng 26:1171–1176

Ti H, Li Q, Zhang R, Zhang M, Deng Y, Wei Z, Chi J, Zhang Y (2014) Free and bound phenolic profiles and antioxidant activity of milled fractions of different *indica* rice varieties cultivated in southern China. Food Chem 159:166–174

Vandeputte GE, Delcour JA (2004) From sucrose to starch granule to starch physical behaviour: a focus on rice starch. Carbohyd Polym 58(3):245–266

Vandeputte GE, Derycke V, Geeroms J, Delcour JA (2003) Rice starches. II. Structural aspects provide insight into swelling and pasting properties. J Cereal Sci 38(1):53–59

Varnamkhasti MG, Mobli H, Jafari A, Keyhani AR, Soltanabadi MH, Rafiee S, Kheiralipour K (2008) Some physical properties of rough rice (*Oryza sativa* L.) grain. J Cereal Sci 47(3):496–501

Wang L, Xie B, Shi J, Xue S, Deng Q, Wei Y, Tian B (2010) Physicochemical properties and structure of starches from Chinese rice cultivars. Food Hydrocoll 24(2):208–216

Wang KM, Wu JG, Li G, Zhang DP, Yang ZW, Shi CH (2011) Distribution of phytic acid and mineral elements in three *indica* rice (*Oryza sativa* L.) cultivars. J Cereal Sci 54(1):116–121

Wani AA, Singh P, Shah MA, Schweiggert-Weisz U, Gul K, Wani IA (2012) Rice starch diversity: effects on structural, morphological, thermal, and physicochemical properties—a review. Compr Rev Food Sci Food Saf 11(5):417–436

Yadav BK, Jindal VK (2007) Dimensional changes in milled rice (*Oryza sativa* L.) kernel during cooking in relation to its physicochemical properties by image analysis. J Food Eng 81(4):710–720

Zavareze EDR, Storck CR, de Castro LAS, Schirmer MA, Dias ARG (2010) Effect of heat-moisture treatment on rice starch of varying amylose content. Food Chem 121(2):358–365

Zhang MW, Zhang RF, Zhang FX, Liu RH (2010) Phenolic profiles and antioxidant activity of black rice bran of different commercially available varieties. J Agric Food Chem 58(13):7580–7587

Zhou Z, Robards K, Helliwell S, Blanchard C (2002) Composition and functional properties of rice. Int J Food Sci Tech 37(8):849–868

Zhu F, Cai YZ, Bao J, Corke H (2010) Effect of γ-irradiation on phenolic compounds in rice grain. Food Chem 120(1):74–77

Chapter 3

Engineering Properties of Brown Rice from Selected Indian Varieties

V.R. Sinija, S. Sulochana, and M.S. Shwetha

Introduction

Rice is the seed of the grass species *Oryza sativa* (Asian rice) or *Oryza glaberrima* (African rice). As a cereal grain, it is the most widely consumed staple food for a large part of the world's human population, especially in Asia. It is the agricultural commodity with the third highest worldwide production, after sugarcane and maize, according to 2015 FAOSTAT data. Rice is the most important grain with regard to human nutrition and caloric intake, providing more than one-fifth of the calories consumed worldwide by humans.

Rice, a monocot, is normally grown as an annual plant, although in tropical areas, it can survive as a perennial and can produce a ratoon crop for up to 30 years. The rice plant can grow to 1–1.8 m (3.3–5.9 ft) tall, occasionally more, depending on the variety and soil fertility. It has long, slender leaves 50–100 cm (20–39 in) long and 2–2.5 cm (0.79–0.98 in) broad. The small wind-pollinated flowers are produced in a branched arching to pendulous inflorescence 30–50 cm (12–20 in) long. The edible seed is a grain (caryopsis) 5–12 mm (0.20–0.47 in) long and 2–3 mm (0.079–0.118 in) thick.

History of Rice in India

India is an important centre for rice cultivation. Historians believe that while the indica variety of rice was first domesticated in the area covering the foothills of the Eastern Himalayas (i.e. north-eastern India), stretching through Burma, Thailand,

V.R. Sinija (✉) • S. Sulochana • M.S. Shwetha
Indian Institute of Food Processing Technology, Thanjavur, Tami Nadu, India
e-mail: sinija@iifpt.edu.in; sulochana@iifpt.edu.in; smshwethashankar@gmail.com

© Springer International Publishing AG 2017
A. Manickavasagan et al. (eds.), *Brown Rice*,
DOI 10.1007/978-3-319-59011-0_3

Laos, Vietnam and Southern China, the japonica variety was domesticated from wild rice in southern China which was introduced to India. Perennial wild rice still grows in Assam and Nepal. It seems to have appeared around 1400 BC in Southern India after its domestication in the northern plains. It then spread to all the fertilized alluvial plains watered by rivers. Some says that the word rice is derived from the Tamil word 'arisi'.

Rice is first mentioned in the Yajur Veda (c. 1500-800 BC) and then is frequently referred to in Sanskrit texts. In India there is a saying that grains of rice should be like two brothers, close but not stuck together. Rice is often directly associated with prosperity and fertility; hence there is the custom of throwing rice at newlyweds.

Brown Rice

Brown rice is whole grain rice, with the inedible outer hull removed, while the white rice is the same grain with the hull, bran layer and cereal germ removed. It is evident that brown rice is better than white, but still majority of consumers typically choose white rice over brown rice because of the difference of appearance. According to a study conducted by the American Journal of Clinical Nutrition, brown rice is the top choice in terms of both nutritional and other inherent healthy benefits.

It becomes white rice when it is milled and polished to remove the outer layers, called the bran and germ. This also removes many naturally occurring nutrients. Brown rice is minimally processed, which makes it nutritionally superior to white rice. The fibre in whole grains can support heart health, digestion and weight management.

Brown rice and white rice have similar amounts of calories and carbohydrates. White rice, unlike brown rice has the bran and germ removed; and has different nutritional content. Brown rice is a whole grain and a good source of magnesium, phosphorus, selenium, thiamine, niacin, vitamin B6, manganese and fibre.

Brown rice has a shelf life of approximately 6 months, but hermetic storage, refrigeration or freezing can significantly extend its lifetime. Freezing, even periodically, can also help control infestations of Indian meal moths. Besides the nutritional benefits of consuming brown rice, there are two economic importance:

1. Forgoing polishing and whitening reduces the power demand of milling by as much as 65%.
2. With the bran and the nutrient-rich embryo intact, and with fewer broken grains, whole grain milling recovery is as much as 10% higher than white rice.

Comparison of calorific and nutrient content of brown and milled rice (http://www.drlam. com/opinion/brown_rice_vs_ white_rice.cfm)

Constituents	Brown rice (1 cup)	White rice (1 cup)
Calories	232.0	223.0
Protein	4.88 g	4.10 g
Carbohydrates	49.7 g	49.6 g
Fat	1.17 g	0.205 g
Dietary fibre	3.32 g	0.74 g
Thiamine (B1)	0.176 g	0.223 g
Riboflavin (B2)	0.039 mg	0.021 mg
Niacin (B3)	2.730 mg	2.050 mg
Vitamin B6	0.294 mg	0.103 mg
Folacin	10 mcg	4.1 mcg
Vitamin E	1.4 mg	0.462 mg
Magnesium	72.2 mg	22.6 mg
Phosphorus	142.0 mg	57.4 mg
Potassium	137.0 mg	57.4 mg
Selenium	26.0 mg	19.0 mg
Zinc	1.05 mg	0.841 mg

Engineering Properties of Brown Rice

Study of engineering properties of food and agricultural products is significant in applications like design of equipment, e.g. seed planting machines, line sorting machines and pumping requirements; development of sensors, e.g. colour and weight sensors in line sorters and in situ viscometry; process design, e.g. thermal treatments; and so on.

Engineering properties of brown rice are those which are useful and necessary for the design and operation of various processing equipments. The physical properties include size, shape, true density, bulk density, porosity, volume, colour and weight of 1000 grains; frictional properties are angle of repose, coefficient of friction and aerodynamic property, namely, terminal velocity. The above engineering properties for brown rice can be determined as per procedures given by Sahay and Singh (1994).

Shape and Size

Shape and size are inseparable in a physical object, and both are generally necessary if the object is to be satisfactorily described. The shape and size of the product is an important parameter that affects conveying characteristics of solid materials by air or water.

Table 3.1 Classification of paddy based on size

Categories	Length (mm)	L:B
Long slender	≥6	≥3
Long bold	≥6	<3
Medium slender	<6	2.5–3.0
	<4.5	2.0–2.5
Short slender	<6	≥3
Short bold	<6	<2.5

The size of the grain can be determined by measuring the longitudinal and lateral diameters using the digital vernier caliper having the least count of 0.01 mm.

In the present study, the L & B of paddy and brown rice were determined by arranging ten randomly selected whole grains end to end for measuring (Adair et al. 1966), and it was measured using the wooden grooved board. Thickness is measured by using Mitutoyo dial caliper.

According to the Ramaiah Committee (1968), the following classification was used for categorizing the paddy varieties (Tables 3.1 and 3.2).

The observations with respect to weight, surface area and hardness are showed in Table 3.3. Results of the analysis showed that in all varieties of different states, obviously paddy weighed more compared to brown rice ranging from 12.87 g (Warangal AP) to 35.56 g (Manipur) for paddy and 10.4 g (Warangal AP) to 29.38 g (Manipur) for brown rice. Surface area of paddy was more when compared to brown rice, and hardness was more in Pondicherry varieties followed by Gujarat and Assam.

Weight of 1000 Grains

Thousand grains, unbroken and sound, from three randomly drawn samples can be weighed and recorded.

1000 grain weight is determined by counting and weighing of 100 grains of paddy and brown rice in triplicate. Surface area can be calculated using the formula (Bhattacharya et al. 1982)

$$S = 3/4 \times 22/7 \times L \times \frac{\sqrt{B^2}}{2} + T^2 \times N \ mm^2/g, \text{where } N \text{ is normalized grain weight.}$$

N = No of grains/g

Hardness of grains has been a subject of interest to millers, livestock feeders and others. Biting or cutting the grain has provided a qualitative evaluation of grain hardness.

Kernel hardness (endosperm texture) affects the processing properties of the grain and the resulting products. Grains with a high proportion of corneous endosperm tend to be more resistant to breakage during decortication (dehulling)

Table 3.2 Percentage of length, breadth and thickness of paddy and brown rice and its categorization

State	Variety	Paddy				Brown rice				Category
		L (mm)	B (mm)	T (mm)	L:B	L (mm)	B (mm)	T (mm)	L:B	As per L:B ratio
Assam	Bahadur	7.9	3.3	1.89	2.43	6.0	2.5	1.69	2.36	Long bold
	Keteki Joha	7.8	3.2	1.60	2.47	6.0	2.0	1.51	3.07	Long slender
	Ranjit	7.9	2.5	1.64	3.23	6.0	2.2	1.56	2.59	Long bold
Manipur	Chakhao phou	7.9	2.8	1.78	2.8	5.9	2.2	1.61	2.7	Medium slender
	Charong phou	9.5	2.7	1.92	3.5	6.4	2.8	1.78	2.3	Long bold
	Drum phou	8.8	3.0	2.29	2.9	6.5	2.8	2.07	2.3	Long bold
	Moirang phou	9.0	3.0	2.02	3.0	6.4	2.5	1.84	2.6	Long bold
	Taothabi phou	8.4	2.8	2.01	3.0	6.0	2.3	1.79	2.6	Long bold
	Tampha phou	8.3	2.8	1.84	3.0	6.0	2.6	1.79	2.3	Long bold
Gujarat	GR 3	10.6	2.3	1.87	4.7	7.8	1.9	1.65	4.1	Long slender
	GR 4	8.1	2.3	1.64	3.6	5.8	1.9	1.48	3.0	Medium slender
	GR 5	8.6	2.7	1.82	3.2	6.5	2.5	1.77	2.6	Long bold
	GR 6	9.9	2.4	1.82	4.1	7.2	2.2	1.71	3.3	Long slender
	GR 7	9.3	2.5	1.77	3.8	6.9	2.2	1.61	3.1	Long slender
	GR 8	7.9	3.0	2.02	2.6	5.8	2.6	1.77	2.2	Medium slender
	GR 11	8.1	2.3	1.57	3.6	6.0	1.9	1.54	3.1	Long slender
	GR 12	8.1	2.0	1.53	4.0	6.2	1.9	1.57	3.2	Long slender
	GR 101	8.5	2.5	1.74	3.4	6.6	2.2	1.52	3.0	Long slender
	GR 102	8.5	3.0	1.80	2.8	6.4	2.0	1.58	3.2	Long slender
	GR 103	8.4	2.5	1.82	3.4	5.9	2.2	1.53	2.7	Medium slender
	GR 104	9.3	2.4	1.72	4.0	7.2	2.1	1.49	3.4	Long slender
	Gurjari	9.2	3.0	1.87	3.1	7.1	2.1	1.71	3.4	Long slender
	IR 22	9.5	2.4	1.77	3.9	7.3	2.1	1.51	3.5	Long slender
	IR 28	9.3	2.4	1.86	4.0	7.1	2.2	1.72	3.2	Long slender
	IR 66	9.8	2.4	1.80	4.2	7.0	2.1	1.59	3.3	Long slender
	Jaya	8.2	2.9	1.93	2.8	6.6	2.5	1.67	2.6	Long bold
	Lal kada	8.2	2.9	1.90	2.8	5.9	2.5	1.75	2.3	Medium slender
	Masuri	8.6	2.2	1.37	3.9	6.8	1.9	1.40	3.6	Long slender
	Narmada	7.6	2.2	1.53	3.4	5.6	1.8	1.38	3.1	Medium slender
	NVSR-20	9.3	2.5	1.73	3.8	6.9	2.2	1.53	3.1	Long slender

(continued)

Table 3.2 (continued)

State	Variety	Paddy				Brown rice				Category
		L (mm)	B (mm)	T (mm)	L:B	L (mm)	B (mm)	T (mm)	L:B	As per L:B ratio
AP	NLR 145	8.83	2.77	1.84	3.19	6.73	2.20	1.66	3.06	Long slender
	NLR 9672	7.47	2.87	1.93	2.60	5.73	2.47	1.86	2.32	Short bold
	NLR9672–96	6.97	3.17	1.96	2.20	6.00	2.77	1.71	2.17	Long bold
	NLR 9674	7.77	2.83	1.87	2.75	5.63	2.47	1.66	2.28	Short bold
	NLR 27999	8.03	3.10	1.97	2.59	5.90	2.57	1.77	2.30	Short bold
	NLR 28523	7.53	2.67	1.95	2.82	5.37	2.43	1.65	2.21	Short bold
	NLR 28600	7.57	3.13	1.91	2.42	5.63	2.43	1.70	2.32	Short bold
	NLR 30491	7.77	2.53	1.78	3.07	5.83	2.37	1.60	2.46	Short bold
	NLR 33057	8.77	2.73	2.01	3.21	6.07	2.53	1.77	2.39	Long bold
	NLR 33358	9.00	2.67	1.82	3.37	6.73	2.23	1.61	3.01	Long slender
	NLR 33359	7.67	3.23	1.91	2.37	6.97	2.17	1.65	3.22	Long slender
	NLR 33641	8.07	2.83	1.84	2.85	6.00	2.47	1.69	2.43	Long bold
	NLR 33654	8.17	2.37	1.87	3.45	5.83	2.07	1.66	2.82	Medium slender
	NLR 33654	8.17	2.57	1.82	3.18	6.07	2.23	1.77	2.72	Long bold
	NLR 33892	7.83	3.17	1.81	2.47	5.83	2.40	1.61	2.43	Short bold
	NLR 34242	7.63	2.80	1.90	2.73	5.57	2.03	1.66	2.74	Medium slender
	NLR 34449	7.87	2.53	1.68	3.11	5.17	2.03	1.57	2.54	Medium slender
	MTU 1001	8.87	2.83	1.94	3.13	6.50	2.47	1.66	2.64	Long bold
	MTU 1010	9.37	2.70	1.92	3.47	7.13	2.17	1.59	3.29	Long slender
Hyderabad –AP	Dhanrasi	8.33	3.2	1.88	2.60	5.8	2.6	1.72	2.23	Short bold
	IET 15420	8.46	2.8	1.94	3.02	5.9	2.7	1.78	2.19	Short bold
	Jaya	8.6	2.86	2.01	3.01	5.9	2.7	1.82	2.19	Short bold
	M.Sugandha	10.13	2.33	1.78	4.35	7.7	2.0	1.64	3.85	Long slender
	Nagarjuna	8.2	2.66	1.95	3.08	6.0	2.3	1.74	2.61	Long bold
	Nidhi	8.53	2.33	1.81	3.66	6.1	2.0	1.67	3.05	Long slender
	Phalguna	9.53	2.6	2.03	3.67	7.6	2.1	1.81	3.62	Long slender
	Rasi	7.73	2.86	1.99	2.70	6.0	2.5	1.8	2.40	Long bold
	Salivahana	7.66	2.8	1.94	2.74	6.0	2.7	1.8	2.22	Long bold
	Shanthi	9.4	2.6	2.0	3.62	7.1	2.2	1.85	3.23	Long slender
	Swarnadhan	7.46	3.13	1.71	2.38	5.6	2.8	1.66	2.00	Short bold
	Thriguna	8.6	2.46	2.07	3.50	7.1	2.1	1.87	3.38	Long slender

(continued)

Table 3.2 (continued)

State	Variety	Paddy				Brown rice				Category
		L (mm)	B (mm)	T (mm)	L:B	L (mm)	B (mm)	T (mm)	L:B	As per L:B ratio
Raghulu, Srikakulam –AP	RGL 1	8.5	2.5	1.76	3.40	6.3	2.1	1.48	3.0	Long slender
	RGL 52I	8.5	2.5	1.69	3.40	6.3	2.3	1.53	2.7	Long bold
	RGL 52II	8.0	2.5	1.73	3.20	6.4	2.5	1.62	2.6	Long bold
	RGL 1880	8.5	2.7	1.62	3.15	6.7	2.4	1.73	2.8	Long bold
	RGL 2332	8.1	2.6	1.82	3.12	6.3	2.2	1.76	2.8	Long bold
	RGL 2537	9.0	2.8	1.83	3.21	6.8	2.4	1.76	2.8	Long bold
	RGL 2538	8.5	2.5	1.77	3.40	6.5	2.3	1.66	2.8	Long bold
	RGL 2624	8.2	2.4	1.68	3.42	6.2	2.0	1.50	3.0	Long slender
	RGL 3996	9.1	2.8	1.86	3.25	6.8	2.5	1.81	2.7	Long bold
	RGL 5613	9.1	2.7	1.88	3.37	6.8	2.4	1.77	2.8	Long bold
	RGL 56513	8.4	2.6	1.83	3.23	6.1	2.4	1.62	2.6	Long bold
	RGL 11414	9.0	2.7	1.89	3.33	6.8	2.4	1.73	2.8	Long bold
Warangal –AP	Sannallu	7.4	2.4	2.65	3.08	5.5	1.8	2.50	3.06	Short slender
	Surekha	9.2	2.8	2.89	3.29	7.4	2.4	2.71	3.08	Long slender
	Varalu	9.2	2.2	2.5	4.18	6.4	2.3	2.55	2.78	Long bold
	Warangal Samba	7.8	2.8	2.04	2.79	3.8	2.0	2.48	2.90	Medium slender
Pondicherry	Puduvai ponni	7.0	3.0	1.91	3.1	5.6	2.5	1.74	2.2	Short bold
	Punithavathy	8.0	2.6	1.78	2.5	5.9	2.1	1.63	2.8	Medium slender
	Bharathidasan	8.7	2.6	1.89	3.3	6.7	2.2	1.75	3.1	Long slender
	Janahar	8.7	2.8	1.85	3.1	6.6	2.6	1.65	2.5	Long bold
	Aravindar	8.1	2.5	1.85	3.2	6.2	2.2	1.64	2.9	Long bold
	Subramaniya Bharathi	7.4	2.6	1.68	2.7	5.6	2.2	1.43	2.6	Medium slender
	Annalakshmi	7.9	2.4	1.78	3.3	5.7	2.1	1.63	2.7	Medium slender

and milling than grain with a high proportion of floury endosperm. During milling hard grains tend to yield proportionally cleaner endosperm of large particle size than soft grains. This is because the corneous endosperm is easily separated from intact starchy endosperm giving a higher yield.

It was measured for randomly selected ten brown rice samples by using Kiya hardness tester.

The percentage of chalkiness of varieties of all states is given in Table 3.4. Observations showed that chalkiness percent was 100 in Manipur varieties.

Chalkiness in rice kernels is an undesirable characteristic because it degrades the visual appearance and cooking quality of milled rice. Head rice yield, defined as the mass percentage of rough rice that remains as head rice (milled kernels that are at least three-fourths of the original kernel length after complete milling, USDA 2005), is the most commonly used indicator of rice milling quality. Chalkiness generally

Table 3.3 1000 grain weight, surface area and hardness of paddy and brown rice

State	Variety	1000 grain weight (g)		Surface area (mm²g)		Hardness (kg)
		Paddy	Brown rice	Paddy	Brown rice	
Assam	Bahadur	23.11	19.47	2166.77	1765.83	6.30
	Keteki Joha	18.64	13.65	2495.31	2684.38	4.40
	Ranjit	18.79	13.76	2095.22	1960.10	9.50
Manipur	Chakhao phou	20.19	15.50	2163.83	1729.61	8.22
	Charong phou	27.95	21.30	1876.91	1661.63	6.31
	Drum phou	35.56	29.38	1556.72	1284.02	6.34
	Moirang phou	28.10	22.88	1930.71	1447.22	8.70
	Taothabi phou	23.64	17.42	2041.33	1673.14	6.38
	Tampha phou	22.83	16.09	2030.24	1961.94	6.42
Gujarat	GR 3	26.27	21.35	1993.58	21.35	9.4
	GR 4	20.12	15.10	1895.48	15.10	7.7
	GR 5	26.17	21.22	1783.48	21.22	8.0
	GR 6	25.33	19.32	1962.15	19.32	9.4
	GR 7	26.55	20.75	1788.37	20.75	8.0
	GR 8	28.70	24.44	1659.31	24.44	6.0
	GR 11	17.13	12.80	2194.76	12.80	6.1
	GR 12	17.09	12.66	1989.25	12.66	5.8
	GR 101	16.79	12.71	2570.14	12.71	6.8
	GR 102	21.29	16.89	2328.12	16.89	6.2
	GR 103	16.29	11.23	2657.74	11.23	5.8
	GR 104	21.25	15.69	2153.84	15.69	7.9
	Gurjari	30.17	25.13	1796.74	25.13	6.5
	IR 22	19.42	15.02	2431.46	15.02	8.1
	IR 28	23.89	18.64	1970.12	18.64	9.4
	IR 66	25.39	20.34	1929.99	20.34	8.0
	Jaya	24.22	19.60	1965.75	19.60	8.0
	Lal kada	25.13	19.24	1885.58	19.24	7.7
	Masuri	21.08	15.29	1762.31	15.29	6.4
	Narmada	25.20	19.94	1347.02	19.94	5.9
	NVSR-20	24.15	18.27	1951.38	18.27	4.9
	Ratna	24.01	18.78	1957.62	18.78	8.0
	Safed kada	25.87	20.53	1862.06	20.53	7.7

(continued)

Table 3.3 (continued)

State	Variety	1000 grain weight (g)		Surface area (mm²g)		Hardness (kg)
		Paddy	Brown rice	Paddy	Brown rice	
AP	NLR 145	23.66	20.00	2025.51	1545.74	6.80
	NLR 9672	19.66	16.66	2133.49	1772.52	7.48
	NLR 9672–96	23.00	19.00	1876.37	1713.40	7.52
	NLR 9674	21.00	17.66	2084.46	1581.32	7.46
	NLR 27999	20.66	18.66	2370.56	1644.53	7.79
	NLR 28523	18.66	15.00	2231.18	1752.64	7.60
	NLR 30491	18.66	15.66	2138.17	1774.36	8.84
	NLR 33057	26.00	22.66	1898.88	1378.58	6.12
	NLR 33358	22.33	19.00	2187.39	1623.82	6.92
	NLR 33359	21.33	18.00	2242.29	1759.41	7.80
	NLR 33641	20.33	18.00	2197.50	1662.77	5.04
	NLR 33654	20.66	16.00	2012.74	1611.47	6.42
	NLR 33654	18.33	16.00	2366.40	1800.27	5.00
	NLR 33892	19.00	15.66	2515.58	1793.27	6.57
	NLR 34242	19.00	16.00	2255.97	1521.56	7.20
	NLR 34449	17.00	13.00	2332.98	1701.07	8.35
	MTU 1001	24.66	19.66	2049.11	1639.95	3.94
	MTU 1010	25.33	21.33	2049.24	1498.82	7.44
Hyderabad –AP	Dhanrasi	24.15	18.19	2133.71	1656.78	7.99
	IET 15420	25.07	20.71	1915.95	1535.59	9.90
	Jaya	26.10	21.55	1919.81	1485.86	10.22
	M.Sugandha	18.32	12.73	2702.31	2607.55	9.02
	Nagarjuna	22.75	18.47	1981.44	1561.54	8.64
	Nidhi	18.51	15.62	2266.20	1695.97	6.30
	Phalguna	28.72	24.22	1824.37	1449.98	6.90
	Rasi	18.35	14.26	2446.35	2160.41	8.50
	Salivahana	22.72	18.61	1914.20	1743.77	7.20
	Shanthi	23.64	19.74	2173.99	1723.21	7.00
	Swarnadhan	22.54	18.48	1967.51	1644.07	6.80
	Thriguna	21.11	17.74	2183.07	1875.77	6.60

(continued)

Table 3.3 (continued)

State	Variety	1000 grain weight (g)		Surface area (mm²g)		Hardness (kg)
		Paddy	Brown rice	Paddy	Brown rice	
Raghulu, Srikakulam –AP	RGL 1	21.0	16.5	2062.63	1634.98	8.4
	RGL 52I	21.7	17.5	1970.14	1657.53	6.9
	RGL 52II	24.3	18.7	1690.33	1484.47	8.5
	RGL 1880	24.8	18.9	1825.89	1672.25	8.0
	RGL 2332	24.1	17.9	1777.89	1652.73	8.5
	RGL 2537	27.2	22.6	1844.75	1492.55	8.3
	RGL 2538	22.8	19.0	1903.37	1617.37	7.6
	RGL 2624	20.9	15.8	1915.77	1635.10	7.9
	RGL 3996	26.9	22.0	1895.36	1590.06	7.0
	RGL 5613	26.3	20.1	1897.40	1681.53	8.5
	RGL 56513	22.6	18.8	1969.67	1565.95	7.2
	RGL 11414	25.4	20.61	1946.42	1626.96	6.5
Warangal –AP	Sannallu	12.87	10.4	3426.35	2715.40	6.20
	Surekha	24.42	19.76	2526.76	2259.54	8.72
	Varalu	17.42	14.12	2931.41	2594.29	7.32
	Warangal Samba	14.56	12.0	3436.18	2566.61	6.36
	Sannallu	12.87	10.4	3426.35	2715.40	6.20
Pondicherry	Puduvai ponni	20.70	16.27	2119.07	1747.39	10.00
	Punithavathy	20.43	15.93	2056.52	1641.05	8.24
	Bharathidasan	22.57	18.33	2065.16	1712.64	7.56
	Janahar	24.20	19.47	2010.91	1739.84	9.86
	Aravindar	20.40	16.40	2058.24	1729.04	7.84
	Subramaniya Bharathi	16.23	12.87	2225.29	1902.96	8.92
	Annalakshmi	16.57	13.74	2374.45	1838.12	6.06

lowers head rice yield (HRY) as chalky kernels tend to be weaker and are more prone to breaking during milling than non-chalky, fully translucent kernels (Webb 1991; Siebenmorgen and Qin 2005). Chalkiness is a major concern in rice (*Oryza sativa* L.) because it is one of the key factors in determining quality and price.

A visual rating of the chalky proportion of the grain is used to measure chalkiness based on the scale presented below:

Scale	% area of chalkiness
0	When area of chalkiness is nil (translucent)
1	1–10
5	10–20
7	20–50
9	More than 50

Calculation for overall chalkiness score and chalky grains %

Scale	Number of grains under different scales
0	60
1	10
5	5
7	15
9	10
Total	100 grains

$$\text{Chalkiness Score} = (60 \times \mathbf{0}) + (10 \times \mathbf{1}) + (5 \times \mathbf{5}) + (15 \times \mathbf{7}) + (10 \times \mathbf{9})$$
$$= 0 + 10 + 25 + 105 + 90 = 230 / 100 = 2.3$$

Total chalky grains = 100; number of translucent grains = 100 − 60 = 40

Density

The density values are used in design of storage bins and silos, separation of desirable materials from impurities, cleaning and grading, etc.

It can be expressed as the ratio of weight of material to the volume of material. The S.I unit of density is kg/m^3.

True Density

True density is used in design of storage bins and in separation of desirable materials from impurities (Tavakoli et al. 2002). It is the density of material calculated from its component densities considering conservation of mass and volume.

The apparatus used for measuring true density of grains consisted of a 100 ml measuring jar and a weighing balance. Fifty ml of toluene should be taken in a measuring jar, and a known weight of grain sample must be poured into the measuring jar. The rise in the toluene level should be recorded as the true volume of the grains without void space. The true density of the grain can be calculated by using the following formula (Mohsenin 1986):

$$\text{True density} \left(kg / m^3 \right) = \frac{\text{weight of grains} \left(kg \right)}{\text{True volume of grains excluding void space} \left(m^3 \right)}$$

Table 3.4 Chalkiness in brown rice

State	Variety	Chalkiness (%)					Chalkiness score	Total chalkiness (%)
		0	1	5	7	9		
Assam	Bahadur	0	32.5	102.5	0.0	0	1.35	43.8
	Keteki Joha	0	22.5	32.5	66.5	0	1.22	35.5
	Ranjit	0	22.5	32.5	66.5	0	1.22	35.5
Manipur	Chakhao phou	0	2	10	0	88	8.5	100
	Charong phou	0	7	15	29	49	7.3	100
	Drum phou	0	0	3	14	83	8.6	100
	Moirang phou	1	9	15	22	52	7.1	99
	Taothabi phou	1	5	10	29	55	7.6	99
	Tampha phou	0	30	29	26	15	4.9	100
	Chakhao phou	0	2	10	0	88	8.5	100
	Charong phou	0	7	15	29	49	7.3	100
Gujarat	GR 3	41	26	8	12	14	2.7	59
	GR 4	36	21	11	23	8	3.2	64
	GR 5	71	25	0	0	4	0.6	29
	GR 6	63	15	4	12	5	1.7	37
	GR 7	53	23	7	10	6	1.9	47
	GR 8	46	18	9	18	8	2.7	54
	GR 11	25	13	4	3	4	1.0	75
	GR 12	50	17	10	10	13	2.5	50
	GR 101	58	16	21	4	0	1.5	42
	GR 102	45	25	13	7	10	2.2	55
	GR 103	60	17	8	9	5	1.7	40
	GR 104	44	9	16	18	12	3.3	56
	Gurjari	13	23	13	25	26	5.0	87
	IR 22	52	15	8	15	10	2.5	48
	IR 28	62	16	6	12	5	1.7	38
	IR 66	47	20	13	7	12	2.5	53
	Jaya	30	26	18	19	8	3.2	70
	Lal kada	50	13	10	18	10	2.7	50
	Masuri	42	19	19	11	9	2.7	58
	Narmada	100	0	0	0	0	0.0	0
	NVSR 20	40	18	6	23	13	3.3	60
	Ratna	50	21	11	12	7	2.2	50
	Safed kada	44	22	17	12	5	2.3	56

(continued)

Table 3.4 (continued)

State	Variety	Chalkiness (%)					Chalkiness score	Total chalkiness (%)
		0	1	5	7	9		
AP	NLR 145	11	31	16	15	26	4.5	89
	NLR 9672	21	57	16	4	2	1.9	79
	NLR 9672–96	16	60	12	8	5	2.1	84
	NLR 9674	22	67	1	6	5	1.5	78
	NLR 27999	8	68	16	4	4	2.1	92
	NLR 28523	9	48	23	11	9	3.2	91
	NLR 28600	17	62	11	6	4	2.0	83
	NLR 30491	34	27	14	13	11	2.9	66
	NLR 33057	6	47	20	14	13	3.6	94
	NLR 33358	20	20	10	15	36	5.0	80
	NLR 33359	9	15	10	16	49	6.2	91
	NLR 33654	8	40	23	15	14	3.8	92
	NLR 33654(A)	41	41	7	10	1	1.5	59
	NLR 33892	10	26	23	19	23	4.8	90
	NLR 34242	13	25	21	17	24	4.6	87
	NLR 34449	24	14	19	24	19	4.5	76
	MTU 1001	21	56	15	4	4	1.9	79
	MTU 1010	30	31	9	11	19	3.2	70
Hyderabad –AP	Dhanrasi	20	30	22	17	11	3.6	80
	IET 15420	15	24	14	12	35	4.9	85
	Jaya	21	21	16	17	24	4.4	79
	M.Sugandha	65	16	8	7	4	1.4	35
	Nagarjuna	32	14	16	21	17	3.9	68
	Nidhi	79	8	7	3	3	0.9	21
	Phalguna	25	32	22	15	6	3.0	75
	Rasi	34	29	10	13	13	2.9	66
	Salivahana	23	29	18	11	19	3.7	77
	Shanthi	76	7	6	6	4	1.2	24
	Swarnadhan	21	7	11	12	49	5.9	79
	Thriguna	34	8	4	16	38	4.9	66
Raghulu, Srikakulam –AP	RGL 1	61	20	8	9	2	1.4	39
	RGL 52I	53	22	12	10	2	1.7	47
	RGL 52II	75	13	8	4	0	0.8	25
	RGL 1880	58	8	15	12	7	2.3	42
	RGL 2332	80	3	10	7	0	1.0	20
	RGL 2537	56	27	17	0	0	1.1	44
	RGL 2538	45	36	7	7	6	1.7	55
	RGL 2624	65	21	6	9	0	1.1	35
	RGL 3996	49	26	14	9	2	1.8	51
	RGL 5613	91	5	0	0	4	0.4	9
	RGL 56513	65	21	6	9	0	1.1	35
	RGL 11414	59	23	12	6	0	1.3	41

(continued)

Table 3.4 (continued)

State	Variety	Chalkiness (%)					Chalkiness score	Total chalkiness (%)
		0	1	5	7	9		
Warangal –AP	Sannallu	70	16	9	5	1	1.0	30
	Surekha	75	11	4	5	5	1.1	25
	Varalu	76	4	7	8	6	1.4	24
	Warangal Samba	98	1	1	0	1	0.1	2
Pondicherry	Puduvai ponni	14	67	9	5	5	1.9	86
	Punithavathy	60	29	0	5	6	1.2	40
	Bharathidasan	60	21	6	3	9	1.5	40
	Janahar	38	47	9	0	6	1.4	62
	Aravindar	37	19	13	12	20	3.4	63
	Subramaniya Bharathi	86	12	0	0	2	0.3	14
	Annalakshmi	39	29	16	9	7	2.3	61

Bulk Density

Bulk density is the density of a material when packed or stacked in bulk. The bulk density can be determined by using wooden box of volume 1000 ml. The grains to be filled into the box and the top should be levelled off. The grains then weighed using a precision electronic balance. The bulk density can be calculated using the formula:

$$\text{Bulk density} \left(kg/m^3 \right) = \frac{\text{weight of grains} \left(kg \right)}{\text{Volume of grains including void space} \left(m^3 \right)}$$

Porosity

All food materials have some porosity (air space) in them. Porosity indicates the volume fraction of void space or air. These internal pores may be closed pores, where the pore space is enclosed from all sides; blind pores, where one end is closed; and through pores where flow can take place. The porosity of the materials can be found out by air comparison pycnometer.

Porosity of grains can be calculated by using the bulk density and true density values by using the following formula:

$$\text{Porosity} \left(\% \right) = 1 - \frac{\text{bulk density}}{\text{True density}} \times 100$$

Frictional Properties

The frictional properties such as coefficient of friction and angle of repose are important in designing of storage bins, hoppers, chutes, pneumatic conveying systems, screw conveyors, forage harvesters, etc.

Angle of Repose

The angle of repose determines the maximum angle of pile of grain in horizontal plane. Flowability of grains is usually measured using the angle of repose value (a measure of the internal friction between grains) that is useful in design of hoppers (Mahmud et al. 2009). The angle of repose is the angle between base and slope of the cone formed on a free vertical fall of grains on to a horizontal plane. It can be determined by following the procedure described by Sahay and Singh (1994). From the height and diameter of grains heaped in natural piles, the angle of repose is calculated by using the following formula:

$$\text{Angle of repose, } \varphi = \tan^{-1}\left(\frac{2H}{D}\right)$$

where
    φ = Angle of repose, degrees
    H = Height of the heap, m
    D = Diameter of the heap, m

Coefficient of Friction

The coefficient of friction is used to determine the angle at which chutes must be positioned in order to achieve consistent flow of materials through the chute (Ghasemi et al. 2007). The coefficient of friction of grains was determined against two material surfaces, namely, cardboard and mild steel, by "inclined surface" method (Mohsenin 1986).The static angle of friction can be determined by measuring the angle of inclination at which the grain placed on it just began to slide of inclined test surface.

$$\text{Coefficient of static friction} = \tan\theta$$

where θ = Angle of friction, degrees

Flow and physicochemical properties like angle of repose, bulk density, EMC and water uptake ratio are given in Table 3.5. Results showed that angle of repose

Table 3.5 Flow and physicochemical properties of brown rice

State	Variety	Angle of repose (degree)	Bulk density kg/m³	EMC-S (%) d.b.	Water uptake 80 °C	Water uptake 98.5 °C	Water uptake ratio 80 °C/ 98 °C
Assam	Bahadur	39.5	–	41.57	0.47	1.44	32.64
	Keteki Joha	35.7	–	53.59	0.41	1.91	21.26
	Ranjit	39.8	–	38.69	0.41	1.80	22.84
Manipur	Chakhao phou	40.7	747.37	48.00	0.87	1.22	71.31
	Charong phou	40.0	743.56	41.95	0.95	1.50	63.00
	Drum phou	39.7	742.36	46.55	0.96	1.53	62.62
	Moirang phou	41.7	743.46	39.77	0.95	1.50	63.55
	Taothabi phou	40.0	732.36	39.08	0.93	1.50	62.00
	Tampha phou	41.0	543.42	35.84	0.27	1.36	19.93
	Chakhao phou	40.7	747.37	48.00	0.87	1.22	71.31
Gujarat	GR 3	35.2	523.10	30.86	0.37	1.74	21.26
	GR 4	35.3	566.26	32.95	0.46	2.16	21.06
	GR 5	36.5	642.80	32.76	0.28	1.95	14.40
	GR 6	38.3	514.06	30.17	0.79	2.08	37.74
	GR 7	34.0	537.10	34.66	0.84	2.20	37.95
	GR 8	35.2	600.95	33.33	0.37	2.47	14.98
	GR 11	37.0	579.60	30.41	0.52	2.08	25.00
	GR 12	35.0	507.26	36.16	0.44	2.34	18.84
	GR 101	36.3	568.65	37.21	1.39	2.26	61.28
	GR 102	45.0	466.35	32.95	0.31	2.21	14.06
	GR 103	35.3	575.00	30.41	0.24	2.39	9.83
	GR 104	35.5	477.95	37.36	1.31	2.40	54.70
	Gurjari	35.5	554.78	32.95	0.52	2.66	19.55
	IR 22	38.8	534.09	34.88	0.62	2.40	25.83
	IR 28	38.7	537.98	34.30	0.76	1.66	45.78
	IR 66	37.7	560.58	34.48	0.78	2.15	36.28
	Jaya	37.2	599.34	38.73	1.03	1.97	52.28
	Lal kada	34.8	602.75	38.32	0.32	1.67	19.16
	Masuri	34.3	530.46	36.87	0.60	2.12	28.30
	Narmada	34.0	527.18	36.99	0.47	2.47	19.03
	NVSR-20	36.2	548.75	31.46	0.40	1.69	23.67
	Ratna	38.3	561.85	33.33	0.49	2.24	21.88
	Safed kada	34.3	605.63	32.22	0.35	2.29	15.28

(continued)

Table 3.5 (continued)

State	Variety	Angle of repose (degree)	Bulk density kg/m³	EMC-S (%) d.b.	Water uptake 80 °C	98.5 °C	Water uptake ratio 80 °C/ 98 °C
AP	NLR 145	37.8	775.49	33.71	–	–	–
	NLR 9672	38.0	802.36	33.71	–	–	–
	NLR 9672–96	37.8	803.50	33.15	–	–	–
	NLR 9674	39.3	801.72	32.39	–	–	–
	NLR 27999	37.7	786.27	35.47	–	–	–
	NLR 28523	36.8	788.69	35.06	–	–	–
	NLR28600	39.3	781.49	43.71	–	–	–
	NLR 30491	40.5	785.33	35.63	–	–	–
	NLR 33057	38.7	779.76	32.39	–	–	–
	NLR 33358	35.8	785.13	32.20	–	–	–
	NLR 33359	37.8	753.43	36.42	–	–	–
	NLR 33641	39.0	793.78	34.66	–	–	–
	NLR 33654	35.7	750.40	33.92	–	–	–
	NLR 33654	37.3	778.66	35.03	–	–	–
	NLR 33892	38.3	790.47	33.33	–	–	–
	NLR 34242	38.3	781.87	34.83	–	–	–
	NLR 34449	40.0	762.21	29.71	–	–	–
	MTU 1001	35.2	777.11	33.14	–	–	–
	MTU 1010	36.2	770.48	33.71	–	–	–
Hyderabad –AP	Dhanrasi	–	–	35.63	0.48	1.22	38.93
	IET 15420	–	–	35.23	0.56	1.49	37.37
	Jaya	–	–	40.80	0.80	2.11	38.00
	M.Sugandha	–	–	39.20	1.52	1.73	88.12
	Nagarjuna	–	–	38.51	0.38	1.74	21.90
	Nidhi	–	–	41.52	1.23	1.81	67.68
	Phalguna	–	–	39.43	1.34	1.83	72.95
	Rasi	–	–	38.07	1.01	1.76	57.55
	Salivahana	–	–	37.50	0.71	1.61	44.10
	Shanthi	–	–	38.51	1.07	1.73	61.85
	Swarnadhan	–	–	31.84	0.62	1.82	34.07
	Thriguna	–	–	35.26	0.62	1.91	32.20

(continued)

Table 3.5 (continued)

State	Variety	Angle of repose (degree)	Bulk density kg/m³	EMC-S (%) d.b.	Water uptake 80 °C	Water uptake 98.5 °C	Water uptake ratio 80 °C/ 98 °C
Raghulu, Srikakulam –AP	RGL 1	39	615.64	36.57	0.51	1.95	26.22
	RGL 52I	37	681.67	37.06	0.52	2.03	25.43
	RGL 52II	–	–	36.21	0.66	2.02	32.67
	RGL 1880	38	688.09	34.86	0.43	1.93	22.34
	RGL 2332	38	730.28	35.09	0.37	1.99	18.39
	RGL 2537	–	–	38.51	0.91	1.54	58.77
	RGL 2538	–	–	33.88	0.76	1.97	38.42
	RGL 2624	37	706.65	34.86	0.67	2.04	32.60
	RGL 3996	–	–	37.93	0.87	1.87	46.26
	RGL 5613	–	–	37.21	0.65	1.73	37.28
	RGL 56513	–	–	35.80	0.72	2.12	33.81
	RGL 11414	–	–	35.63	0.73	1.80	40.28
Warangal –AP	Sannallu	33.70	0.41	2.13	19.06	33.70	0.41
	Surekha	69.93	0.39	2.43	16.05	69.93	0.39
	Varalu	35.33	0.86	2.23	38.65	35.33	0.86
	Warangal Samba	31.15	0.51	2.53	20.40	31.15	0.51
Pondicherry	Puduvai ponni	38.3	616.31	35.84	0.63	1.48	42.23
	Punithavathy	37.5	621.44	33.90	0.43	1.45	30.10
	Bharathidasan	37.2	614.43	34.66	0.48	1.41	33.69
	Janahar	38.3	564.43	19.54	0.34	1.36	25.09
	Aravindar	36.3	617.34	35.63	0.51	1.56	32.69
	Subramaniya Bharathi	36.7	605.36	38.92	0.27	1.55	17.10
	Annalakshmi	36.7	603.12	35.63	0.54	1.43	37.54

was more in Gujarat and Manipur varieties followed by other places. Bulk density was more in AP varieties followed by Manipur. Water uptake ratio was higher in Manipur varieties.

Cracked Grains/Suncheck

Using the paddy grain crack detector, the number of cracked grains in a 100 grain sample was analysed.

Equilibrium Moisture Content upon Soaking (EMC-S) and Water Uptake Ratio

Equilibrium moisture content upon soaking (EMC-S) of polished rice at room temperature (Indhudhara Swamy, Y.M, et al., 1971) has been reported as one of the indices of ageing factors, which also influences the cooking quality of rice. EMC-S ratios of water uptake at 80 °C to that at 96 °C were assessed (Bhattacharya and Sowbhagya 1971).

Colour

Colour of agricultural materials is valuable physical characteristics for selective separation in the field as subsequent handling and processing and also helps in anticipating end product colour quality. Colour is the overall visual perception of the colour of the grain as viewed by the naked eye.

The colour measurement of brown rice can be made using Minolta chromameter (Make: Minolta Instrument Co., Japan; Model-CRD 200). It is a lightweight, compact tristimulus colour analyser for measuring reflected-light colour of a sample. It combines advanced electronics and optical technology to provide high accuracy and complete portability of data. Using an 8 mm diameter (measuring area) diffused illumination and a 0° viewing angle, the chromameter takes accurate colour measurements instantaneously, and the readings are displayed. A Pulsed Xenon Arc (PXA) lamp in a mixing chamber provides diffused lighting over the surface of sample. Six high-sensitivity silicon photocells, filtered to match the CIE (Commission International IEclairage) Standard Observer Response, were used by meter's double feedback system to measure both incident and reflected lights. The chromameter detects any slight variation in the spectral power distribution of the PXA lamp and compensates automatically. Chromaticity may be measured in either Yxy (CIE, 1931) or L*a*b* (CIE, 1976) coordinates, and the colour difference could be in terms of Δ (Yxy), Δ (L*a*b*) or Δ (E*ab). Data generated can be converted between coordinate systems between chromaticity and colour difference measuring modes by the meter. The instrument also offers a choice of either CIE illuminate C or D65 lighting conditions, and in the present experiment, the CIE illuminate C was used. After initial calibration of the meter with a white colour standard, the sample was placed on the measuring head of the meter, and the sample colour was measured in terms of Yxy, L*a*b* and ΔE (Hunterlab 2001).

Aerodynamic Properties

Aerodynamic properties of agricultural products are important and required for design of air conveying systems and the separation equipment (Sahay and Singh 1994).

Terminal Velocity

Terminal velocity is required to decide the velocity of winnowing air blown to separate a lighter material (Sahay and Singh 1994). Terminal velocity is equal to air velocity at which the particle remains in suspended state in a vertical pipe. Terminal velocity of brown rice can be measured using an air column. For each test, a sample was dropped into the air stream from the top of the air column, and air was blown up the column to suspend the material in the air stream. The air velocity near the location of the sample suspension was measured by digital anemometer having a least count of 0.1 m/s (Gharibzahedi et al. 2010).

Milling Properties

Milling in general refers to reduction of food grain into various end products. With respect to rice, it refers to overall operations like cleaning, dehusking, separation, etc. During rice milling processes, in which rough rice hull is removed from brown rice, the occurrence of mechanical damage due to intensive forces and stresses cannot be neglected. The extent of these stresses could be induced by changes in material properties such as moisture and texture. If the stresses exceed the rupture strength of the material, it will lead to cracks or breakage. The most important difference between rice and other cereal is the economic and qualitative aspects of rice production. In contrast with other cereals, rice is preferably consumed as whole grains. Therefore, the percentage of whole and unbroken kernels is an important quality criterion in rice trade (Siebenmorgen 1994). In addition, the breakage of rice grains does adversely affect the seed germination, storability and cooking quality (Li et al. 1999).

Conclusion

Engineering properties of brown rice are very important to analyse since many design relay on it. Results of the analysis showed that most of the varieties in all states were long bold and long slender as per L:B ratio. Surface area of paddy was more when compared to brown rice, and hardness was more in Pondicherry varieties followed by Gujarat. Percentage of chalkiness was found more in Manipur varieties. Flow and physicochemical properties like angle of repose were more in Gujarat and Manipur varieties, bulk density was more in AP varieties, and water uptake ratio was higher in Manipur varieties.

References

Adair CR, Beachell HM, Jodon NE, Johnston TH, Thysell JR Green VE, Jr., Webb BD, Atkins JG, (1966) Rice in the United states: Varieties and production. Agric. Handbook no.289, Agric. Res. Service, U.S. Dept. Agric, 19

Anonymous (Ramaiah Committee) (1968) Report on Classification of rice. Ministry of Food, Agriculture, Community Development and Cooperation, Govt. of India, New Delhi

Bhattacharya KR, Sowbhagya CM (1971) Water uptake by rice during cooking. Cereal Sci Today 16:420–424

Bhattacharya KR, Ramesh BS, Sowbhagya CM (1982) Dimensional classification of rice for marketing. J Agric Eng XIX(4):69–76

FAOSTAT (2015) Rice production statistics. Retrieved April 5, 2015, from http://faostat.fao.org

Gharibzahedi SMT, Etemad V, Mirarab RJ, Foshat M (2010) Study on some engineering attributes of pine nut (*Pinus pinea*) to the design of processing equipment. Res Agric Eng 56(3):99–106

Ghasemi V, Mobli MH, Jafari A, Rafiee S, Heidary Soltanabadi M, Kheiralipour K (2007) Some engineering properties of paddy (var. Sazandegi). Int. J. Agric and Biol., properties of chick pea seeds. Biosyst Eng 5:763–766

Hunterlab (2001) The basics of colour perception and measurement. Version 1.4. pp 1–.124. http://www.hunterlab.com

Indudhara Swamy YM, Ali SZ, Bhattacharya KR (1971) Hydration of raw and parboiled rice and paddy at room temperature. J Food Sci Technol 8:20–22

Li YB, Cao CW, Yu QL, Zhong QX (1999) Study on rough rice fissuring during intermittent drying. J Drying Tech 17(9):1779–1793

Mahmud TK, Zhu K, Issoufou A, Fatmata T, Zhou H (2009) Functionality, in vitro digestibility and physicochemical properties of two varieties of defatted foxtail millet protein concentrates. Int J Mol Sci 10:5224–5238

Mohsenin NN (1986) Physical properties of plant and animal materials. Gordon and Breach Science Publisher, New York

Sahay KM, Singh KK (1994) Unit operations in agricultural processing. Vikas Publishing House Pvt. Ltd, New Delhi, pp 154–194

Siebenmorgen TJ (1994) Role of moisture content in affecting head rice yield. Rice Sci Tech 15:341–380

Siebenmorgen TJ, Qin G (2005) Relating rice kernel breaking force distributions to milling quality. Trans of the ASAE 48(4):223–228

Tavakoli T, Payman MH, Alizadeh M, Khoshtaghaza MH, (2002) Effect of moisture content and linear speed difference on paddy dehulling quality. International Conference on Agricultural Engineering. Budapest, pp 68–69

USDA (2005) Standards for rice, revised; Federal Grain Inspection Service. U.S. GPO, Washington, DC

Webb BD (1991) Rice quality and grades. In: Luh BS (ed) Rice utilization. Van Nostrand Reinhold, New York, pp 89–119

Chapter 4
Brown Rice Flour Rheology

Shumaila Jan, H.A. Pushpadass, D.C. Saxena, and R.P. Kingsly Ambrose

Introduction

Rice (*Oryza sativa* L.) is the staple food of about half the world's human population and particularly for the people in Asia. Brown rice is the dehulled rice obtained from paddy grains, with the bran and germ still intact. It consists of roughly 6–7% (w/w) of bran, 2–3% (w/w) of embryo, and 90% (w/w) of endosperm. The germ and bran layers are the nutrient-rich components in brown rice. The brown rice is normally subjected to abrasion to remove the bran layers from the endosperm and obtain white rice as the latter form is preferred by the consumers. The extent of removal of the bran layers is termed as degree of milling, which determines the whiteness of rice. During milling of brown rice, considerable amount of proteins and minerals are lost. In contrast, the brown rice with germ and bran layers is richer in nutrients such as proteins, lipids, fibres, vitamins, and minerals (Chen et al. 1998; Lamberts et al. 2007).

Rice and brown rice flours are used as primary ingredients in the preparation of many traditional and unique food products. For example, rice flour is used in the preparation of foods such as noodles, breakfast cereals, unleavened breads, snack food items, crackers, candies, and baby foods (Bao and Bergman 2004). In addition,

S. Jan • D.C. Saxena
Department of Food Engineering & Technology, Sant Longowal Institute of Engineering & Technology, Longowal, Punjab, India
e-mail: shumailanissar@gmail.com; dcsaxena@sliet.ac.in

H.A. Pushpadass
ICAR-National Dairy Research Institute, Bangalore, India
e-mail: heartwin1@gmail.com

R.P. Kingsly Ambrose (✉)
Department of Agricultural and Biological Engineering, Purdue University, West Lafayette, IN, USA
e-mail: rambrose@purdue.edu

© Springer International Publishing AG 2017
A. Manickavasagan et al. (eds.), *Brown Rice*,
DOI 10.1007/978-3-319-59011-0_4

as a gluten-free ingredient, rice flours are being increasingly used in the preparation of baked products. Three methods, viz. dry-, semidry-, and wet-grinding procedures, are used in the preparation of rice flour (Chiang and Yeh 2002). Semidry grinding involves soaking the rice kernels in water for few hours, drying to remove excess moisture, and then grinding them. For wet grinding, the rice kernels are soaked in water for specific duration and then ground at high moisture content.

The method of grinding depends on the final use of the flour. For example, wet-grinding method is used to prepare flour to make traditional Indian food products such as *idli* and *dosa*. For easy storability, for handling, and for the convenience of using to prepare different products, dry grinding of low-moisture kernels is a popular method to prepare rice flours. In addition, dry grinding consumes less energy and does not use water. For dry grinding of rice, hammer mill, roller mill, pin mill, and disc-type stone mills are being commonly used. The grinding procedure, to make brown rice flour, depends on the locality and availability of milling machines. The grinding method has a considerable influence on the physicochemical characteristics of the rice flours produced (Chen et al. 1999). Similarly, the chemical and particulate characteristics also depend on the flour milling procedure. Yeh (2004) stated that the model and design of the milling equipment affected the grinding performance as well as the particle size of the flour.

Particle and powder properties affect the handling behaviour of flour during processing, flow through hoppers and silos, mixing, and packaging. From grinding till the end use in bakeries or in other industrial processes, flour handling involves pneumatic conveying, blending, aeration, compaction, and storage in bins or hoppers (Bian et al. 2015). Undesirable bulk powder phenomena such as agglomeration and caking prevent free flow of rice flour through the processing/handling equipment (Emami and Tabil 2007). Bulk powders subjected to a series of static and dynamic loads during handling, transport, processing, and storage undergo change in their physical and flow properties (Bargale et al. 1995). Furthermore, poor flow could lead to downtime due to the clogging of conveying lines, improper discharge from bins, and process downtime.

Flow of powders is defined as the relative movement of bulk particles among neighbouring particles or along the container wall surface. Flowability of powders is inherently complex as there are many factors that influence the response of powders when handled (Freeman 2007). Bulk properties such as moisture content, density, shape, particle size distribution, composition, glass transition temperature, relative humidity (RH), and storage time influence the flow properties of particulate materials (Teunou et al. 1999). In addition, Freeman (2007) reported that the powder flowability was a consequence of the combined effects of various engineering, chemical, and environmental variables. Being hygroscopic, rice flour exhibits poor flow properties and higher cohesiveness (Neel and Hoseney 1984; Kuakpetoon et al. 2001). Under favourable conditions, particulate materials can stick together due to intermolecular forces such as van der Waal's and electrostatic interactions and form stable structures, preventing flow. The magnitude of intermolecular forces depends on the composition, surface characteristics, particle size, and shape. In addition, the bulk porousness of the powders influences the interparticle interaction which consequently impacts the bulk flow properties.

Different techniques have been used to characterize flow properties of particulates, namely, through measurement of flow indicators, compressibility, shear tests, etc. (Abu-Hardan and Hill 2010; Ambrose et al. 2016). The test methods and apparatus that have been used in powder classification can be divided into three types: those used for non-compacted powders (e.g. angle of repose, bulk density), powders that are tapped (tapped density, Hausner ratio, and Carr index), or consolidated (e.g. shear tests). The accuracy of first two methods is operator dependent, including the repeatability. Results of shear test, which analyses the bulk characteristics such as compressibility, cohesion, unconfined yield strength (UYS), and wall friction, are more quantitative, reliable, and directly applicable in design of conveying and storage systems.

With high variability in rice flours due to their differences in physical and chemical characteristics, many traditional powder flow characterization techniques result in poor repeatability and do not simulate dynamic processing conditions. There are no extensive studies on comparing the flow properties of rice flours processed and ground using different mills and at non-compacted, tapped, and consolidated conditions. This chapter describes the flow properties of brown rice flours milled using a traditional stone mill and a modern hammer mill.

Characterization of Brown Rice Flour Rheology

Two grinding principles, namely, impact (using a hammer mill) and abrasion (using a stone mill), were used to grind the rice kernels. The hammer mill (Fab-Tech Engineers Pvt. Ltd., Ahmedabad, Gujarat, India) had seven hammers rotating at 3200 rpm and with a 100 mesh size screen. In each run, 5 kg of cleaned brown rice grains was ground. The stone mill (S.V. Industries Pvt. Ltd., Ajmer, India) had discs rotating at 4200 rpm in concurrent direction. This mill had a capacity of 45 kg per batch.

Particle size analysis of brown rice flour was carried out using a laser diffraction particle size analyser (model SLAD-2300, Shimadzu Corporation, Kyoto, Japan). The laser diffraction and laser scattering intensity pattern with a measuring range of 0.017–2500 µm was used. A suspension was prepared with 0.5 g of powder and 1 mL of ethanol (refractive index: 1.36) by continuous stirring on a magnetic stirrer for 3–5 min at room temperature. Results were expressed as d (v, 0.1), d (v, 0.5), and d (v, 0.9), which were particle sizes in microns at which 10%, 50%, and 90% of the sample volume diameter is smaller. The term "v" indicates that the particle size determination was based on the diameter of a sphere that had the same volume of the measured particle. Particle size distribution was determined as a span, a measure of width of the size distribution.

The various particle diameters (d), namely, d (10%), d (50%), and d (90%), of the stone mill flour were 8.75 ± 0.06, 58.89 ± 0.2, and 172.48 ± 0.77 µm, respectively

Table 4.1 Particle size of brown rice flour ground using different mills[a]

Mill type	d (10%) (μm)	d (50%) (μm)	d (90%) (μm)	Span (d 90–d 10)/d 50
Stone mill	8.75 ± 0.06	58.89 ± 0.20	172.48 ± 0.77	2.78
Hammer mill	14.55 ± 0.84	59.96 ± 0.01	106.69 ± 1.40	1.53

[a]d-particle diameter

(Table 4.1). The corresponding d (10%), d (50%), and d (90%) particle diameters of the hammer mill flour were 14.55 ± 0.84, 59.96 ± 0.01, and 106.69 ± 1.40 μm. It is evident from Table 4.1 that the span of particle size of the flour ground using hammer mill was much less (1.53) when compared with the value of 2.78 for flour ground using stone mill. Thus, the grinding of brown rice in hammer mill resulted in narrow particle size distribution compared to stone milled flour.

Particle, Aerated, and Tapped Density of Rice Flour

Aerated (loose/bulk), tapped, and particle densities are being used to assess the powder bulk characteristics. Aerated density measures the overall packing of powders in a specific volume, while the tapped density is the density after vibration or tapping. Tapped density estimates the compaction characteristics of powders under no-load conditions. True density is the density of the solids without intergranular spaces and voids. The aerated bulk density of rice flour was determined by carefully filling a standard graduated cylinder of 100 mL volume (Vankle's design, Standard Instrument Corporation, Patiala, India) with the sample. The aerated bulk density was calculated from the weight and volume of powders Eq. (4.1). The cylinder was then tapped on a flat surface for about ten times to allow the material to settle and pack. The cylinder was further filled with material to make up the volume to 100 mL. The process was repeated until there was no change in volume. From this compacted volume and weight, the tapped density was calculated Eq. (4.2). The particle density (true density) was determined using the gas multi-pycnometer (Quantachrome Corporation, Boynton Beach, FL, USA) fitted with small cell of 2.74 × 10 × 5 cm dimension.

$$\rho_l = \frac{W_l}{V_l} \tag{4.1}$$

$$\rho_t = \frac{W_t}{V_t} \tag{4.2}$$

where ρ_l and ρ_t are the loose and tapped densities (kg.m^{-3}), W_l and W_t are the weight of flour (kg) under loose and tapped conditions, and V_l and V_t are the volume of flour (m^3) under the same conditions, respectively.

Table 4.2 Bulk properties of brown rice flour milled from different mills

Mill type	Particle density (kg/m^3)	Aerated bulk density (kg/m^3)	Tapped density (kg/m^3)	Hausner ratio	Carr index
Stone mill	1380 ± 20	620 ± 40	810 ± 60	1.31 ± 0.04	23.46 ± 0.11
Hammer mill	1310 ± 50	550 ± 30	770 ± 100	1.40 ± 0.09	28.57 ± 0.09

Hausner ratio (HR) and Carr index (CI) are the measures the propensity of a powder to be compressed (Ambrose et al. 2016). The bulk and tapped density values were used to calculate HR Eq. (4.3) and CI Eq. (4.4) as given below:

$$HR = \frac{\rho_t}{\rho_l} \tag{4.3}$$

$$CI = \frac{100 \times \left(\rho_t - \rho_l \right)}{\rho_t} \tag{4.4}$$

The mean particle density of brown rice flour was in the range of 1310–1380 kg/m^3, while the aerated bulk and tapped densities were in the range of 550–620 and 770–810 kg/m^3, respectively (Table 4.2). The tapped density was higher than bulk density because tapping enabled the smaller particles to occupy the voids between larger particulates and attain a dense packing condition. From Table 4.2, it is also evident that the particle and bulk densities of the two flours were quite similar. The HR and CI of brown rice flour ranged from 1.31 to 1.40 and 23.46 to 28.57, respectively. According to the powder flow classification proposed by Lebrun et al. (2012) based on HR and CI values, brown rice flour ground using both mills was found to have "passable" to "poor" flowability.

Angle of Repose and Coefficient of Friction

A hollow cylinder was placed over a smooth flat surface, and rice flour was filled from the top. The samples were tapped to obtain uniform packing and to minimize the wall effect if any. The tube was slowly raised above the floor so that the whole sample could slide and form a natural slope. The height of heap above the floor and the diameter of the heap at its base were measured, and the angle of repose (φ) was measured using the following relationship:

$$\text{Angle of repose}\left(\phi \right) = \tan^{-1}\left(\frac{2h}{D} \right) \tag{4.5}$$

Table 4.3 Physical properties of brown rice flour from different mills

Mill type	Angle of repose (deg.)	Static coefficient of friction		
		Glass	Galvanized iron	Plywood
Stone mill	58.41 ± 0.04	0.56 ± 0.04	0.62 ± 0.03	0.70 ± 0.02
Hammer mill	53.67 ± 0.06	0.52 ± 0.02	0.59 ± 0.01	0.66 ± 0.04

where ϕ was the angle of repose in degrees, "h" was the height of the pile in mm, and "D" was the diameter of the pile in mm.

The static coefficient of friction (μ) was determined on three structural materials, namely, glass, galvanized iron sheet, and plywood. A hollow plastic cylinder of 30 mm diameter and 35 mm height was placed on an adjustable tilting flat plate and filled with the sample. The cylinder was raised slightly so as not to touch the surface. The structural surface with the cylinder resting on it was inclined gradually, until the cylinder started to slide. The angle of tilt (α) was noted using a graduated scale, and the angle of inclination was calculated as:

$$\mu = \tan \alpha \tag{4.6}$$

The angle of repose of rice flour ground using stone mill was significantly higher than that of the flour ground using the hammer mill (Table 4.3). This could be due to the wider particle size distribution of flour ground using stone mill. The span of stone mill ground flours was almost double (2.78) as compared to 1.53 for the hammer mill ground flours. The wider particle size distribution allowed the intergranular voids between the larger particles to be occupied by the smaller particles, resulting in closer packing of the solids and a higher angle of repose. Carr (1965) stated that powders with an angle of repose less than 40° flow easily but those with greater than 45° will probably not flow well. Based on this classification, brown rice flour ground using both the mills is not easily flowable. For both the flours, the static coefficient of friction was the least on glass surface and the highest on plywood surface. This was expected because plywood has the roughest and most abrasive surface of the three tested materials, which allowed the particles to form a more stable heap without sliding. Also, it is evident from Table 4.3 that the flour ground using hammer mill had marginally lower coefficient of friction across all surfaces as compared to that of stone mill ground flour.

Dynamic Flow Properties

Powders with good dynamic flow characteristics pass through hoppers and processing machines without assistance. In contrast, cohesive powders require mechanical agitation, vibration, or some kind of external assistance to make the bulk powder flow. The flow properties of brown rice flour ground using the two milling machines were evaluated using the FT4 powder rheometer (Freeman Technologies,

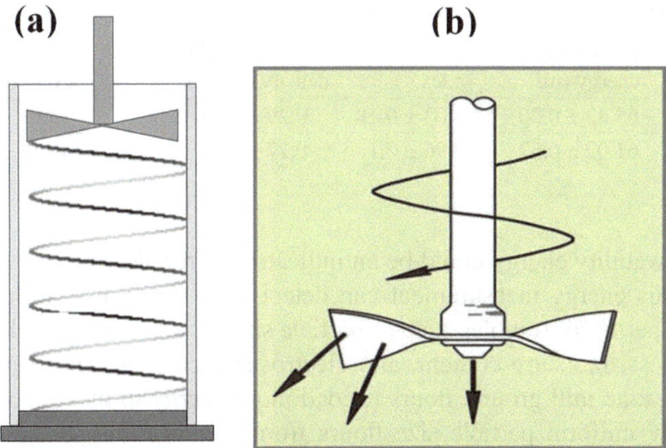

Fig. 4.1 (**a**) FT4 powder rheometer set up showing the helical movement of the blade in the sample holder; (**b**) blade movement within the sample holding cell during testing

Gloucestershire, UK). The vertical glass sample container (120 mm height; 50 mm internal diameter) (Fig. 4.1a) and a rotating blade (48 mm diameter; 10 mm height) that slices through the sample up and down either in clockwise or anticlockwise direction (Fig. 4.1b) were used for testing the flour. Detailed descriptions of this equipment and its use in powder characterization can be found in Freeman (2007). From the forces causing deformation and flow of the flour, which were imposed by the blade moving in a controlled manner, the flow properties were calculated. Basic flowability energy (BFE), stability index (SI), conditioned bulk density (CBD), compressibility, and shear properties were measured as described below.

Basic Flowability Energy (BFE) BFE is the energy required for establishing a particular flow pattern in known volume of conditioned powder. BFE corresponds to the stabilized flow energy needed to displace a conditioned powder sample during downward movement of the probe at 5° helix and at a blade tip speed of 100 mm/s in testing. It can be used as a tool to measure the effect of milling, moisture, and agglomeration on bulk flow properties. The BFE of noncohesive powders, which are easy flowing with low shear strength, is higher than the BFE of cohesive powders.

Stability Index (SI) It is the factor by which the measured flow energy changes during repeated testing and assesses if the powder is affected by being made to flow. The structure of this stability test programme includes seven test cycles of energy measurement. Stable samples would have similar energy values during cycles of measurements. 100 mm/s blade tip speed was used for all test cycles with the blades moving transversely down the vessel. The stability index (SI) was calculated using the following relationship:

$$\text{Stability index}\ (SI) = \frac{\text{Total energy consumed } at \text{ test number } 7, mJ}{\text{Total energy consumed } at \text{ test number } 1, mJ} \tag{4.7}$$

Table 4.4 Flow properties of brown rice flour samples

Mill type	Basic flow energy (mJ)	Stability index	Conditioned bulk density (g/mL)	Compressibility (%)
Stone mill	65.42 ± 0.20	1.03 ± 0.18	0.29 ± 0.01	33.98 ± 2.55
Hammer mill	61.02 ± 0.67	0.96 ± 0.1	0.27 ± 0.01	48.13 ± 0.68

Basic flowability energy could be an indicator of free-flowing characteristics of powders. This energy measurement can detect even very small variations in the physical properties of samples such as particle size distribution, particle shape, surface roughness, moisture content, and electrostatic charges. The densely packed particles of stone mill ground flours needed more energy to move as compared to the relatively uniform particle size flours from hammer mill in the forced flow regime (Table 4.4). It is well known that powder flow properties are dependent on particle size and its distribution (Fu et al. 2012). The presence of smaller particles tends to reduce flowability as the particle surface area per unit mass increases, providing a greater surface area for surface cohesive forces to interact. Another probable explanation for this behaviour was that the relatively cohesive stone mill flour had a much large flowing zone ahead of and around the blade in which shearing occurred because of the high transmissibility of forces from particle to particle in the system. Thus, during the test, a high proportion of sample volume was moved; more work was done in this sample. However, these energy figures were much lower than the values of 680 and 713 mJ reported by Bian et al. (2015) for hard and soft wheat flours, respectively. The stability index of both the flours was close to "1" (Table 4.4), indicating that the bulk powders did not segregate or agglomerate when made to flow. Brown rice flour had conditioned bulk density in the range of 0.27–0.29 g/mL.

Compressibility

Compressibility characteristics of the rice flours were determined by measuring the change in volume (or density) of the powder sample as a function of applied normal stress. During the test, the conditioning blade was replaced by a vented porous piston which was programmed to apply increased stepwise pressure of 1, 2, 4, 6, 8, 10, 12, and 15 kPa normal stresses across the cross section of the sample. The ratio between the density and bulk density at each normal pressure was calculated. The distance travelled by the piston is measured, and the compressibility is calculated from the percentage change in volume. Though compressibility is not a direct measure of flowability, it can be used to understand the behaviour of powders in various process or handling environments. Compressibility would reveal the mechanical compaction that occurs during handling, storage, and transportation. This bulk property is influenced by particle size distribution, cohesivity, particle stiffness, particle shape, and particle surface texture.

The compressibility test is a measurement that also reveals the change in density as a function of applied normal stress. For example, free-flowing powders tend to be less sensitive to direct compression. Irrespective of the milling method, brown rice flours showed an increase in compressibility, in the range of 19–21%, with applied normal stress (Fig. 4.2). However, the wider particle size distribution in stone mill ground flour with relatively closer packing of the particulates did not favour further cohesion and compression, and thus, the values were significantly lower than that of hammer mill ground flours in the range of applied normal stress tested.

Shear Tests

The shear test provides an indication of how easily a powder will move from a static condition to dynamic flow. A rotational shear head attachment, which induced both vertical and rotational stresses, was used in a 50 mm sample holder vessel to conduct the test. As the powder bulk resists the rotation of the shear head, the shear stress increases until the bed shears, at which time a maximum shear stress was observed. The normal stress is maintained constant throughout the measurement. Mohr circle analysis (Fig. 4.3) was conducted on the yield loci (relationship between applied and shear stress) to calculate the cohesion (best fit line intercepting y-axis), unconfined yield strength (UYS, the greater of the two values at which the smaller Mohr circle intercepts the x-axis), flow function (FF, major principal stress/unconfined yield strength), and angle of internal friction (AIF).

For shear analysis, the samples were pre-consolidated at 6 kPa normal load using shear head attachment. AIF is the slope of the line passing through the origin of

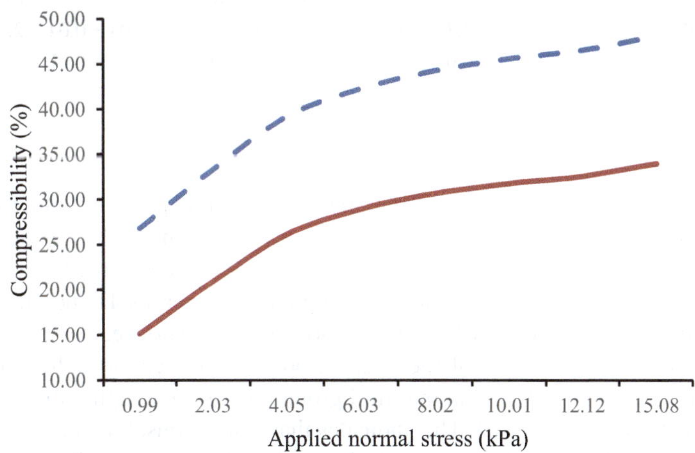

Fig. 4.2 Compressibility of brown rice flour (━ ━ Hammer mill; ━━━ Stone mill)

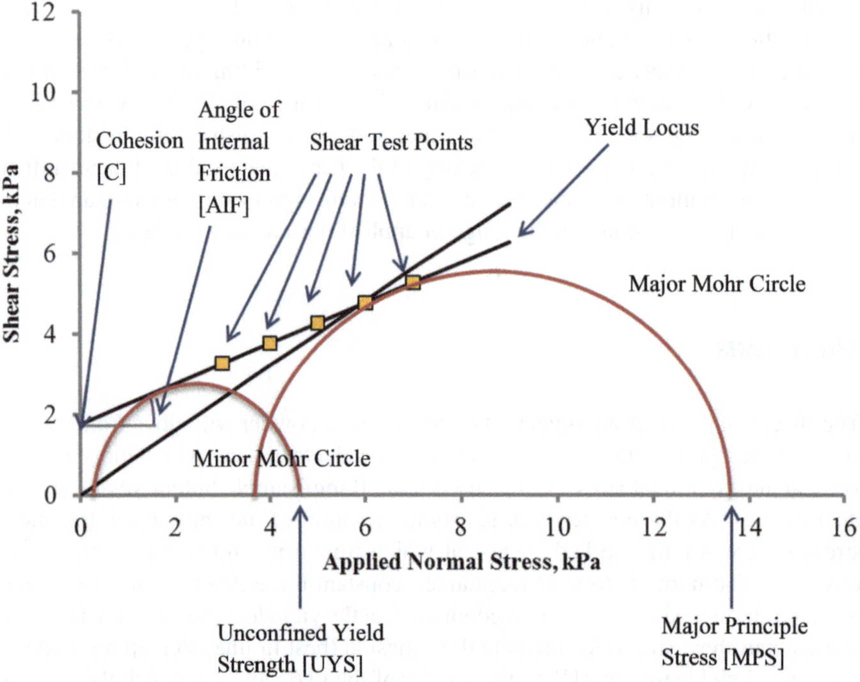

Fig. 4.3 Typical Mohr circle plot to analyse shear test results

Table 4.5 Shear flow properties of brown rice flour samples

Mill type	Major principal stress (kPa)	Unconfined yield stress (kPa)	Angle of internal friction (deg.)	Cohesion (kPa)	Flow function
Stone mill	11.28 ± 0.15	3.79 ± 0.20	32.58 ± 0.34	1.04 ± 0.04	2.98 ± 0.06
Hammer mill	11.74 ± 0.16	6.62 ± 0.28	34.10 ± 0.09	1.76 ± 0.07	1.77 ± 0.08

normal and shear stress plot and tangent to the Mohr circle passing through the pre-shearing point. The flow function is a measure of stress needed to make an arch collapse and make the material flow, while the effective AIF is a measure of friction between particles during steady-state flow of granular materials. Higher the AIF, the more difficult it is for the powder to move along the wall surface.

Shear test is an indicator of the stress a powder can support before it yields. Cohesion, UYS, and AIF of hammer mill ground flours were higher than the stone mill ground flours (Table 4.5). This indicates that, under consolidation, flow behaviour of stone mill ground flours would be better than the hammer mill ground flours. AIF is the measure of the internal friction at steady-state flow of granular materials.

The higher the AIF, the more difficult it is for the powder to move along the wall surface. The AIF of brown rice flour was in the range of 32.58–34.10°. The mean FF values of stone mill and hammer mill flours were 2.98 and 1.77, respectively. Lower FF value suggests easy flow of materials (Fitzpatrick et al. 2004).

Conclusions

Brown rice flour is widely used for various food and non-food applications. Characterization of flow of brown rice flour is important for understanding the handling and transport behaviour during processing operations. Static, dynamic, and shear flow analyses were found to be good indicators of the flowability of differently milled brown rice flour. The particle size distribution of flours depended on the milling technique. The hammer mill ground flours had relatively better flow characteristics compared to the stone mill ground flours. Under consolidation, stone mill ground flours had better flowability. Further studies on the effect of environmental variables such as temperature and relative humidity are essential to understand the significant influence of particle properties on the flow characteristics of rice flour.

References

Abu-Hardan M, Hill SE (2010) Handling properties of cereal materials in the presence of moisture and oil. Powder Technol 198:16–24

Ambrose RPK, Jan S, Siliveru K (2016) A review on flow characterization methods for cereal grain-based powders. J Sci Food Agric 96:259–364

Bao J, Bergman CJ (2004) The functionality of rice starch. In: Eliasson A (ed) Starch in food: structure, functional and application. CRC Press, Cambridge, pp 258–292

Bargale PC, Irudayaraj J, Marquis B (1995) Studies on rheological behaviour of canola and wheat. J Agric Eng Res 61:267–274

Bian Q, Sittipod S, Garg A, Ambrose RPK (2015) Bulk flow properties of hard and soft wheat flours. J Cereal Sci 63:88–94

Carr RL (1965) Classifying flow properties of solids. Chem Eng 72:69–72

Chen H, Siebenmorgen TJ, Griffin K (1998) Quality characteristics of long-grain rice milled in two commercial systems. Cereal Chem 75:560–565

Chen JJ, Lu S, Lii CY (1999) Effect of milling on the physicochemical characteristics of waxy rice in Taiwan. Cereal Chem 76:796–799

Chiang PY, Yeh AI (2002) Effect of soaking on wet-milling of rice. J Cereal Sci 35:85–94

Emami S, Tabil LG (2007) Friction and compression characteristics of chickpea flour and components. Powder Technol 175:14–21

Fitzpatrick JJ, Barringer SA, Iqbal T (2004) Flow property measurement of food powders and sensitivity of Jenike's hopper design methodology to the measured values. J Food Eng 61:399–405

Freeman R (2007) Measuring the flow properties of consolidated, conditioned and aerated powders. A comparative study using a powder rheometer and a rotational shear cell. Powder Technol 174:25–33

Fu X, Huck D, Makein L, Armstrong B, Willen U, Freeman T (2012) Effect of particle shape and size on flow properties of lactose powders. Particuology 10(2012):203–208

Kuakpetoon D, Flores RA, Milliken GA (2001) Dry mixing of wheat flours: effect of particle properties and blending ratio. Lebensm Wiss Technol 34:183–193

Lamberts L, De Bie E, Vandeputte GE, Veraverbeke WS, Derycke V, De Man W, Delcour JA (2007) Effect of milling on colour and nutritional properties of rice. Food Chem 100:1496–1503

Lebrun P, Krier F, Mantanus J, Grohganz H, Yang M, Rozet E, Boulanger B, Evrard B, Rantanen J, Hubert P (2012) Design space approach in the optimization of spray-drying process. Eur J Pharm Biopharm 80:226–234

Neel DV, Hoseney RC (1984) Factors affecting flowability of hard and soft wheat flours. Cereal Chem 61:262–266

Teunou E, Fitzpatrick JJ, Synnott EC (1999) Characterisation of food powder flowability. St.Paul, MN, USA, J Food Eng 39:31–37

Yeh A-I (2004) Preparation and applications of rice flour. In: Champagne ET (ed) Rice: chemistry and technology, 3rd edn. The American Association of Cereal Chemists, Inc., St.Paul, MN, USA, p 495–539

Chapter 5
Physicochemical Characteristics of Rice Bran

M.N. Lavanya, N. Venkatachalapathy, and Annamalai Manickavasagan

Rice bran, which contributes about 10% of whole grain weight, is obtained during milling as a by-product. Every year production of rice bran is increasing as the production of paddy increases; the world rice bran production in 2014–2015 is about 23.80 million tonnes, and during 2015–2016, it is estimated to be 38.50 million tonnes (SEA, India). Interest in the field of rice bran is increasing due to its amazing nutrient content and its health benefits.

Raw rice bran has characteristics of bland flavor, slightly bitter, and sweet in taste. Rice bran is nutritionally rich in protein, fiber, lipids, and antioxidants; it also contains anti-nutritional factors, such as trypsin inhibitor and lectins. Bran contains less soluble fiber and cholesterol-lowering and other health-enhancing properties. These include rice bran oil, plant sterols, tocopherols, oryzanol, and β-sitosterol (Sairam et al. 2011). Although the nutritional and food potential of rice bran have been recognized, the consumption of rice bran in human foods has been limited to a very small quantity due to quick rancidity of the bran. Stabilization process will inactivate the rancid-causing enzyme and anti-nutritional factors (Sharma et al. 2004). As bran deteriorates due to the presence of lipase enzyme, the flavor of bran also will change to incipient rancid, musty, and sour.

Physicochemical properties are very important to know how rice bran will behave during food processing. Physicochemical properties include physical properties (bulk density, color, particle size, water activity), functional properties (water absorption capacity, water solubility, and fat absorption), and chemical properties (nutritional content, antioxidant property). These properties will help to utilize maximum rice bran in food applications. In this chapter, the physicochemical properties of rice

M.N. Lavanya (✉) • N. Venkatachalapathy
Indian Institute of Food Processing Technology, Thanjavur, Tami Nadu, India
e-mail: lavanya1993mn@gmail.com; venkat@iicpt.edu.in

A. Manickavasagan
School of Engineering, University of Guelph, Guelph, ON, Canada
e-mail: manicks33@hotmail.com

© Springer International Publishing AG 2017
A. Manickavasagan et al. (eds.), *Brown Rice*,
DOI 10.1007/978-3-319-59011-0_5

"

bran and its importance are discussed. The physicochemical properties vary according to variety and treatment given to the bran. All raw brans are stabilized before using commercially. The process of stabilization also affects the properties of bran.

Rancidity-Causing Enzymes in Rice Bran

The primary reason for the failure of using rice bran as a food source is the presence of rancid-causing enzymes, which are activated soon after the milling. If oil is not extracted soon after the milling, the oil is hydrolyzed into free fatty acids (FFA) and glycerol. The resulting fatty acids increase bran acidity and reduce pH; produce off-flavor and soapy taste; and change the functional properties of the bran. In grain, the enzymes are present in the testa-cross layer of the grains, but oil vacuoles are present in aleurone and sub-aleurone layers and also in germ. Upon milling, oil is exposed to lipases, causing rapid breakdown to free fatty acids at the rate of 5–7% of the weight of oil per day. The rate of FFA formation is highly dependent on environmental conditions. Up to 70% FFA has been reported for a single month of bran storage (Thanonkaew et al. 2012). Rice bran oil contains 2–4% FFA at the time of milling. Bran oil with an excess of 10% FFA is unfit for human consumption and less than 5% FFA is desirable for producing rice bran oil because high FFA results in high refining losses (Tao et al. 1993).

Rice bran, germ, and the outer layers of the caryopsis have higher enzyme activities. The enzymes like α-amylase, β-amylase, ascorbic acid oxidase, catalase, cytochrome oxidase, dehydrogenase, deoxyribonuclease, esterase, flavin oxidase, α- and β-glycosidase, invertase, lecithinase, lipase, lipoxygenase, pectinase, peroxidase, phosphatase, phytase, proteinase, and succinate dehydrogenase are present (Orthoefer 2005) in the bran. Particularly, lipase, lipoxygenase, and peroxidase are the most important enzymes because they affect the keeping quality and shelf life of rice bran commercially.

Several types of lipase are present in rice bran that are site specific and cleave the 1,3-site of triglycerols (Orthoefer 2005). Depending on the type of lipases present in the bran, storage conditions, and packaging methods, spoilage due to lipase continues. The rancid oil and their related products are toxic and therefore are considered as anti-nutritional factors that reduce the shelf life of the bran.

Stabilization of Rice Bran

It is necessary to stabilize the rice bran by using suitable techniques for controlling undesirable reactions to avoid rancidity, off-flavor, refining loss, and deterioration of nutrients. Stabilization is aimed at destruction or inhibition of lipase, the enzyme that causes the development of FFA. This is done to reduce refining losses, which are directly proportional to the FFA content. To process bran into a food grade product of good keeping quality and high industrial value, all the components which cause deterioration must be removed or their activity should be arrested. Important in this respect is that inactivation of enzymes must be complete and irreversible. At the same time, the valuable nutrients should be preserved. Bran, after proper stabilization, can serve as a good source of protein, essential unsaturated fatty acids, calories, and nutrients such as tocopherols and ferulic acid derivatives.

Two types of rancidity will occur in rice bran such as oxidative rancidity and hydrolytic rancidity. Oxidative rancidity is caused by the oxidation of the double bonds of the fatty acids, while hydrolytic rancidity is caused by the removal of fatty acids from the glycerol molecule (da Silva et al. 2006).

Several thermal methods are used for rice bran stabilization (to inhibit lipase activity). Heat treatment is the most common method to stabilize rice bran. High temperatures above 120°C denature the enzyme responsible for lipid degradation in rice bran without destroying the nutritional value of the rice bran (Thanonkaew et al. 2012). Most of the processes involve dry or moist heat treatment. It is suggested that moist heat treatment may be more effective than dry heat, but few processes that use steam have achieved satisfactory results. To achieve proper stabilization, every discrete bran particle must have a proper moisture content, depending upon the time and temperature of the treatment.

Various stabilization methods, applied to protect rice bran degradation, have been reported such as steaming (Juliano and Bechtel 1985), microwave heating (Lakkakula et al. 2004; Zigoneanu et al. 2008), ohmic heating (Loypimai et al. 2009; Ramezanzadeh et al. 2000), extrusion (Randall et al. 1985), refrigeration, and pH lowering (Amarasinghe et al. 2004). Heat stabilization is accomplished commercially by wet or dry-heating methods, i.e., hot air, drum drying, dry extrusion, and microwave (Narisullah and Krishnamurthy 1989). Although hot air drying is an effective method of stabilization, the nonuniform heating of the material in the tray dryers limits its application.

Extrusion cooking for bran stabilization has been shown to be effective but requires large capital investment (Malekian et al. 2000). Operating and equipment maintenance costs make the process uneconomical. The activity of lipase is dependent on temperature, moisture content, and pH value. The optimum acting condition for the lipase is 30–40°C and pH 7.5–8.0; thus, the basic principal for bran stabilization is inactivating the lipase enzyme by controlling these conditions (Qingci et al. 1999). Hydrochloric acid, acetic acid, and propanol were used in chemical method to inactivate the enzyme (Prakash and Ramaswamy 1996).

Ohmic heating was found to be an effective method for rice bran stabilization with moisture addition. Furthermore, it was observed that free fatty acid concentration increased more slowly than the control for raw bran samples subjected to ohmic heating with no corresponding temperature rise, indicating that electricity has a nonthermal effect on lipase activity. Bioactive compounds of rice bran yielded the highest levels of phenolic compound, α-tocopherol, γ-oryzanol, and antioxidant activity at electric field strengths in the range of 140–225 V/cm (Lakkakula et al. 2004).

Infrared radiation has been used recently for inactivation of enzymes. FFA content of rice bran stabilized with a shortwave infrared radiation remained almost constant for 6 months (Yılmaz et al. 2014). Besides, infrared (IR) radiation offers many advantages including versatility, simplicity of the required equipment, fast response of heating, easy installation, and low capital cost. It mentioned as a low-cost approach mainly because of the radiative energy, is directly transferred from the heating element to the material without heating the surrounding air. The preference of the wavelength of IR radiation is considerably important in food processes because not only it affects the temperature and the emissivity of the emitter, but also it affects the absorption intensity of the radiative energy by food components. For instance, the level of radiation, which reflected back by the materials with a rough surface, is 50% for NIR and less than 10% for FIR region (Chua and Chou 2003).

The use of microwave heat for stabilization of rice bran is considered as one of the effective methods in controlling deterioration of bran (Wu and Salunkhe 1977; Lee and Yoon 1984). Compared with other heat treatments, microwave heating is efficient, economically superior, and shorter in processing time; has little effect on the nutritional value of the bran; and has little or no effect on the original color of the bran (Tao et al. 1993). Microwave is having these advantages and effective on lipase enzyme in rice bran. Bran stabilized at 850 W for 3 min showed 16 weeks of storage life, and the microwave-stabilized rice bran yields maximum oil as compared to unstabilized rice bran.

Recently, researchers focused on subcritical water as an environmentally friendly technique for decomposition. The eco-friendly treatment technique for rice bran oil stabilization under subcritical water condition results indicated that subcritical water could effectively inactivate the enzyme and total free fatty acids concentration remained constant in the treated oils, whereas it considerably increased from 5.6% to 36.0% in untreated samples (Pourali et al. 2009).

Physical Properties

Color

Raw rice bran has light tan or yellowish brown color, and it varies according to the variety and the processing conditions. Most of the bran is stabilized before using commercially. In heat processing, the color of the bran changes to brown by losing its yellowish light color, and the bran stabilized by dry heat changes its color to brown and extruded bran dark brown (Garcia et al. 2012). Upon heating, change in the color may be attributed to the formatting of Maillard reaction compounds and partly to the color of pigments of raw bran.

Rice bran is composed of beta-carotene and lycopene, which is considered as carotenoid, which gave its reddish brown appearance. Lycopene and beta-carotene are precursors of vitamin A, and both of them are antioxidants in food and biological system (Lamberts et al. 2006). Carotenoids act as photoprotective agents by absorbing the potentially harmful light energy or by quenching singlet oxygen and trapping free peroxyl radicals.

Water Activity(A_w)

Water activity of any material is very important to avoid spoilage and to extend shelf life. In the case of bran also, it is important to know the water activity to avoid hydrolysis. When A_w below 0.30 reaches primary adsorption zone, where water molecules linked to carboxyl (COOH) group, which links to other molecules by hydrogen bonds, this water layer would not dissolve the food components, but it would cover the food, which could lead to an acceleration of lipid oxidation. The raw bran had A_w around 0.54 ± 0.03, and upon dry heat stabilization process, it reduces to 0.25 ± 0.0 (Garcia et al. 2012).

Particle Size

The particle size of the bran varies according to the type of polisher used to remove bran from rice kernel and the processing conditions which are used for stabilization. The commercially available rice brans are sieved in different mesh size to get the proper particle size for specified use of the bran. For example, to get maximum oil yield after stabilization, the pellets will be formed so course bran is needed, and to use in food systems, fine flour is required (Table 5.1).

Bulk Density

The bulk density of stabilized bran would be more when compared to raw bran of 0.500 g/ml increase in bulk density after stabilization which is observed to increase percolation rate of the solvent during extraction of oil and has better increase in oil extraction (Sharma et al. 2004). After extraction of oil also, the bulk density would decrease to 0.34 g/ml, and the bulk density of different stabilized bran is given in Table 5.2.

Table 5.1 Rice bran particle size distribution, %

Mesh	Particle size (μm)	Raw bran	Moist heat stabilized bran
18	>1000	0	0
18–30	1000–595	2.4	18.6
30–50	595–297	30.0	32.7
50–80	297–177	12.2	18.5
80–100	177–149	8.5	10.8
<100	<149	46.7	19.4

Luh et al. (1991)

Table 5.2 Functional properties of rice bran

	Bulk density (g/ml)	Water absorption (g/100g)	Water solubility (g/100g)	Fat absorption (g/100g)
Raw bran	0.500	143.68	7.57	96.32
Commercially defatted rice bran	0.340	240.00	–	210.00
Extruded stabilized bran	0.593	170.93	9.33	72.80
Dry heat stabilized bran	0.548	156.91	9.34	81.28

Sairam et al. (2011), Sharma et al. (2004)

Functional Properties

To incorporate bran in any food, there is a need to know its functional characteristics like water absorption capacity, water solubility, and fat absorption capacity. These functional properties will vary according to the bran source and processing conditions. Bran is rich in protein and fiber; with fiber having high water-holding capacity, hence bran will absorb the maximum amount of water around 143.68 ml/100 g, and as bran stabilized the bran, the moisture content is less than that of the raw bran, and water absorption capacity is increased. Also in defatted bran, the water absorption is more due to change in the porosity of the bran. During bran heat stabilization, the starch molecules absorb water and get solubilized which cause water solubility of the bran. During the heat treatment, there is a change in the bed porosity and pore size of the bran due to the rupture of fatty vacuoles, and fat leads to leach out from the tissue, which causes decrease in fat absorption.

Because of the high water-holding capacity of rice bran, it maintains moisture and freshness in baked foods. Raw rice bran has the ability to stay stable in emulsified layer of food even after 30 min of heating. It shows the possibility of using bran as fat emulsifiers in foods (Tang et al. 2003). The forming capacity aids air incorporation, leavening and maintaining texture in baked foods and whipped toppings. Defatted rice bran absorbs more water and has good foaming capacity and stability in food. Raw rice bran has high foaming capacity as well as emulsification. The extruded rice bran with 115.5% foaming value could be the best bran for use in food systems (Table 5.3).

Chemical Properties

Chemical properties include proximate and antioxidant properties of the bran. For any food material, we need to know the composition of what we are eating, and how much we are eating is very important for the betterment of health. As mentioned earlier bran is a rich source of protein, fiber, and fat and other nutrients, having unique antioxidants like oryzanol and tocopherol, which makes bran more interesting and passionate to incorporate into the foodstuff. The nutrients are distributed

Table 5.3 Emulsification and foaming percentage of bran

	Emulsification (%)	Foaming (%)
Raw rice bran	46	9
Microwave-stabilized bran	41	5
Extrusion stabilized bran	22	4

Tang et al. (2003)

throughout the aleurone layer in varying quantity. The nutrient content varies according to the source of bran and processing conditions; stabilized bran has the ability to retain maximum nutrients and the percentage of fat also increases. The bran is also a rich source of silica due to the presence of rice hull fragments. Commercial bran is a fine, floury material made up of the outer layer of brown rice and pulverized germ with some hull fragments and some endosperm (Orthoefer 2005). The particle size distribution varies according to polishing percentage, i.e., the degree of milling, milling conditions, and type of milling. According to these conditions, the composition of bran also will vary. Generally, a low degree of milling is practiced.

The amounts of starch and other nutrients in the bran depend on the degree of milling and extent of kernel endosperm breakage during milling. Whitening is the process in which brown rice is subjected to removal of bran and germ giving white rice by applying pressure to grain which generates heat causing cracking and brokens which result in less head yield. Therefore, to avoid the overheating of grains, three to four whitening machines are provided which are collectively known as a multibreak system. The multibreak system will provide the resting time while milling, to avoid the reduction in the production of heat due to friction in polishing and also serve as a port to the separation of the bran which has been removed from the grain (Yilmaz 2016). The amount of bran removed from the whitening machine is known as bran fraction of that particular multibreak pass, and mixture of outputs from these passes is known as composite rice bran (Table 5.4).

Nutrients are not uniformly distributed throughout the bran layers, and the removal of bran is due to friction and abrasion of the kernel and the surface of polisher. The removal of bran is not layer by layer; abrasion will occur in a zigzag way which may remove endosperm layer along with the bran. In some portion, the bran will not be removed completely as it may have attached to the kernel, and in some other portion, there will be complete removal of bran, which causes variations in compositions according to polisher and degree of milling.

As mentioned earlier the stabilization techniques change the properties of bran, and it retains the maximum amount of nutrients by inactivating the rancid-causing enzymes especially lipases. The proximate composition of different brans is given in Table 5.5. Oil quality will be improved by inactivating enzyme, increasing shelf-stability of the oil by reducing FFA of the bran. The FFA of raw bran is more as compared to the stabilized one over the period of bran storage. The high amount of FFA indicates the bran rancid, and it is no longer able to add in the human diet.

Rice bran contains 15–20% oil with three main fatty acids, i.e., palmitic acid (12–18%), oleic acid (40–50%), and linoleic acid (30–35%). The bran is a rich

Table 5.4 Composition of rice bran during degree of milling

	Degree milling (%)	Protein	Fat	Fiber	Ash
1st cone	0–3	17.0	17.7	10.5	9.8
2nd cone	3–6	17.6	17.5	10.3	9.4
3rd cone	6–9	17.0	16.5	5.7	8.4
4th cone	9–10	16.7	14.2	5.7	7.5

Danforth and Orthoefer (1989)

Table 5.5 The proximate composition of different bran g/100g

	Raw bran	CDRB	Microwave
Moisture	11.23	11.1	8.4
Protein	14.63	17.2	17.5
Fat	16.4	0.66	17.5
Total dietary fiber	24.58	11.44	28.98
Ash	7.4	14.65	7.6
Carbohydrate	51.7	43.6	48.9

Sairam et al. (2011)
CDRB: commercially defatted rice bran

source of minerals also like iron, calcium, and phosphorus. As told earlier the bran has unique compounds like oryzanol and tocopherol (Orthoefer 2005).

The rice bran contains a unique complex of naturally occurring antioxidant compounds like oryzanols, tocopherols, tocotrienols, and phytosterols groups. Rice bran oil contains a high amount of antioxidants like tocopherol approximately 1.0% (v/v) compared to other oil seeds and tocotrienol around 1.7% (v/v). Crude rice bran oil contains $\leq 2\%$ (v/v) of oryzanol that is a mixture of sterol esters of ferulic acid (Lloyd et al. 2000). The concentration of these antioxidants also depends on the degree of milling; as milling progresses these compounds got reduced and also stabilization technique which is used to inactivate lipolytic enzymes.

Uses of Rice Bran

Rice bran is finding enormous applications in food industries for increasing the nutritional quality of processed foods. Rice bran being high in dietary fiber and in view of its therapeutic potential, its addition can contribute to the development of value-added foods or functional foods that currently are in high demand. Supplementation of rice bran has been successfully carried in various foods like bread, cakes, noodles, pasta, and ice creams without significantly affecting the functional and textural properties.

The primary use of rice bran as an additive in food is due to its high fiber content, which mildly promotes stool regularity. From a marketing point, the most commonly available rice bran-derived product is the oil (Prasad et al. 2012). Rice bran oil has an impressive nutritional quality that makes it suitable for nutraceutical products.

Table 5.6 Potential application of rice bran as food

Product enriched	Purpose of addition	Inference	Reference
Pizza with stabilized rice bran flour	Effect on chemical and functional properties of storage frozen pizzas	5% rice bran incorporated pizza dough stable for 60 days at −18°C	De Delahaye et al. (2005)
Pasta	Effect on textural and antioxidant properties	Pasta supplemented with rice bran was highly acceptable up to 4 months of storage	Kong et al. (2012)
Bread	Effect of replacing wheat flour by infrared stabilized rice bran on minerals and B vitamins	Increase in te amount of B vitamins and minerals, especially niacin and phytic acid	Tuncel et al. (2014)
Bread enriched with full-fatted rice bran (FFRB) and defatted rice bran (DFRB)	Effect on functional properties of bread	Loaf volume increased with 2% of FFRB and decreased with 6% DFRB Hardness, gumminess, and chewiness increased with increase in level of DFRB than FFRB	Lima et al. (2002)
Bread with rice bran hemicellulose	Effect on chemical and functional properties	Enrichment resulted in higher water-binding and swelling capacity Significantly reduced loaf volume and increased firmness of bread	Hu et al. (2009)
Pork meatballs enriched with bran	Effect on sensory and physicochemical properties	Protein, fat, and white index decreased as bran level increased <10% bran incorporated meatballs showed good textural properties	Huang et al. 2005
Cookies	Fiber and mineral enrichment	Supplementation improved dietary fiber content and mineral profile Defatted rice bran can be substituted up to 20% in wheat flour	Sharif et al. (2009)

It also has the potential to be used as an additive to improve the storage stability in food due to its antioxidant properties. Rice bran oil has industrial potential, especially in the preparation of snack food due to the great stability of frying, whereas rice bran fiber can be used as both a nutritional and functional ingredient. Chicken coated with stabilized rice bran fiber tends to absorb less fat during frying, and the small amount of fat present naturally in rice bran fiber can act as a carrier of flavors (Mohd Esa et al. 2013).

The nutritional and functional properties of rice bran are suitable for baked products, namely, cookies, muffins, bread, crackers, pastries, and pancakes. The addition of rice bran into the wheat flour further increased the protein, lysine, and dietary fiber contents in bread and cookies. The color, flavor, protein extractability, and solubility of bran, as well as other properties, such as water and fat absorption and emulsifying and foaming capacity, have demonstrated improvements that further enlighten us on the potential use of bran in foods (Tuncel et al. 2014). Due to its naturally occurring enzymatic activity (lipases) and subsequent hydrolytic rancidity, rice bran needs to be stabilized to control these undesirable reactions.

The antioxidants present in bran destroy the fungi, bacteria, and insect infestations, hence enhancing the shelf life of rice bran. The stabilized rice bran was successfully incorporated in up to 20% of the production of yeast bread because the hygroscopicity of the rice bran may improve its moisture retention in the baked products, while its ability to foam improved the air incorporation and leavening processes (Gul et al. 2015). Defatted rice bran can be used to substitute for up to 10–20% of the wheat flour used for making cookies without adversely affecting the quality. Biscuits prepared with broken rice powder were highly acceptable in terms of taste and feel in the mouth.

Besides oil, rice bran also has 10–15% protein content, consisting of 37% water-soluble, 31% salt-soluble, 2% alcohol-soluble, and 27% alkali-soluble storage proteins. Rice bran proteins have been found to be of high quality and application in food and pharmaceutical industries. Its unique properties, hypoallergenicity, and anticancer effects make it a superior cereal protein with a wide range of possible applications (Gul et al. 2015; Tuncel et al. 2014). However, as of now, commercial rice bran protein is still unavailable on the market (Table 5.6).

References

Amarasinghe BMWPK, Gangodavilage NC (2004) Rice bran oil extraction in Sri Lanka: data for process equipment design. Food Bioprod Process 82:54–59

Chua KJ, Chou SK (2003) Low-cost drying methods for developing countries. Trends Food Sci Technol 14(12):519–528

da Silva MA, Sanches C, Amante ER (2006) Prevention of hydrolytic rancidity in rice bran. J Food Eng 75(4):487–491

Danforth S, Orthoefer FT (1989) Presented at 64th Tri-state Oil Mill Assoc. Mtg., Biloxi

DeDelahaye EP, Jimenez P, Perez E (2005) Effect of enrichment with high content dietary fiber stabilized rice bran flour on chemical and functional properties of storage frozen pizzas. J Food Eng 68(1):1–7

Garcia MC, Benassi MDT, SoaresJúnior MS (2012) Physicochemical and sensory profile of rice bran roasted in microwave. Food Sci Technol (Campinas) 32(4):754–761

Gul K, Yousuf B, Singh AK, Singh P, Wani AA (2015) Rice bran: Nutritional values and its emerging potential for development of functional food – a review. Bioact Carbohydr Diet Fibre 15:30002–30004

Hu G, Huang S, Cao S, Ma Z (2009) Effect of enrichment with hemicellulose from rice bran on chemical and functional properties of bread. Food Chem 115(3):839–842

Huang SC, Shiau CY, Liu TE, Chu CL, Hwang DF (2005) Effects of rice bran on sensory and physico-chemical properties of emulsified pork meatballs. Meat Sci 70(4):613–619

Juliano BO, Bechtel DB (1985) The rice grain and its gross composition. In: Juliano BO (ed) Rice chemistry and technology, 2nd edn. American Association Cereal Chemist, St Paul, pp 17–57

Kong S, Kim DJ, Oh SK, Choi IS, Jeong HS, Lee J (2012) Black rice bran as an ingredient in noodles: chemical and functional evaluation. J Food Sci 77(3):C303–C307

Lakkakula NR, Lima M, Walker T (2004) Rice bran stabilization and rice bran oil extraction using ohmic heating. Bioresour Technol 92(2):157–161

Lamberts L, Brijs K, Mohamed R, Verhelst N, Delcour JA (2006) Impact of browning reactions and bran pigments on color of parboiled rice. J Agric Food Chem 54(26):9924–9929

Lee JS, Yoon HN (1984) Stabilization of rice bran by microwave energy. Korean J Food Technol (Korea R.) 16(1):113–119

Lima I, Guraya H, Champagne E (2002) The functional effectiveness of reprocessed rice bran as an ingredient in bakery products. Nahrung/Food 46(2):112–117

Lloyd BJ, Siebenmorgen TJ, Beers KW (2000) Effects of commercial processing on antioxidants in rice bran 1. Cereal Chem 77(5):551–555

Loypimai P, Moonggarm A, Chottanom P (2009) Effects of ohmic heating on lipase activity, bioactive compounds and antioxidant activity of rice bran. Aust J Basic Appl Sci 3(4):3642–3652

Luh BS, Barber S, de Barger CB (1991) In: Luh BS (ed) Rice utilization, vol II. Avi Book, Van Nostrand Reinhold, New York

Malekian F, Rao RM, Prinyawiwarkul W, Marshall WF, Windhauser M, Ahmedna M (2000) Lipase and lipoxygenase activity, functionality and nutrient losses in rice bran during storage, Bulletin number, 879. Louisiana Agricultural Experiment Station, LSU Agricultural Center, Baton Rouge

Mohd Esa N, Tan BL, Loh SP (2013) By-products of rice processing: an overview of health benefits and applications. J Rice Res 1:1–11

Narisullah HW, Krishnamurthy MN (1989) Effect of stabilization on the quality characteristics of rice bran oil. J Am Oil Chem Soc 66:661–663

Orthoefer FT (2005) Rice bran oil. In: Shahidi F (ed) Bailey's industrial oil and fat products, Edible oil and fat products: chemistry, properties, and health effects, vol 1, 6th edn. Memorial University of Newfoundland, St. John's

Pourali O, Salak FA, Yoshida H (2009) A rapid and ecofriendly treatment technique for rice bran oil stabilization and extraction under sub-critical water condition. In: Proceedings of the world congress on engineering and computer science, vol 1

Prakash J, Ramaswamy HS (1996) Rice bran proteins: properties and food uses. Crit Rev Food Sci Nutr 36(6):537–552

Prasad N, Sanjay KR, Vismaya MN (2012) Health benefits of rice bran-a review. J Nutr Food Sci (2011) 1(3):1–7

Qingci H, Well H, Yong Z, Chongyr C (1999) Experimental study on the storage of heat-stabilized rice bran. In: Proceedings of the 7th international working conference on stored-product protection, vol 2. Sichuan Publishing House of Science and Technology, Chengdu, pp 1685–1688 (ISBN 7536440987)

Ramezanzadeh FM, Rao RM, Prinyawiwatkul W, Marshall WE, Windhauser M (2000) Effects of microwave heat, packaging, and storage temperature on fatty acid and proximate compositions in rice bran. J Agric Food Chem 48(2):464–467

Randall JM, Sayre RN, Schultz WG, Fong RY, Mossman AP, Tribelhorn RE, Saunders RM (1985) Rice bran stabilization by extrusion cooking for extraction of edible oil. J Food Sci 50(2):361–364

Sairam S, Krishna AG, Urooj A (2011) Physico-chemical characteristics of defatted rice bran and its utilization in a bakery product. J Food Sci Technol 48(4):478–483

Sharif MK, Butt MS, Anjum FM, Nawaz H (2009) Preparation of fiber and mineral enriched defatted rice bran supplemented cookies. Pak J Nutr 8(5):571–577

Sharma HR, Chauhan GS, Agrawal K (2004) Physico-chemical characteristics of rice bran processed by dry heating and extrusion cooking. Int J Food Prop 7(3):603–614

Tang S, Hettiarachchy NS, Horax R, Eswaranandam S (2003) Physicochemical properties and functionality of rice bran protein hydrolyzate prepared from heat-stabilized defatted rice bran with the aid of enzymes. J Food Sci 68(1):152–157

Tao J, Rao R, Liuzzo J (1993) Microwave heating for rice bran stabilization. J Microw Power Electromagn Energy 28:156–156

Thanonkaew A, Wongyai S, McClements DJ, Decker EA (2012) Effect of stabilization of rice bran by domestic heating on mechanical extraction yield, quality, and antioxidant properties of cold-pressed rice bran oil (Oryza saltiva L.) LWT-Food Sci Technol 48(2):231–236

Tuncel NB, Yılmaz N, Kocabıyık H, Uygur A (2014) The effect of infrared stabilized rice bran substitution on B vitamins, minerals and phytic acid content of pan breads: part II. J Cereal Sci 59(2):162–166

Wu MT, Salunkhe DK (1977) Extending shelf-life of fresh soybean curds by in-package microwave treatments. J Food Sci 42(6):1448–1450

Yılmaz N (2016) Middle infrared stabilization of individual rice bran milling fractions. Food Chem 190:179–185

Yılmaz N, Tuncel NB, Kocabıyık H (2014) Infrared stabilization of rice bran and its effects on γ-oryzanol content, tocopherols and fatty acid composition. J Sci Food Agric 94(8):1568–1576

Zigoneanu IG, Williams L, Xu Z, Sabliov CM (2008) Determination of antioxidant components in rice bran oil extracted by microwave-assisted method. Bioresour Technol 99(11):4910–4918

Part III
Nutritional and Health Benefits

Chapter 6
Chemical and Nutritional Properties of Brown Rice

Sila Bhattacharya

Introduction

Rice, one of the oldest food crops, is the staple food for about half of the world population. The annual world production of paddy (rough rice) is about 740 MT at present (Anon 2016). Most of the rice is produced in the Asian countries mainly in China, India, Indonesia, Vietnam, Thailand, Bangladesh, etc. Rice is a good source of easily digestible starch and good quality protein due to the presence of high content of lysine compared to other cereal grains. Rice, in its whole form, is a good source of vitamins like thiamine and niacin and minerals like iron, phosphorus, and magnesium (Juliano 1985). However, table rice is usually eaten as its milled form, where the husk as well as the bran is removed during milling. Thus, the concept of brown rice has emerged; it can be described as the dehusked or dehulled whole grain rice with its bran and germ. Since it contains the bran and germ layers, it is believed to be nutritionally superior to milled rice; it is the only form of grain which contains vitamin E and offers a cholesterol-lowering effect. Brown rice has a mild nutty flavor but exhibits fat-degraded rancid flavor because of the presence of bran and germ. In some countries, the milled rice is fortified with thiamine, niacin, and essential minerals to enrich the grain. However, in many countries where rice constitutes the main diet, enrichment is not the common practice. A diet containing brown rice diet is better than that of milled rice considering nutritional status. The photographs of paddy (rough rice), brown rice, and milled (polished) rice are shown in Fig. 6.1. The brown rice has a dull brown or light yellow color compared to corresponding polished rice.

Thus, there is an increasing inclination of consuming brown rice although white rice is the most popular form among the rice eaters throughout the world. Brown

S. Bhattacharya (✉)
Grain Science & Technology Department, CSIR-Central Food Technological Research Institute, Mysore, India
e-mail: silabhat@yahoo.com

© Springer International Publishing AG 2017
A. Manickavasagan et al. (eds.), *Brown Rice*,
DOI 10.1007/978-3-319-59011-0_6

Fig. 6.1 Photographs of (**a**) paddy, (**b**) brown rice, and (**c**) milled rice

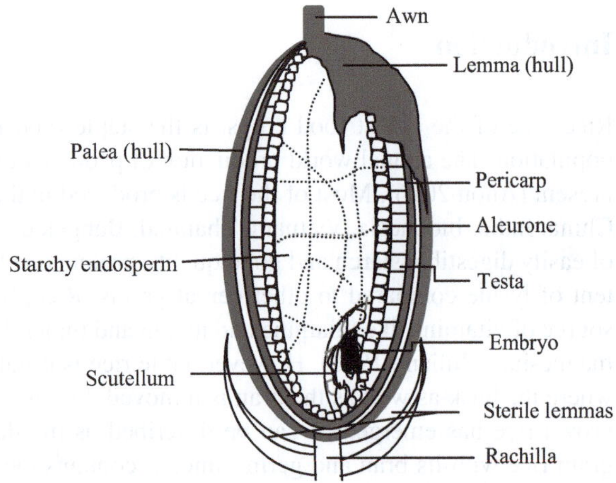

Fig. 6.2 Cross-sectional view of rice kernel

rice has a shelf life of about 6 months at the ambient condition but can be extended further by hermetic storage or by refrigeration. Some of the brown rice-based products have been developed recently and possess a good scope for commercial exploitation. These products include expanded brown rice, popped rice, brown rice flakes, brown rice chips, brown rice vermicelli, brown rice noodles, brown rice instant semolina, etc.

Structure and Milling Quality of Rice

Whole rice grain with its husk, bran, and endosperm is called rough rice or paddy. Rice is covered by inedible two thick glumes known as lemma and palea which are removed by shelling. Two glumes are joined by interlocking. The typical structure of rice is shown in Fig. 6.2. The firm interlocking of lemma and palea gives a good

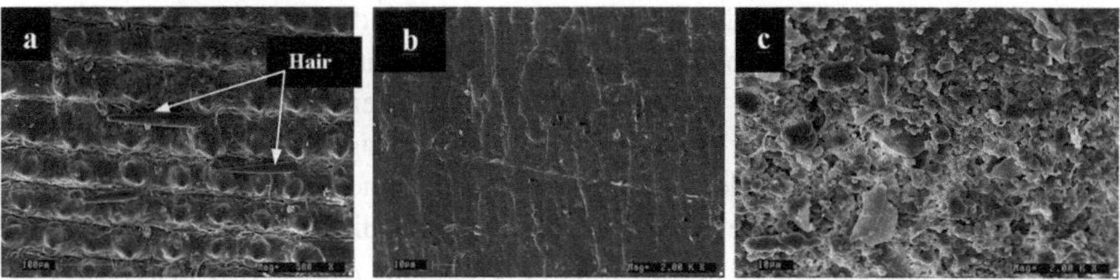

Fig. 6.3 Photomicrographs of the outer surface of (**a**) paddy, (**b**) brown rice, and (**c**) milled rice at magnifications of 500×, 2000×, and 2000×, respectively

storage life and puffing characteristics in the paddy but makes the shelling process difficult and tricky. Moreover, the presence of minute hairs (Fig. 6.3a) on the rough surface glume causes more wear and tear of machinery employed for polishing/shelling. During the grain development stages, due to the pressure of husk, the longitudinal grooves and ridges (Fig. 6.3b) are formed on the surface of the brown rice. Varieties having deep grooves require more extent of polishing and thus offer higher nutritional losses. Hence, milling quality is markedly influenced by the structure of the grain. Dehusked whole grain, called brown rice, is covered by the caryopsis coat consisting of pericarp, seed coat or testa, and aleurone layer. Milled rice endosperm is composed of starch, protein, and fat (Fig. 6.3c).

The embryo or germ is located on the ventral side of the caryopsis and is attached to the endosperm portion by the scutellum. The inedible husk contributes 18–28% of the rough rice, and the remaining 72–82% is brown rice. Brown rice constitutes about 1–2% pericarp, 2–3% aleurone, 4–6% seed coat, and 2–3% germ, and the remaining is the endosperm. The cell walls of the pericarp are 2 μm in thickness and contain protein, hemicellulose, and cellulose (Juliano 1972). Next to pericarp, there exists a layer of seed coat or tegmen. Beneath the tegmen, the aleurone layer of a thickness of about 2 μm encloses the endosperm and the embryo. The aleurone layer is composed of protein, cellulose, and hemicellulose. The embryo is present on the outer side of the aleurone layer. For obtaining brown rice from rough rice, two rubber rolls with a rubber belt operating against a ribbed steel roll can be used for effective milling. The abrasive hulls damage the rubber rolls and thus frequent replacement of them is necessary. Since rubber rolls are less bruising on the surface of the brown rice, the produced brown rice is more stable to lipase attack.

Influencing Factor for Composition

Environmental factors such as wider spacing of plants during cultivation, the use of high nitrogen-containing fertilizer, and ambient temperature during the growth of plants and ripening increase the protein and ash contents of brown rice but possess a nonsignificant effect on the fat content of the rice kernel. Protein content tends to increase with the short growth duration of plant and due to cloudy weather during grain development (Liu et al. 2014). Stresses such as drought condition, salinity, and alkalinity of water and soil; higher or lower temperature of environment; occurrence of diseases; and attack by pests increase the protein content of rice grain but decrease the starch content. The contents of total nitrogen and minerals (such as calcium, copper, and molybdenum) and the total chlorine content of soil also affect the protein content of the rice grain. Environmental conditions thus affect the rate of starch and protein deposition during grain development. An increase in protein content reduces the starch content in the grain. An increase in the duration of plant growth also decreases the protein content of brown rice. The mineral content of the grain is obviously affected by the mineral content of the soil as well as water used for irrigation. Sulfur deficiency in the soil reduces grain yield without having any adverse effect on the cysteine and methionine contents of the rice protein.

Nutritional Composition of Rice

Nutritional composition of rice is highly varietal as well as environmental dependent. The same variety of rice may show variations in the composition due to different climatic and soil conditions. The proximate composition (as dry basis) of brown rice indicates that it contains 2.4–3.9% fat, 1.5–2.1% ash, 0.8–2.6% fiber, and 7.3–15.4% protein (N × 5.95); the energy content is 1520–1610 kJ/100 g, while the bulk density is 676–683 kg/m^3 (Juliano et al. 1964; Matz 2014). Carbohydrate or starch is distributed mostly in the endosperm cells of brown rice. The linear fraction of the starch (called amylose) constitutes 8–37% of its starch content in non-waxy rice, whereas the branched fraction (called amylopectin) is the major fraction of starch in the waxy type (glutinous) of rice. Rice starch shows the characteristic A-type X-ray diffraction pattern of cereal starches and has a similar granule size and gelatinization temperature (Wani et al. 2012). Gelatinization temperature is a range of temperature when starch granules start to swell in hot water and lose its birefringence. Gelatinization temperature may vary from 55°C to 80°C depending on the variety and environmental conditions such as the differences in the temperature during the development of the grain. Differences in the cooking time of rice also may be observed due to the differences in gelatinization temperature.

The superior protein quality of rice compared to other cereals is due to the presence of high content of the limiting amino acid such as lysine. Further, the major rice protein fraction is called glutelin which is soluble in dilute acid or alkali. Since the contents of protein and amino acid vary in different fractions of the grain, there is a considerable variation in the total protein content as well as the amino acid sequence of the milled fractions of the brown rice. Further, a high degree of milling adversely affects the nutritive value of the milled rice. The process of milling removes the pericarp, seed coat, aleurone layer, and embryo to yield milled rice. Hence, there is a loss in the contents of fat, protein, fiber, minerals, thiamine, riboflavin, niacin, and α-tocopherol. However, the high content of phytic acid and dietary fiber present in the bran layers hinders the bioavailability of the minerals in the brown rice.

Carbohydrate

Carbohydrate is the major component of any cereal grain. The bran layers of brown rice contain the maximum content of hemicellulose (about 1.4–2.1% pentosans) of which 43% is distributed in the bran layers and 8% in germ (Juliano et al. 1964). The water-soluble hemicelluloses of bran have an arabinose/xylose ratio of 1.8. The bran layer also contains some galactose and protein, whereas alkali-soluble hemicelluloses contain about 37% arabinose, 34% xylose, and 11% galactose with a trace amount of glucose along with protein and uronic acid. Brown rice also contains a high amount of cellulose (about 62% in the bran) due to the presence of seed coat, aleurone layer, and thick pericarp cell walls. Starch is the major constituent of rice. The amylopectin constitutes 25–50% by number and 30–60% by weight of amylose. The iodine affinity of amylose from rice is 20–21% by weight, while for amylopectin it is 0.4–0.9% in low- and intermediate-amylose rice, whereas it is 2–3% in high-amylose rice. Based on the starch-iodine color absorption, rice is classified as waxy (1–2%), very low-amylose (2–12%), low-amylose (12–20%), intermediate-amylose (20–25%), and high-amylose (25–33%) rice. The waxy endosperm is opaque due to the presence of air spaces among the starch granules, which have a lower density than the non-waxy granules. Viscoamylograph characteristics of waxy and non-waxy rice flour show that waxy rice has the lower peak viscosity than the non-waxy rice flour possibly due to the presence of non-starch constituents, mainly water-insoluble rice proteins in waxy rice which hinder the swelling of the starch granules. Common rice starch has an amylose/amylopectin ratio of about 20:80 (Xie et al. 2008). These varieties have small granule sizes in the range of 3–8 μm; irregularly shaped polygons are present in the waxy rice to exhibit compound granules (Hegenbart 1996). Non-starch polysaccharides consist of water-soluble polysaccharides and insoluble dietary fiber (Juliano 1985) which form complexes with starch and may have a hypocholesterolemic effect.

In brown rice, the total free sugars present are about 0.8–1.4% of which 0.1–0.13% are the reducing sugars mainly as glucose and a trace amount of fructose. Brown rice contains about 0.2% of phytin or myoinositol hexaphosphate, an important constituent of the bran layers. Ribonucleic acid is the main nucleic acid in brown rice and constitutes about 0.2–0.3% of the rice and 2–3% of brown rice protein; deoxyribonucleic acid constitutes about 0.01% of brown rice (Juliano et al. 1964).

Protein

Brown rice contains about 8–15% (dry weight basis) protein of which 14% is distributed in bran and about 80% in the endosperm. There are discrepancies of the reported amino acid content due to the differences in the methodological conditions followed by the researchers. A wide range of reported lysine content values is also noticed indicating the existence of high lysine varieties. The bran and germ fractions in the brown rice contain higher levels of lysine and lower levels of glutamic acid than the starchy endosperm. Both waxy and non-waxy rice have the similar pattern of amino acid distribution. The brown rice and milled rice have the same pattern of amino acid content for most of the cases; however, some amino acids particularly the lysine content vary negatively and tyrosine and phenylalanine contents positively with the total protein content of the brown rice sample. Protein solubility fractions are unevenly distributed in brown rice and milled rice fractions. Rice protein is mainly alkali-soluble protein glutelin (78–79%), salt-soluble globulin (10–11%), and water-soluble albumin (7–8%), and the alcohol-soluble fraction is the least (3%). Albumin and globulin contents are higher in brown rice than milled rice as they are more concentrated in the germ and aleurone layers. The proportion of albumin and globulin in protein is high at the periphery of the kernel but gradually decreases toward the center, whereas glutelin content is distributed inversely. Albumin distribution is 51% in bran, whereas globulin distribution is 40% in bran. Globulin is composed of 43% glutamic acid and arginine, 4–9% cysteine, and 6.6% methionine. Prolamin can be extracted by 70% ethanol followed by acetone precipitation from ethanolic extraction. Prolamin distribution is about 21% in bran, whereas glutelin distribution is only 5% in bran. Glutelin has the highest molecular weight among the rice protein fractions. Aspartic and glutamic acids are the major amino acids present in glutelin. The soluble protein fractions vary with an increase in protein content; glutelin and prolamin contents also increase. Protein bodies of rice are composed of about 60% protein, 10–28% lipid, and 12–29% carbohydrates along with a small amount of ash, ribonucleic acid, phospholipid, phytic acid, and niacin (Tanaka et al. 1980).

Amino acid compositions of brown rice, reported as g/16 g of N (mentioned by Anon (1969) and cited by Juliano (1972)), are alanine, 5.5–6.5; arginine, 7.9–9.5; aspartic acid, 9.0–10.5; cysteine, 1.2–2.1; glutamic acid, 16.9–19.9; glycine, 4.5–5.4; histidine, 2.1–2.9; isoleucine, 4.1–4.8; leucine, 7.9–8.9; lysine, 3.5–4.6; methionine, 1.9–2.9; phenylalanine, 5.3–6.0; proline, 4.4–5.5; serine, 4.6–5.9; threonine, 3.6–4.4; tryptophan, 0.9–1.6; tyrosine, 4.4–5.4; and valine, 5.9–7.0. Free amino acids constitute about 0.7% by weight of brown rice protein of which the major free amino acids are aspartic and glutamic acids. Amino acid composition of the four protein fractions of brown rice reveals that albumin has the highest lysine content followed by glutelin, globulin, and prolamin. If the albumin content is higher, the higher is the lysine and the lower is the glutamic acid content which reflects the better protein quality in germ and bran in rice. Among the cereal proteins, rice protein offers the best nutritional status due to its high lysine content, although lysine is still the limiting amino acid followed by threonine. It is observed that there is a further increase in the nitrogen balance of the food for children if rice is fortified with methionine, threonine, and tryptophan instead of lysine alone. For brown rice, a wide range of protein efficiency ratio (PER) has been reported because of the differences of the experimental conditions employed; the reported PER values are between 1.73 and 1.93, and biological values are 67–89 (Juliano 1972). Improved nutritional status is noticed for high protein rice samples containing higher levels of all essential amino acids; this fact encourages the breeders to provide more effort on the breeding of high protein containing rice varieties rather than on improving their protein quality.

Vitamins and Minerals

Brown rice is rich in vitamins like thiamine (0.29–0.61), riboflavin (0.04–0.14), niacin (3.5–5.3), and tocopherol (0.90–2.50) and minerals like calcium (10–50), phosphorus (170–430), iron (0.2–5.2), and zinc (0.6–2.8); values are shown here as mg/100 g of flour (Juliano 1985; Pedersen and Eggum 1983). Vitamins are mostly concentrated in the aleurone layers of the brown rice. The overall composition of rice is not affected by storage, but vitamin content decreases progressively. Juliano (1972) has indicated that brown rice is an important source of thiamine and niacin and also tocopherol which is about 0.9–2.3 mg/100 g of brown rice. Since the B vitamins are more concentrated in the bran layer, the major nutritional advantage of brown rice over milled rice is its high content of vitamin B. About 50% of the total thiamine is in the scutellum portion of the grain, and 80–85% of the niacin is present in the pericarp and aleurone layers; the embryo accounts for more than 95% of total tocopherols. The thiamine of brown rice is mostly concentrated in the bran (65%), and 22% is present in the milled rice fraction (Juliano and Bechtel 1985). About 39% riboflavin and 54% niacin are present in the bran, respectively. The minerals are also concentrated mostly in the outer layers of brown rice. A major portion (90%) of the phosphorus in bran is phytin phosphorus. Studies on the same subjects have shown lower apparent absorption rates for sodium, potassium, and phosphorus; a lower phosphorus balance has been reported for the brown rice diet when the protein intake is low (Miyoshi et al. 1987). In case of a standard protein intake, even if the potassium, phosphorus, calcium, and magnesium levels are higher in the brown rice diet, the absorption rates of potassium and phosphorus are significantly lower for the brown rice diet (Miyoshi et al. 1987). The important factor is the high phytate level in the bran fraction (aleurone and germ). Pigmented brown rice shows higher riboflavin but similar thiamine contents compared to nonpigmented IR rice (Deepa et al. 2008). The selenium content of brown rice, grown in Japan, is reported to be 30–40 mg/g (Miller and Engel 2006); 13% Se is present in the hull, 15% in bran, and 72% in rice kernel.

Lipid

In brown rice, 80% of lipid is present in the aleurone layer and bran, specifically as lipid bodies or spherosomes. The lipid characteristics for brown rice, bran, and germ are similar. The unsaponifiable matter of bran oil is composed of 42% sterols, 24% higher alcohols, 20% ferulic acid, and 10% hydrocarbons (Juliano 1972). Oryzanol, a ferulate ester of unsaturated triterpenoid alcohols, is a potent antioxidant present in the bran oil at a level of 0.96–2.89%. Brown rice also contains another potent antioxidant called tocopherols at 5% level of which 47% is the principal tocopherol, i.e., α-tocopherol which is present at a level of 0.005–0.015% of the brown rice lipids. Other two tocopherols, such as β-tocopherol and γ-tocopherol, are present at about 26% levels. The wax content of rice bran oil is 3–9%. Brown rice contains significantly higher levels of linoleic, palmitic, and oleic acids but has lower contents of myristic, palmitic, palmitoleic, and stearic acids compared to the milled rice. About 43% oleic, 28% palmitic, and 25% linoleic acids (Herting and Drury 1969) are the major fatty acid components in brown rice and are concentrated mostly in the bran and germ portions of the grain. Free fatty acids and mono- and diglycerides mainly comprise palmitic, oleic, and linoleic acids. Waxy and non-waxy rice have a similar fatty acid composition.

Anti-nutritional Factors

Rice contains the least amount of anti-nutritional and allergenic substances among all other cereals. Rice is the best among the cereals in most of the aspects of nutritional point of view, though the bran layer in rice contains some anti-nutritional factors like phytates, trypsin inhibitor, oryza cystatin, hemagglutinin-lectin, etc. which are mostly concentrated in the embryo and aleurone layer (Juliano 1985). These are protein in nature and except phytate, others can be denatured by heat treatment. Phytin phosphates readily form a complex with the essential minerals like Fe, Ca, and Zn and also with protein and make these minerals unavailable to the system (Miyoshi et al. 1987). Phytates can be inactivated by soaking and cooking of brown rice; the cooking process also inactivates the hemagglutinin-lectin and trypsin inhibitors. Wet heat treatment is more effective to inactivate trypsin inhibitor than the dry heat treatment of the embryo. Hemagglutinin-lectins can bind the specific carbohydrate receptor sites in the intestinal mucosal cells and thus interfere the nutrient absorption across the intestinal wall. Inhibitors and lectins are mostly located in the embryo. Oryzacystatin is another protease inhibitor present in brown rice but decreases its activity when subjected to a high temperature (45% at 120°C). Oryzacystatin can effectively inhibit cysteine proteinases such as ficin, papain, chymopapain, and cathepsin C but hardly affects the serine proteinases (subtilisin, trypsin, and chymotrypsin) or carboxyl proteinase (pepsin).

Digestibility Characteristics

Starch digestibility of brown rice is significantly lower than of milled white rice (Juliano 1972). It indicates that the rate at which brown rice starch is digested and the appearance of glucose in blood are also significantly lower compared to milled rice sample. Insulin response of cooked brown rice is much lower than cooked milled rice which has 70–90% starch digestibility (Miller et al. 1992). Brown rice shows approximately the same protein digestibility as that of milled rice in their cooked form (Bradbury et al. 1984). A slightly lower true digestibility for the brown rice protein is observed in rats compared to milled rice. On the other hand, both rice show similar biological value (BV, a measure of the proportion of absorbed protein from the food) and net protein utilization (NPU, the ratio of amino acid converted to proteins to the ratio of supplied amino acids) (Eggum et al. 1982). Brown rice has a true nitrogen digestibility of 99.7% and biological value of about 74%, and net protein utilization is 96.3% (Eggum 1979).

Stabilization

In spite of nutritional advantages, consumption of brown rice is limited. A short shelf life of only 3–6 months of brown rice is one of the major causes restricting its mass consumption. Fat-degraded off-flavors and off-odors have limited the commercial production and consumption of it. During dehusking, the outer bran layer of rice gets damaged and thus gets exposed to hydrolytic and oxidative deterioration, leading to the hydrolysis of the oil in rice. Hence, for stabilizing brown rice, three approaches are conventionally followed such as (1) inactivation and denaturation of lipases present in rice kernel by heat treatment, (2) free oil removal by employing an organic solvent, and (3) inactivation and denaturation of lipase and lipase-producing bacteria and mold by ethanol extraction. However, the oxidative degradation is rather difficult to prevent completely. However, by maintaining a low oxygen level (below 1%) in the packaging system, storing of brown rice at a low temperature can reduce oxidative rancidity. Storage in a dark environment under modified atmospheres or under vacuum can also control it but cannot completely prevent oxidative changes.

Constraint

Although brown rice is believed to be nutritionally superior, it has been recently challenged due to concerns over arsenic levels. Brown rice has been reported to possess about 80% more inorganic arsenic on average than that of white rice of the same type. Rice readily absorbs more arsenic than many other plants. If consumed over time, the presence of arsenic may lead to cancer and skin lesions. Researchers now are in a hope to obtain rice plants that express increased levels of *Oryza sativa* C-type ATP-binding cassette (ABC) transporter (OsABCC) family, OsABCC1, or to genetically engineer a rice sample to overexpress the transporter. This approach may solve the problem of arsenic contamination of rice and the rice-based products in a cost-effective manner (Song et al. 2014). Some semi-dwarf *indica* rice in Japan has shown a high cadmium level; major causes of epidemic of "itai-itai" disease in Japan are due to the high cadmium content in rice (Kasuya et al. 1992).

Trends in Brown Rice Research

Though brown rice is an old concept emphasizing a nutritious product, there have been a few recent researches that focus on the nutritious status of brown rice (Table 6.1). Anthocyanin content in different colored rice has been identified and quantified to evaluate its potential application as the functional food ingredients (Abdel-Aal et al. 2006). The total anthocyanin pigment content varies widely. However, red and black rice contain a few pigments of which the most abundant anthocyanin is cyanidin 3-glucoside in black and red rice.

Colored rice cultivars show stronger antioxidant activities and free-radical scavenging activities than that of white (polished) rice. The antioxidant properties are mainly due to the presence of phenolic compounds other than anthocyanin pigments. The antioxidant capacity results mainly from the outer seed coat, and not from the endosperm (Chen et al. 2012). This phenomenon may help the future breeding researchers to have a beneficial effect on rice milling process with distinct desirable colors of brown rice.

Massaretto et al. (2011) have investigated the inhibitory effect of phenolic compounds on the activity of angiotensin I-converting enzyme (ACE); the effect of cooking on phenolics and their inhibitory activities for several pigmented and nonpigmented rice varieties is also studied. Pigmented rice shows significantly higher inhibitory effect than that of nonpigmented rice on ACE. Further, cooking significantly reduces the content of total phenolics and ACE inhibition.

Investigations of the antioxidant activity and the lipophilic and hydrophilic components of total phenolic contents for some cereal grains from China reveal that these cereals possess diverse antioxidant capacities. Phenolic compounds such as gallic acid, kaempferol, quercetin, galangin, and cyanidin 3-glucoside are widely found in those cereals (Deng et al. 2012). Further, it has been claimed that the pigmented or colored cereals, such as black rice, red rice, and purple rice, can be important sources of natural antioxidants for health promotion and reduction in disease risk as they possess the highest antioxidant capacities and total phenolic contents among other cereals.

Finocchiaro et al. (2007) have investigated on the total antioxidant capacities and chemical constituents of the antioxidants for dehulled red rice and dehulled white rice. Dehulled red rice possesses three times more antioxidant capacity than dehulled white rice, and the antioxidants present are the proanthocyanidins and phenolic compounds. During milling of red rice, bran gets removed causing a significant loss of antioxidants. Further, during the cooking, additional loss of antioxidants occurs, although cooking in limited water can minimize the loss to a small extent. Thus, from a nutritional point of view, cooking of brown rice or partially milled rice in limited water is a preferred choice for consumption.

Table 6.1 Studies on the composition and nutritional status of brown rice

Area of research	Important finding	References
Aroma compounds in four varieties of cooked brown rice employing aroma extract dilution analyses (AEDA)	Forty one odor-active compounds have been found of which three major aroma compounds like 2-amino acetophenone, 2-acetyl-1-pyrroline, and 3-hydroxy-4,5-dimethyl-2(5H)-furanone are the prominent flavor components	Jezussek et al. (2002)
Extraction of soluble and insoluble phenolic compounds from white rice, brown rice, and germinated brown rice	Phenolic compounds are present in higher quantities in brown rice or germinated brown rice compared to that of white rice. Ferulic acid content increases to one-and-half times, and sinapinic acid content increases to ten times in germinated brown rice compared to that of non-germinated brown rice	Tian et al. (2004)
Nutraceutical qualities of the fat-soluble nutraceuticals such as oryzanol, tocopherols, and tocotrienols contents of three varieties of brown rice and their corresponding milled rice	Highest tocopherol and tocotrienol contents are observed in the parboiled brown rice of *Basmati* variety. The higher amount of saturated fatty acids is in the milled rice oil than that of brown rice. The total lipid content is about four times higher in brown rice compared to that of the milled rice	Khatoon and Gopalakrishna (2004)
Identification and quantification of anthocyanin contents in different colored rice	Red and black rice contain a few pigments of which the most abundant anthocyanin is cyanidin 3-glucoside	Abdel-Aal et al. (2006)
Effect of nutraceutical lipid content during milling of brown rice	An increase in the degree of milling of brown rice significantly decreases the lipid content and also a simultaneous decrease in tocopherols. An increase in the degree of milling of brown rice markedly decreases the concentrations of γ-oryzanol, squalene, and octacosanol	Ha et al. (2006)
Content of γ-oryzanol and steryl ferulates of 30 different varieties of brown rice samples of European origin	Cycloartenyl ferulate and 24-methylenecycloartanyl ferulate are the major components of the γ-oryzanol	Miller and Engel (2006)
Dehulled red and white rice analyzed for their total antioxidant capacities and chemical constituents of the antioxidants	Dehulled red rice possesses three times more antioxidant capacity than dehulled white rice, and the contributed antioxidants are due to the proanthocyanidins and phenolic compounds	Finocchiaro et al. (2007)

(continued)

Table 6.1 (continued)

Area of research	Important finding	References
Fat-soluble phytochemicals such as tocotrienols and γ-oryzanol in 32 rice genotypes including the subspecies of *japonica* and *indica* varieties	A significant variation in the contents of vitamin E isomers and the γ-oryzanol in the rice genotypes have been observed. The α-tocopherol, α-tocotrienol, and γ-tocotrienol are the most abundant compounds present in *japonica* rice, while γ-tocotrienol, α-tocopherol, and α-tocotrienol are in *indica* rice	Heinemann et al. (2008)
Investigation on the inhibitory and cooking effects of phenolic compounds for several pigmented and nonpigmented rice varieties	Pigmented rice shows significantly higher inhibitory effect than that of nonpigmented rice on angiotensin I-converting enzyme (ACE)	Massaretto et al. (2011)
Determination of the contents of γ-oryzanol, total phenolics, individual phenolic acid profile, and the antioxidant activity of different rice milling fractions	About 94% of γ-oryzanol content is reduced in milled rice compared to the brown rice sample	Tuncel and Yilmaz (2011)
Investigation on the antioxidant properties of colored rice cultivars	Colored rice cultivars exhibit stronger antioxidant activities and free-radical scavenging activities than that of white rice	Chen et al. (2012)
Investigation of the antioxidant activity and the lipophilic and hydrophilic components of total phenolic contents for some cereal grains from China	These cereals possess diverse antioxidant capacities. Phenolic compounds such as gallic acid, kaempferol, quercetin, galangin, and cyanidin 3-glucoside are widely found in those cereals. Black, red, and purple rice can be the important sources of natural antioxidants	Deng et al. (2012)
Investigation of several *indica* and *japonica* rice varieties for their composition and distribution of vitamin E	The contents of vitamin E and total tocopherol are much higher in *japonica* rice than in *indica* rice. The γ-tocotrienol is the most abundant component in *indica* rice, while α-tocopherol is the major component in the *japonica* rice	Zhang et al. (2012)
Investigation on traditional red and brown rice for their antioxidant properties and phenolic contents	Proanthocyanidin-containing traditional red rice possesses markedly higher antioxidant properties and phenolic contents than light brown rice varieties	Gunaratne et al. (2013)
Study on the nutritional status of rice having red- or purple-colored bran	The presence of the intact bran layer of whole grain rice or brown rice makes it nutrient dense	Bett-Garber et al. (2013)

(continued)

Table 6.1 (continued)

Area of research	Important finding	References
Investigation on the effects of parboiling, storage, and cooking of brown rice on their biological activities	The γ-tocotrienol is the major constituent of the total tocols, whereas α-tocopherol, α-tocotrienol, and γ-tocopherol are present in small quantities. The parboiled rice, after 6 months of storage followed by cooking, shows about 90% losses of tocopherols	Pascual et al. (2013)
Effect of the anti-colitis effects of brown rice and identification of the anti-oxidative and inhibitory effects	The inhibitory effects of pro-inflammatory cytokines, myeloperoxidase activity, neutrophil infiltration in the colonic mucosa, and activation of nuclear factor kappa B for brown rice are determined because these are the major factors responsible for anti-colitis effects	Shizuma (2014)

Proanthocyanidin-containing traditional red rice (Sri Lankan variety) has been shown to possess over sevenfold higher antioxidant properties and phenolic contents than that of light brown rice varieties. Further, it is observed that these traditional red varieties also contain a significant amount of protein, well-balanced amino acids, and a higher content of fat, fiber, and vitamin E compared to the new types (Gunaratne et al. 2013).

The cooking quality of aromatic pigmented and nonpigmented rice on the physical and physicochemical properties, color, and viscosity profile and the effect of cooking on phytochemical contents and antioxidant capacities have been investigated by Saikia et al. (2012); these examined samples have different sizes and shapes. Pigmented varieties contain the highest amount of total phenolics, total flavonoids, and antioxidant properties. A drastic reduction has been reported for all these contents and also their antioxidant properties due to cooking.

Zhang et al. (2012) have investigated several *indica* and *japonica* rice varieties for their composition and the distribution of vitamin E in the rice kernel. The content of tocopherols or tocotrienols in brown rice of these varieties has been determined using a reverse phase HPLC method. Results reveal that the contents of vitamin E between these two types differ significantly; the contents of vitamin E and total tocopherol are much higher in *japonica* rice than in *indica* rice. The γ-tocotrienol is the most abundant component in *indica* rice, while α-tocopherol is the major component in the *japonica* rice.

Shizuma (2014) has investigated the anti-colitis effects of brown rice by employing experimental colitis model and identified the anti-oxidative and inhibitory effects of pro-inflammatory cytokines, myeloperoxidase activity, neutrophil infiltration in the colonic mucosa and activation of nuclear factor kappa B; these are the major factors responsible for anti-colitis effects. In addition, dysbiosis may have some relation to the anti-colitis effects.

Pascual et al. (2013) have studied the effects of parboiling, storage, and cooking of brown rice on their biological activities especially hypocholesterolemic, anti-inflammatory, and antioxidant activities. The γ-tocotrienol is the major constituents (~75%) of the total tocols, whereas α-tocopherol, α-tocotrienol, and γ-tocopherol are present in small quantities. The parboiled rice, after 6 months of storage followed by cooking, shows about 90% losses in tocopherols. However, parboiled rice, after 6 months of storage, shows about 60% retention of γ-oryzanol, which is stable after cooking.

Tian et al. (2004) have extracted the soluble and insoluble phenolic compounds from white rice, brown rice, and germinated brown rice; it is concluded that all these phenolic compounds are present in higher quantities in brown rice or germinated brown rice compared to that of white rice. However, in all these cases, the insoluble phenolic content is significantly higher compared to the soluble phenolic compounds. Ferulic acid content increases to one-and-half times, and sinapinic acid content increases to ten times in germinated brown rice compared to that of non-germinated brown rice samples. Hence, these researchers have suggested that brown rice consumption is a good practice, but germinated brown rice may offer even higher health benefits.

Tiwari and Cummins (2009) have reviewed the importance of tocopherols in human health and nutrition that are mostly concentrated in the germ and bran fractions in the grains. The content of tocopherols in a product depends on the unit operations employed during food processing including the milling steps. Hence, the by-products obtained from milling operations may be incorporated in different formulations to develop nutritious brown rice-based functional foods.

Nutraceutical qualities of the fat-soluble nutraceuticals such as oryzanol, tocopherols, and tocotrienols contents of three varieties of brown rice and their corresponding milled rice have been investigated by Khatoon and Gopalakrishna (2004). The total lipid content is about four times higher in brown rice compared to that of the milled rice. However, the parboiling process reduces the tocopherol and tocotrienol contents significantly, while oryzanol content is not affected. Highest tocopherol and tocotrienol contents are observed in the parboiled brown rice of *Basmati* variety. The higher amount of saturated fatty acids is present in the milled rice oil than that of brown rice. The cause of lowering of the ratio of saturated to monounsaturated fatty acid in milled rice may be due to the change in the fatty acid composition of the brown and milled rice.

Ha et al. (2006) have studied the effect of nutraceutical lipid content during milling of brown rice. An increase in the degree of milling of brown rice significantly decreases the lipid content accompanied by a simultaneous decrease in the content of tocopherols. An increase in the degree of milling of brown rice markedly decreases the concentrations of γ-oryzanol, squalene, and octacosanol in the milled rice.

The content of γ-oryzanol and steryl ferulates of 30 different varieties of brown rice samples of European origin has been analyzed for their composition (Miller and Engel 2006). Cycloartenyl ferulate and 24-methylenecycloartanyl ferulate are the major components of the γ-oryzanol. The content of γ-oryzanol ranges from 26 to 63 mg/100 g of the brown rice sample. The environmental conditions influence the variation in the γ-oryzanol content and steryl ferulates composition. However, the degree of maturity of the grains does not affect these components.

Tuncel and Yilmaz (2011) have determined the γ-oryzanol content and total phenolics, individual phenolic acid profile, and the antioxidant activity of the free and bound extracts of the different rice milling fractions. About 94% reduction in γ-oryzanol content occurs in the milled rice compared to the brown rice. It is worth mentioning here that the rice bran fraction is a rich source of γ-oryzanol, phenolics and antioxidants.

Heinemann et al. (2008) have analyzed the fat-soluble phytochemicals such as tocotrienols and γ-oryzanol in 32 rice genotypes including the subspecies of *japonica* and *indica*. A significant variation in the contents of vitamin E isomers and the γ-oryzanol in the rice genotypes has been observed. The α-tocopherol, α-tocotrienol, and γ-tocotrienol are the most abundant compounds present in *japonica* rice, while γ-tocotrienol, α-tocopherol, and α-tocotrienol are in *indica* rice. Total vitamin E content in *japonica* is reported to be about 24 mg/kg which is significantly higher than that of *indica* rice (about 17 mg/kg).

Bett-Garber et al. (2013) have indicated that the presence of the intact bran layer of whole grain rice or brown rice makes it nutrient dense. Further, rice having red- or purple-colored bran shows higher contents of phenols and flavonoids, and it is also associated with some flavor attributes. The variation of the mass of bran and its thickness is attributed due to the proportions of bran, hay, and straw, whereas amylose content negatively correlates well with the sweet taste of the bran. The hardness of the kernel is attributed to the kernel density and bran thickness which in turn also varies with the cooking time of rice.

Jezussek et al. (2002) have found 41 odor-active compounds when aroma compounds are analyzed in four varieties of cooked brown rice employing the aroma extract dilution analyses (AEDA). Three major aroma compounds like 2-amino acetophenone (medicinal, phenolic like), 2-acetyl-1-pyrroline (popcorn like aroma), and 3-hydroxy-4,5-dimethyl-2(5H)-furanone (seasoning like) are the prominent flavor components reported for these three varieties of brown rice. The *indica* variety differs most in its overall aroma from these three Asian brown rice samples.

Quality Factors and Standards

There are different sets of the US standards for rice practiced for commercial purposes to serve as the foundation for purchasing specifications. The US standard for brown rice is defined as "Rice (*Oryza sativa* L.), which consists of more than 50% of kernels of brown rice, and which is intended for processing to milled rice" (Anon 2009). As per the percent content of paddy kernels, red rice, damaged kernels, objectionable seeds, chalky kernels, broken kernels, well-milled kernels, etc., there are five numbered grades for each of long grain, medium grain, and short grain and mixed type of rice grains.

Codex standard for rice (Anon 1995) defines husked rice (brown rice or cargo rice) is the paddy rice from which the husk is only removed. Similarly, Indian Standard (Anon 1999) indicates that brown rice is the paddy from which husk only has been removed. The process of husking and handling may result in some loss of bran. The products covered by the provisions of this standard shall be free from heavy metals in amounts which may pose a health hazard to human. The pesticide residues for rice shall comply with those maximum residue limits established by the Codex Alimentarius Commission for this commodity.

Conclusion

Brown rice is the dehusked or dehulled whole grain rice possessing the bran and germ. The presence of bran and germ makes the brown rice to contain a significant amount of dietary fiber, important vitamins, and minerals which are present in negligible quantities in the milled white rice. The composition and nutritional characteristics of brown rice indicate the health-benefitting aspects for human consumption.

References

Abdel-Aal ESM, Young JC, Rabalski I (2006) Anthocyanin composition in black, blue, pink, purple, and red cereal grains. J Agric Food Chem 54:4696–4704

Anon (1969) International Rice Research Institute, Los Banos, Laguna, Philippines 1970, p 27

Anon (1995) *Standard for rice*, Codex Standard 198–1995. Codex Alimentarius, International Food Standards, WHO, Food and Agriculture Organisation of the United Nations

Anon (1999) Rice grader–specification (*Second Revision*) ICS 67.060; 53.1000 BIS 1999, *Bureau of Indian Standards*, Manak Bhavan, 9 Bahadur Shah Zafar Marg, New Delhi 110002

Anon (2009) United States Standards for brown rice for processing. Federal Food, Drug, and Cosmetic Act Effective 27 Nov 2009, 42 FR 40869, 12 Aug 1977; 42 FR 64356, 23 Dec 1977

Anon (2016) Food and Agriculture Organization of the United Nations. Statistics Division. Retrieved 22 July 2016. http://faostat3.fao.org/

Bett-Garber KL, Lea JM, McClung AM, Chen MH (2013) Correlation of sensory, cooking, physical, and chemical properties of whole grain rice with diverse bran color. Cereal Chem 90:521–528

Bradbury JH, Collins JG, Pyliotis NA (1984) Digestibility of proteins of the histological components of cooked and raw rice. British J Nutr 52:507–513

Chen XQ, Nagao N, Itani T, Irifune K (2012) Anti-oxidative analysis, and identification and quantification of anthocyanin pigments in different coloured rice. Food Chem 135:2783–2788

Deepa G, Singh V, Naidu KA (2008) Nutrient composition and physicochemical properties of Indian medicinal rice–Njavara. Food Chem 106:165–171

Deng GF, Xu XR, Guo YJ (2012) Determination of antioxidant property and their lipophilic and hydrophilic phenolic contents in cereal grains. J Funct Foods 4:906–914

Eggum BO (1979) The nutritional value of rice in comparison with other cereals. In: Proceedings of workshop on chemical aspects of rice grain quality, IRRI, Los Baños, Laguna, Philippines, pp 91–111

Eggum BO, Juliano BO, Maningat CC (1982) Protein and energy utilization of rice milling fractions by rats. Plant Foods Hum Nutr 31:371–376

Finocchiaro F, Ferrari B, Gianinetti A (2007) Characterization of antioxidant compounds of red and white rice and changes in total antioxidant capacity during processing. Mol Nutr Food Res 51:1006–1019

Gunaratne A, Wu K, Li D (2013) Antioxidant activity and nutritional quality of traditional red-grained rice varieties containing proanthocyanidins. Food Chem 138:1153–1161

Ha TY, Ko SN, Lee SM (2006) Changes in nutraceutical lipid components of rice at different degrees of milling. Eur J Lipid Sci Technol 108:175–181

Hegenbart S (1996) Food product design. Understanding Starch Functionality. 1 Jan 1996, Weeks Publishing Company. Retrieved 22 June 2016. www.foodproductdesign.com

Heinemann RJB, Xu Z, Godber JS, Lanfer-Marquez UM (2008) Tocopherols, tocotrienols and γ-oryzanol contents in *japonica* and *indica* subspecies of rice (*Oryza sativa* L.) cultivated in Brazil. Cereal Chem 85:243–247

Herting DC, Drury EJ (1969) Alpha-tocopherol content of cereal grains and processed cereals. J Agric Food Chem 17:785–790

Jezussek M, Juliano BO, Schieberle P (2002) Comparison of key aroma compounds in cooked brown rice varieties based on aroma extract dilution analyses. J Agric Food Chem 50:1101–1105

Juliano BO (1972) The rice caryopsis and its composition. In: Houston DF (ed) Rice: chemistry and technology. Am Assoc Cereal Chem, St Paul, pp 16–74

Juliano BO (1985) Rice: chemistry and technology, 2nd edn. Am Assoc Cereal Chem, St Paul, p 774

Juliano BO, Bechtel DB (1985) The rice grain and its gross composition. In: Juliano BO (ed) Rice: chemistry and technology, 2nd edn. Am Assoc Cereal Chem, St Paul, pp 17–57

Juliano BO, Bautista GM, Lugay JC, Reyes AC (1964) Rice quality, studies on physicochemical properties of rice. J Agric Food Chem 12:131–138

Kasuya M, Teranishi H, Aoshima K, Katoh T, Horiguchi H, Morikawa Y, Iwata K (1992) Water pollution by cadmium and the onset of Itai-itai disease. Water Sci Technol 25:149–156

Khatoon S, Gopalakrishna AG (2004) Fat-soluble nutraceuticals and fatty acid composition of selected Indian rice varieties. J Am Oil Chem Soc 81:939–943

Liu Q-H, Wu X, Chen B-C, Ma J-Q, Gao J (2014) Effects of low light on agronomic and physiological characteristics of rice including grain yield and quality. Rice Sci 21:243–251

Massaretto IL, Alves MFM, De Mira NVM (2011) Phenolic compounds in raw and cooked rice (*Oryza sativa* L.) and their inhibitory effect on the activity of angiotensin I-converting enzyme. J Cereal Sci 54:236–240

Matz SA (2014) The chemistry and technology of cereals as food & feed, 2nd edn. Sci Inter Pvt. Ltd., New Delhi, p 751

Miller A, Engel KH (2006) Content of γ-oryzanol and composition of steryl ferulates in brown rice (*Oryza sativa* L.) of European origin. J Agric Food Chem 54:8127–8133

Miller JB, Pang E, Bramall L (1992) Rice: a high or low glycemic index food? Am J Clin Nutr 56:1034–1036

Miyoshi H, Okuda T, Okuda K, Koishi H (1987) Effects of brown rice on apparent digestibility and balance of nutrients in young men on low protein diets. J Nutr Sci Vitaminol 33:207–218

Pascual CSCI, Massaretto IL, Kawassaki F (2013) Effects of parboiling, storage and cooking on the levels of tocopherols, tocotrienols and γ-oryzanol in brown rice (*Oryza sativa* L.) Food Res Int 50:676–681

Pedersen B, Eggum BO (1983) The influence of milling on the nutritive value of flour from cereal grains. IV. Rice. Qual. Plant. Plant Foods Hum Nutr 33:267–278

Saikia S, Dutta H, Saikia D, Mahanta CL (2012) Quality characterisation and estimation of phytochemicals content and antioxidant capacity of aromatic pigmented and non-pigmented rice varieties. Food Res Int 46:334–340

Shizuma T (2014) Anti-colitis effects of brown rice reported by experimental studies. J Rice Res 2:1–3

Song W, Yamaki T, Yamaji N (2014) A rice ABC transporter, OsABCC1, reduces arsenic accumulation in the grain. PNAS 111:15699–15704

Tanaka K, Sugimoto T, Ogawa M, Kasai Z (1980) Isolation and characterization of two types of protein bodies in the rice endosperm. Agric Biol Chem 44:1633–1639

Tian S, Nakamura K, Kayahara H (2004) Analysis of phenolic compounds in white rice, brown rice, and germinated brown rice. J Agric Food Chem 52:4808–4813

Tiwari U, Cummins E (2009) Nutritional importance and effect of processing on tocols in cereals. Trends Food Sci Technol 20:511–520

Tuncel NB, Yilmaz N (2011) Gamma-oryzanol content, phenolic acid profiles and antioxidant activity of rice milling fractions. Eur Food Res Technol 233:577–585

Xie L, Chen N, Duan B, Zhu Z, Liao X (2008) Impact of proteins on pasting and cooking properties of waxy and non-waxy rice. J Cereal Sci 47:372–379

Wani AA, Singh P, Shah MA, Schweiggert-Weisz U, Gul K, Wani IA (2012) Rice starch diversity: effects on structural, morphological, thermal, and physicochemical properties – a review. Compr Rev Food Sci Food Saf 11:417–436

Zhang GY, Liu R-R, Zhang P (2012) Variation and distribution of vitamin E and composition in seeds among different rice varieties. Acta Agron Sin 38:55–61

Chapter 7
Medicinal and Health Benefits of Brown Rice

Shruti Pandey, K.R. Lijini, and A. Jayadeep

Introduction

Rice is a staple food and primary crop grown all over the world. There are different types of rice cultivated in the globe which includes white, red, purple, black, sticky, long grain basmati, etc. Brown rice (BR) or whole rice is unpolished grain obtained after removal of hull from the paddy. BR consists of bran layer, germ, and inner endosperm. It may vary in color such as light brown, reddish, purplish, or black based on bran color. BR is one of the healthiest and most-studied types of rice. BR is rich in nutrients as the bran layer is intact in it and is not removed as in the case of milled rice or polished rice. Polishing of BR to obtain white rice removes the bran layer and results in the loss of nutrients and health beneficial polyphenols, sterols, and tocopherols (Finocchiaro et al. 2007; Ha et al. 2006). Although white rice is consumed as a staple food, its composition (about 90% starch in dry solids) may lead to a nutritional imbalance. On the other hand, BR contains the bran and the embryo which makes it rich in many nutrients like protein and dietary fiber as shown in Table 7.1 and vitamin B and E as well as minerals as shown in Table 7.2. BR also contains bioactive components which include functional lipids, amino acids, phytosterols, phenolic compounds, dietary fiber, and gamma-aminobutyric acid (GABA) (Cho and Lim 2016). However, consumption of BR is limited because of its rough sensory attributes (Ohtsubo et al. 2005).

S. Pandey • K.R. Lijini • A. Jayadeep (✉)
Department of Grain Science and Technology, CSIR-Central Food
Technological Research Institute (CFTRI), Mysore 570020, Karnataka, India
e-mail: shruti@cftri.res.in; raj.lijini@gmail.com; jayadeep@cftri.res.in

© Springer International Publishing AG 2017
A. Manickavasagan et al. (eds.), *Brown Rice*,
DOI 10.1007/978-3-319-59011-0_7

Table 7.1 Proximate composition of brown and white rice (at 14% moisture)

Rice fraction	Protein (g N×5.95)	Fat (g)	Fiber (g)	Ash (g)	Carbohydrate (g)	Energy (Calorie)
Brown rice	7.1–8.3	1.6–2.8	0.6–1.0	1.0–1.5	73–87	363–385
White rice	6.3–7.1	0.3–0.5	0.2–0.5	0.3–0.8	77–89	349–373

Sources: Juliano 1985a; Eggum et al. 1982; Pedersen and Eggum 1983

Table 7.2 Vitamin and minerals in brown and white rice (at 14% moisture)

Rice fraction	Thiamine (mg)	Riboflavin (mg)	Niacin (mg)	α-tocopherol (mg)	Calcium (mg)	Phosphorus (g)	Phytin (g)	Iron (mg)	Zinc (mg)
Brown rice	0.29–0.61	0.04–0.14	3.5–5.3	0.90–2.50	10–50	0.17–0.43	0.13–0.27	0.2–5.2	0.6–2.8
white rice	0.02–0.11	0.02–0.06	1.3–2.4	75–0.30	10–50	0.08–0.15	0.02–0.07	0.2–2.8	0.6–2.3

Sources: Juliano 1985b; Pedersen and Eggum 1983

Phenolic Components in Brown Rice

India is a land of rice diversity (Fig. 7.1). Directorate of Rice Research (Hyderabad, India) revealed, based on an evaluation program for biotic stresses, that 28.31% of the 12,750 entries were colored rice. Of these, 10.48%, 9.41%, and 8.40% had red, brown, and purple pericarps (Anonymous 1998). In India, red rice varieties were prevalent in the south and the hilly tracts of the northeast and west, whereas black rice is in Manipur, eastern part of India and in some areas of South India only.

Njavara, medicinal rice with red pigmented bran endemic to Kerala, is unique grain plant in India. Murthy (2001) has reported that descriptions of the medicinal qualities of brown rice forms of *Njavara* can be found in various ancient histories of Ayurveda such as *Ashtanga Hrudayam* (Vagbhatta, circa 400–500 A.D.). Unique properties of *Njavara* are also available in the web site (http://www.njavara.com). It has applications in treatment, prevention, and cure of degeneration of muscles, tuberculosis, for children with Anemia, for women during lactation, in certain ulcers, and skin diseases. *Njavara* showed higher chemical indices, phytate content, antioxidant activity, and anti-inflammatory effect compared to corresponding rice and bran of other staple varieties, "Sujatha" (nonpigmented) and "Palakkadan Matta" (red pigmented rice) (Smitha et al. 2013). Deepa et al. (2012) concluded that pigmented rice varieties *Njavara* and non-medicinal red rice *Jyothi* had more total phenol content than nonpigmented IR 64 rice. Reddy et al. (1995) reported that the reddish-colored testa in red rice is associated with the presence of a class of polymeric compounds, the proanthocyanidins.

Black rice that is having dark purple-pigmented bran, broadly known as enriched rice, is reported to have number of medicinal effects. Brown rice forms of it has number of nutritional advantages over common rice such as a higher content of protein, vitamins and minerals, oryzanols, tocopherols, and anthocyanins (Suzuki et al. 2004; Yoshida et al. 2010). In addition, black rice has other beneficial compo-

Fig. 7.1   Paddy, brown rice, and polished rice of few Indian rice varieties

Table 7.3   Phenolic and flavonoid contents in brown rice of few genotypes in Greece

Rice	Phenolic content (mg GAE/100 g)			Flavonoid content (mg CATE/100 g)		
	Free	Bound	Total	Free	Bound	Total
Nonpigmented	30.4 ± 8.5	43.9 ± 8.0	74.3 ± 13.2	34.9 ± 6.8	16.5 ± 2.9	51.3 ± 6.0
Pigmented	185.4 ± 53.0	89.2 ± 7.2	274.6 ± 51.0	123.6 ± 15.4	66.2 ± 10.5	189.7 ± 18.3

Source: Irakli et al. 2016
GAE gallic acid equivalent, *CATE* catechin equivalent

nents, including polyphenolics, flavonoids, vitamin E, phytic acid. Pereira-Caro et al. (2013) characterized the phytochemical profile of Japanese black-purple rice and found to have higher anthocyanin and carotenoids.

In a study on nonpigmented and pigmented rice genotypes in Greece, Irakli et al. (2016) reported (Table 7.3) that the content of total phenolics and flavonoids contents of pigmented brown rice were fourfold greater than nonpigmented ones.

Phenolic Acids

Phenolic acids are aromatic compounds with one benzene ring and one or more hydroxyl group. Many of the phenolic acids are either derivatives of benzoic or cinnamic acid, and they can be subdivided into two major groups, hydroxybenzoic acid and hydroxycinnamic acid. Hydroxybenzoic acid derivatives include hydroxybenzoic, protocatechuic, vanillic, syringic, and gallic acids. These components are generally present in the bound form and are typically components of complex structures such as lignins and hydrolysable tannins. Hydroxycinnamic acid derivatives include p-coumaric, caffeic, ferulic, and sinapic acids. They are generally present in the bound form, linked to cell wall structural components such as cellulose, lignin, and proteins through ester bounds (Balasundram et al. 2006).

The common phenolic acids found in rice include ferulic acid, p-coumaric acid, vanillic acid, caffeic acid, syringic acid, protocatechuic acid, etc. Studies on brown rice from Greece (Irakli et al. 2016) reported that compounds in the free fraction of pigmented rice had higher antioxidant capacity relative to those in the bound form, whereas the nonpigmented rice cultivars exhibited the opposite trend. Ferulic acid was the main phenolic acid of all rice genotypes, whereas black rice contained protocatechuic and vanillic acids in higher contents than red rice and nonpigmented rice genotypes. Phenolic acids exert their health effect by acting as antioxidant which neutralizes free radicals (superoxide, nitric oxide, and hydroxyl radical) which could cause oxidative damage of cell membranes and DNA. Chlorogenic acids such as ferulic acid, p-coumaric acid, and caffeic acid are reported to have cardioprotective effect like lowering of blood pressure (Onakpoya et al. 2014). Ferulic acid has the potential to decrease the levels of some inflammatory mediators, e.g., prostaglandin E2 and tumor necrosis factor-alpha (Ou et al. 2003).

Proanthocyanidins

Proanthocyanidins are condensed tannins which are oligomers or polymers of flavan-3-ols which can be present in some cereals and legume seeds. The variety of phenolic hydroxyl groups (especially ortho-hydroxyl groups) in their structure contributed to the strongest antioxidant power.

Oki et al. (2002) indicated that Japanese red bran rice cultivars contained significant amounts of (+)-catechin and/or (−)-epicatechin derivatives and whose degree of polymerization (DP) ranged from 1 to 38. It is quite interesting to note that proanthocyanidins in methanol extract of the seeds of an Italian dehulled black rice, Venere, were undetectable, while that of another dehulled black rice, Artemide, were detectable as studied by Finocchiaro et al. (2010). In general, proanthocyanidins were typically observed in the red but not in black rice varieties (Finocchiaro et al. 2007). The bioavailability studies showed that gastrointestinal absorption of proanthocyanidins was possible after the decomposition of the oligomers (DP 3–6) in the acid environment of the stomach (Spencer et al. 2000) and of polymers (DP > 10) by the microflora in the colon thereby providing the health beneficial effects. Another study by Smitha et al. (2013) reported that bran of *Njavara* black-hulled variety had higher content (0.98 mg catechin equivalent/g dry weight) of proanthocyanidin than its rice which had 0.07 mg catechin equivalent/g dry weight of sample in the methanolic extract of defatted material determined by vanillin assay.

Catechins, one of the prominent proanthocyanidin and low molecular weight proan-thocyanidins, have received considerable attention owing to their different biological activities, in particular their effects on arteriosclerosis and their oxygen free radical scavenging ability (Sun et al. 1998). So, it may be concluded that the red rice varieties, especially *Njavara* medicinal rice consumption, can provide additional health benefits because of these bioactive components.

Flavonoids

Flavonoids have the general structure of a 15-carbon skeleton, which consists of two phenyl rings and heterocyclic ring. The most common flavonoids are flavones, flavonols, and their glycosides (Bravo 1998). This group includes flavones (luteolin, apigenin), flavanones (myricetin, naringin, hesperetin, naringenin), the flavonols (quercetin, kaempferol), flavan-3-ols (catechin, epicatechin, gallocatechin), the anthocyanins (e.g., cyanidin, pelargonidin, petunidin), and the isoflavones (genistein, daidzein).

Shen et al. (2009) has reported that the flavonoid content of black rice was higher than that of red and white rice. Coloration of rice may be derived from accumulation of anthocyanins (Furukawa et al. 2007), and it may be the reason for high content of flavonoid in black varieties. Flavonoids have potent antioxidant and anticancer activities (Dykes and Rooney 2007; Hu et al. 2003). Cardioprotective effects of flavonoids are reported in rat models (Lara Testai et al. 2013) and human hyperlipidemia and atherosclerosis (Cappello et al. 2015).

Anthocyanins

Anthocyanins as mentioned earlier are a flavonoid formed from anthocyanidins by glycosylation and are water-soluble vacuolar pigments. Studies have shown substantial variation in anthocyanin content and composition between grains. A similar trend was observed by Finocchiaro et al. (2010) in which there is 34% difference in total anthocyanin between two black varieties, and the presence of very low amount was observed in red rice. Chen et al. (2006) also reported that black rice was rich in anthocyanin (79.5–473.7 mg/100 g) and red rice had low content of (7.9–34.4 mg/100 g) of anthocyanin. Earlier studies have shown that cyanidin-3-glucoside (C3G) is the major anthocyanin in black rice, and minor ones reported were either malvidin-3-glucoside (Yoon et al. 1995) or peonidin-3-glucoside (Pt3G) (Choi et al. 1994).

Anthocyanins have been recognized as health-promoting functional food ingredients due to their antioxidant activity (Satue-Gracia et al. 1997), anticancer (Zhao et al. 2004), hypoglycemic (Tsuda et al. 2003), and anti-inflammatory effects (Tsuda et al. 2002), and these functions provide synergic effects with various nutrients in vivo.

Medicinal and Health Benefits of Brown Rice

Antioxidant Properties

Antioxidant properties of BR extracts of different varieties have been reported based on DPPH radical scavenging assay and ferric reducing power assay. DPPH is commercially available nitrogen-centered stable free radical and is widely used to evaluate the radical scavenging activity of antioxidant compounds (Huang et al. 2005). The ability to act as donors of hydrogen atoms in the transformation of the DPPH radical to its reduced form (DPPH•-H) was measured by the loss of deep purple color of DPPH after reaction with extracts of different BR. A study by Oki et al. (2002) reported that red-hulled rice cultivars exhibited higher free radical scavenging activity than black- and white-hulled rice cultivars. The reducing power of the soluble extracts of different pigmented and nonpigmented rice varieties was determined by direct electron donation in the reduction of ferricyanide $[Fe(CN)_6]^{3-}$ to ferrocyanide $[Fe(CN)_6]^{4-}$. The product was visualized by addition of free Fe^{3+} ions after the reduction reaction, by forming the intense Prussian blue color complex, $(Fe^{3+})_4[Fe^{2+}(CN^-)_6]_3$, and quantified by absorbance measurement at 700 nm (Ribeiro et al. 2008). Deepa et al. (2012) reported that medicinal rice *Njavara* has high reducing activity compared to red rice *Jyothi*. In another study, Laokuldilok et al. (2011) reported that pigmented rice varieties have higher reducing power.

Phenolic compounds act as antioxidants which include reduction of oxygen concentration, termination of free radicals, decomposition of primary products of oxidation to nonradical species, and prevention of continued hydrogen abstraction from substrate and chelators of metal ions that lead to the formation of peroxidation compounds (Shahidi and Naczk 2004).These antioxidant compounds eliminate reactive oxygen species (ROS) such as lipid peroxide and superoxide anion radicals and also lower cholesterol content (Ichikawa et al. 2001; Nam et al. 2008). Free radical scavenging activity analyzed by DPPH (1, 1-diphenyl-2-picrylhydrazyl) showed higher activity in red rice's than in black and white rice which is correlated with polyphenols and proanthocyanidin content (Oki et al. 2002). There are reports for in vitro study of antioxidants and antioxidant activity of some Chinese and Thai rice varieties with three extracts (Chanida et al. 2013). Compared to white rice, studies on black rice is limited to anthocyanins, phenolics, and antioxidant activity (Goffman and Bergman 2004; Chung and Shin 2007).

Prevention of Atherosclerosis

Chen et al. (2000) found that black and red rice are effective in reducing atherosclerotic plaques on the aorta of rabbits fed a cholesterol-enriched diet. They found that effectiveness was related to the high level of serum HDL-C and ApoA1. Supplementation of black rice anthocyanins in dyslipidemic rats on a high-fat diet has been shown to have a reduction in platelet hyperactivity and hyperglyceridemia and facilitates the maintenance of optimal platelet function (Yang et al. 2011). BR also helps to raise blood levels of nitric oxide, a small molecule known to develop blood vessel dilation, and to inhibit oxidative (free radical) damage of cholesterol and the adhesion of white cells to the vascular wall, which are vital steps in the development of atherosclerotic plaques (Hu et al. 2003). Pigmented rice extracts have also been reported to effectively decrease oxidative stress, inflammation, as well as atherosclerotic lesions (Xia et al. 2003).

Lowering of Type 2 Diabetes Risk

Research by Van Dam et al. (2006) suggests that regular consumption of whole grains lessens the risk of type 2 diabetes. In an 8-year trial, involving 41,186 participants of the Black Women's Health Study, confirms that there is some inverse associations between magnesium, calcium, and the other major food sources in relation to type 2 diabetes that had already been reported mainly in white populations (Willett et al. 2002). In addition, a low-glycemic index diet with a higher amount of fiber and minimally processed whole grain products reduces glycemic and insulinemic responses and lowers the risk of type 2 diabetes (Hu et al. 2001). Supplementation of brown rice of raw and parboiled of white rice variety was reported to ameliorate the diabetic complications in experimental rat (Hameeda et al. 2016). Anthocyanin was shown to ameliorate the insulin resistance, hyperglycemia, and hyperlipidemia in high-fat-fed mice (Tsuda et al. 2003; Jayaprakasam et al. 2006). An in vitro study reported that phenolic compounds isolated from the pigmented rice showed significant inhibitory activity against aldose reductase in the following decreasing order: cyanidin-3-glucoside >quercetin > ferulic acid > peonidin-3-glucoside > tocopherol (Yawadio et al. 2007).

Cholesterol Lowering Effect

Studies on BR have shown that it contains oil which lowers cholesterol (Most et al. 2005). Consumption of rice bran oil lowered cholesterol by 7%. Researchers suggested that the unsaponifiables existing in rice bran oil could become important functional foods for cardiovascular health. Studies by Erkkila et al. (2005) revealed that eating of whole grains, such as BR, is good for postmenopausal women with high cholesterol, high blood pressure, or any other signs of CVD. Experiments revealed that intake of at least six servings of whole grains by women each week experienced both slower progression of atherosclerosis, the buildup of plaque that constricts the blood vessels, and less progression in stenosis, the narrowing of the arterial passageways. The health benefits of BR are partly related with its fiber, which has been already proven to reduce high cholesterol levels, one more way BR helps prevent atherosclerosis (MacDougall et al. 2008).

Prevention of Cancer

Chen et al. (2006) observed that anthocyanins extracted from black rice exerted an inhibitory effect of cell invasion on various cancer cells. The fiber in BR helps to protect against colon cancer since fiber binds to cancer-causing chemicals, protecting them away from the cells lining the colon (MacLean et al. 2007). In addition, it can help stabilize bowel function, reducing constipation. It was observed that a diet rich in fiber from whole grains, such as BR, and some fruit offered significant protection counter to breast cancer for premenopausal women (Cade et al. 2007). Premenopausal women eating the maximum whole grain fiber (at least 13 g/day) had a 41% reduced risk of breast cancer, compared to those with the lowermost whole grain fiber intake (4 g or less per day) (Reid 2002). Cheng (1993) reported that rice bran contains large amount of dietary fiber (24–34% of total bran solids) which corresponds to 1.4–3.3% in total BR. Fiber supplied by whole grain gives most protection.

Prevention of Gallstones

Consumption of foods high in insoluble fiber, such as BR, can help women evade gallstones, as reported by Tsai et al. (2004). The insoluble fiber not only speeds intestinal transit time (how quickly food moves through the intestines) but reduces the secretion of bile acids (excessive amounts contribute to gallstone formation), increases insulin sensitivity, and lowers triglycerides (blood fats) (Liu 2007). Studying the overall fiber intake and types of fiber consumed, taken over a 16-year period by above 69,000 women in the Nurses' Health Study, researchers found that those consuming the most whole overall fiber (both soluble and insoluble) had a 13% lesser risk of developing gallstones compared to women consuming the smaller quantity fiber-rich foods. Those consuming the most foods rich in insoluble fiber gained even extra protection against gallstones, a 17–19% lower risk compared to women eating the least. The protection was dose related; a 5-gram increase in insoluble fiber consumption dropped risk 10% (Liu et al. 2003). Investigators suggest that insoluble fiber not only speeds intestinal transit time but decreases the secretion of bile acids, increases insulin sensitivity, and in return lowers triglycerides.

Conclusion

Brown rice or dehusked rice with bran and germ is rich in nutrients and health beneficial nutraceuticals. A number of varieties of rice like white, red, black, medicinal, etc. are there. In addition to the matrix of nutrients like fiber, vitamins, and minerals, the whole grain arsenal includes a wide variety of phytochemicals that reduce the risk of diseases. Compounds in whole grains that have disease-lowering effects include phenolic acids, proanthocyanidins, flavonoids, anthocyanins, etc. Brown rice is also important dietary sources of water-soluble, fat-soluble, and insoluble antioxidants. In many studies, eating whole grains, such as brown rice, has been linked to protection against oxidative stress, atherosclerosis, diabetes, insulin resistance, hyperlipidemia, obesity, ischemic stroke, cancer, and premature death. Considering the beneficial effects of brown rice, more research on identification of bioactive components and bioactivities of brown rice as well as development of value-added convenience health food products which can prevent present-day chronic diseases needs to be undertaken.

Acknowledgment Authors wish to acknowledge the funding from the project WELFO-BSC 210.

References

Anonymous (1998) Evaluation of rice germplasm for biotic stresses. ICAR Ad-Hoc Research Network Programme, 1993–1998. Directorate of Rice Research, Hyderabad, p 284

Balasundram N, Sundram K, Samman S (2006) Phenolic compounds in plants and agri-industrial by-products: antioxidant activity, occurrence, and potential uses. Food Chem 99:191–203

Bravo L (1998) Polyphenols: chemistry, dietary sources, metabolism, and nutritional significance. Nutr Rev 56:317–333

Cade JE, Burley VJ, Greenwood DC (2007) Dietary fiber and risk of breast cancer in the UK Women's Cohort Study. Int J Epidemiol 23:58–62

Cappello AR, Dolce V, Iacopetta D, Martello M, Fiorillo M, Curcio R, Muto L, Dhanyalayam D (2015) Bergamot (*Citrus bergamia* Risso) flavonoids and their potential benefits in human hyperlipidemia and atherosclerosis: an overview. Mini-Rev Med Chem 15:1–11

Chanida S, Liu Z, Huang J, Gong Y (2013) Anti-oxidative biochemical properties of extracts from some Chinese and Thai rice varieties. Afr J Food Sci 7:300–305

Chen Q, Ling WH, Ma J, Mei J (2000) Effects of black and red rice on the formation of aortic plaques and blood lipids in rabbits. J Hyg Res 29:170–172

Chen PN, Kuo WH, Chiang CL, Chiou HL, Hsieh YS, Chu SC (2006) Black rice anthocyanins inhibit cancer cells invasion via repressions of MMPs and u-PA expression. Chem Biol Interact 163:218–229

Cheng HH (1993) Total dietary fiber content of polished brown and bran types of japonica and indica rice in Taiwan, resulting physiological effects of consumption. Nutr Res 13:93–101

Cho DH, Lim ST (2016) Germinated brown rice and its biofunctional compounds. Food Chem 196:259–271

Choi SW, Kang WW, Osawa T (1994) Isolation and identification of anthocyanin pigments in black rice. Food Biotechnol 3:131–136

Chung HS, Shin JC (2007) Characterization of antioxidant alkaloids and phenolic acids from anthocyanin-pigmented rice (*Oryza sativa* cv. Heugjinjubyeo). J Food Chem 104:1670–1677

Deepa G, Singh V, Akhilender Naidu K (2012) Characterization of antioxidant compounds and antioxidant activity of Indian rice varieties. J Herbs Spices Med Plants 18:18–33

Dykes L, Rooney LW (2007) Phenolic compounds in cereal grains and their health benefits. Cereal Foods World 52:105–111

Eggum BO, Juliano BO, Maniñgat CC (1982) Protein and energy utilization of rice milling fractions by rats. Qual Plant Plant Food Hum Nutr 31:371–376

Erkkila AT, Herrington DM, Mozaffarian D, Lichtenstein AH (2005) Cereal fiber and whole-grain intake are associated with reduced progression of coronary-artery atherosclerosis in postmenopausal women with coronary artery disease. Am Heart J 150:94–101

Finocchiaro F, Ferrari B, Gianinetti A, Dall'Asta C, Galaverna G, Scazzina F (2007) Characterization of antioxidant compounds of red and white rice and changes in total antioxidant capacity during processing. Mol Nutr Food Res 51:1006–1019

Finocchiaro F, Ferrari B, Glaninetti A (2010) A study of biodiversity of flavonoid content in the rice caryopsis evidencing simultaneous accumulation of anthocyanins and proanthocyanidins in a black-grained genotype. J Cereal Sci 51:28–34

Furukawa T, Maekawa M, Oki T, Suda I, Lida S, Shimada H, Takamure I, Kadowaki KI (2007) The Rc and Rd genes are involved in proanthocyanidin synthesis in rice pericarp. Plant J 49:91–102

Goffman FD, Bergman CJ (2004) Rice kernel phenolic content and its relationship with antiradical efficiency. J Sci Food Agr 84:1235–1240

Ha T-Y, Ko S-N, Lee S-M, Kim H-R, Chung S-H, Kim S-R, Yoon H-H, Kim I-H (2006) Changes in nutraceutical lipid components of rice at different degrees of milling. Eur J Lipid Sci Technol 108:175–181

Hameeda BNI, Pradeep SR, Singh V, Srinivasan K, Jayadeep A (2016) Beneficial influence of phosphorylated parboiled dehusked red rice (*Oryza sativa* L.) in streptozotocin-induced diabetic rats. Starch/Stärke 68:568–580

Hu FB, Van Dam RM, Liu S (2001) Diet & risk of type II diabetes: the role of types of fat and carbohydrate. Diabetologia 44:805–817

Hu C, Zawistowski J, Ling W, Kitts D (2003) Black rice pigmented fraction suppresses both reactive oxygen species and nitric oxide & biological model systems. J Agric Food Chem 51:5271–5177

Huang D, Ou B, Prior RL (2005) The chemistry behind antioxidant capacity assays. J Agric Food Resour 53:1841–1856

Ichikawa H, Ichiyanagi T, Xu B, Yoshii Y, Nakajima M, Konishi T (2001) Antioxidant activity of anthocyanin extract from purple black rice. J Med Food 4:211–218

Irakli MN, Samanidou VF, Katsantonis DN, Biliaderis CG, Papadoyannis IN (2016) Phytochemical profiles and antioxidant capacity of pigmented and non-pigmented genotypes of rice (*Oryza sativa* L.) Cereal Res Commun 44:98–110

Jayaprakasam B, Olson LK, Schutzki RE, Tai MH, Nair MG (2006) Amelioration of obesity and glucose intolerance in high-fat-fed C57BL/6 mice by anthocyanins and ursolic acid in cornelian cherry (*Cornus mas*). J Agric Food Chem 54:243–248

Juliano BO (1985a) Factors affecting nutritional properties of rice protein. Trans Nat Acad Sci Technol 7:205–216

Juliano BO (ed) (1985b) Rice: chemistry and technology, vol 7, 2nd edn. American Association of Cereal Chemistry, St Paul, pp 774–778

Laokuldilok T, Shoemaker CF, Jongkaewwattana S (2011) Antioxidants and antioxidant activity of several pigmented rice brans. J Agric Food Chem 59:193–199

Liu RH (2007) Whole grain phytochemicals and health. J Cereal Sci 46:207–219

Liu S, Willett WC, Manson JE, Hu FB, Rosner B, Colditz G (2003) Relation between changes in intakes of dietary fiber and grain products and changes in weight and development of obesity among middle-aged women. Am J Clin Nutr 78:920–927

MacDougall C, Gottschall-Pass K, Neto C (2008) Dietary whole cranberry (*Vaccinium macrocarpon*) modulates plasma lipid and cytokine profiles, and prevents liver toxicity in response to cholesterol-feeding in the JCR-LA-cp corpulent rat model. FASEB J 22:702–706

MacLean MA, Matchett MD, Amoroso J (2007) Cranberry (*Vaccinium macrocarpon*) flavonoids inhibit matrix metalloproteinases (MMPs) in human prostate cancer cells. FASEB J 21:A1000

Most MM, Tulley R, Morales S, Lefevre M (2005) Rice bran oil, not fiber, lowers cholesterol in humans. Am J Clin Nutr 81:64–68

Murthy KRS (2001) Vagbatta's Ashtanga Hrdayam (Text, English translation, notes, appendix, indices), 5th edn. Krishna Das Academy, Varanasi

Nam YJ, Nam SH, Kang MY (2008) Cholesterol-lowering efficacy of unrefined bran oil from the pigmented black rice (*Oryza sativa* L cv. Suwon 415) in hypercholesterolemic rats. Food Sci Biotechnol 17:457–463

Ohtsubo K, Suzuki K, Yasui Y, Kasumi K (2005) Biofunctional components in the processed pre-germinated brown rice by a twin extruder. J Food Compos Anal 18:303–316

Oki T, Masuda M, Kobayashi M, Nishiba Y (2002) Polymeric procyanidins as radical-scavenging components in red hulled rice. J Agric Food Chem 50:7524–7529

Onakpoya IJ, Spencer EA, Thompson MJ, Heneghan CJ (2014) The effect of chlorogenic acid on blood pressure: a systematic review and meta-analysis of randomized clinical trials. J Hum Hypertens 29:77–81

Ou L, Kong LY, Zhang XM, Niwa M (2003) Oxidation of ferulic acid by *Momordica charantia* peroxidase and related anti-inflammation activity changes. Biol Pharm Bull 26:1511–1516

Pedersen B, Eggum BO (1983) The influence of milling on the nutritive value of flour from cereal grains IV rice. Qual Plant Plant Food Hum Nutr 33:267–278

Pereira-Caro G, Watanabe S, Crozier A, Fujimura T, Yokota T, Ashihara H (2013) Phytochemical profile of a Japanese black–purple rice. Food Chem 141:2821–2827

Reddy VS, Dash S, Reddy AR (1995) Anthocyanin pathway in rice (*Oryza sativa* L.): identification of a mutant showing dominant inhibition of anthocyanins in leaf and accumulation of proanthocyanidins in pericarp. Theor Appl Genet 91:301–312

Reid G (2002) The role of cranberry and probiotics in intestinal and urogenital tract health. Crit Rev Food Sci Nutr 42:293–300

Ribeiro RP, Flemming JS, Bacila AR (2008) Yeast usage (*Saccharomyces cerevisiae*), yeast cell wall (SSCW), organic acids and avilamycin in broiler feeding. Arch Vet Sci 13:210 217

Satue-Gracia M, Heinonen IM, Frankel EN (1997) Anthocyanins as antioxidants on human low-density lipoprotein and lecithin-liposome system. J Agric Food Chem 45:3362–3367

Shahidi F, Naczk M (2004) Phenolics in food and nutraceutical. CRC Press, London

Shen Y, Jin L, Xiao P, Lu Y, Bao J (2009) Total phenolic, flavonoids, antioxidant capacity in rice grain and their relations to grain colour, size and weight. J Cereal Sci 49:106–111

Smitha MR, Parvathy R, Shalini V, Ratheesh M, Helen A, Jayalekshmy A (2013) Chemical indices, antioxidant activity and anti-inflammatory effect of extracts of the medicinal rice "*njavara*" and staple varieties: a comparative study. J Food Biochem 37:369–380

Spencer H, Brouder A-M, Mitullah W (2000) Food safety requirements and food experts from developing countries: the case of fish exports from Kenya to the European union. Am J Agric Econ 82(5):1159–1169

Sun S, Zhang X, Tough DF, Sprent J (1998) Type I interferon- mediated stimulation of T cells by CPG DNA. J Exp Med 188:2335–2342

Suzuki M, Kimura T, Yamagishi K, Shinmoto H, Yamaki K (2004) Comparison of mineral contents in 8 cultivars of pigmented brown rice. Nippon Shokuhin Kagaku Kogaku Kaishi 51(8):424–427

Testai L, Martelli A, Cristofaro M, Breschi MC, Calderone V (2013) Cardioprotective effects of different flavonoids against myocardial ischaemia/reperfusion injury in Langendorff-perfused rat hearts. J Pharm Pharmacol 65(5):750–756

Tsai CJ, Leitzmann MF, Willett WC, Giovannucci EL (2004) Long-term intake of dietary fiber and decreased risk of cholecystectomy in women. Am J Gastroenterol 99:1364–1370

Tsuda T, Horio F, Osawa T (2002) Cyanidin 3-O-β-glucoside suppresses nitric oxide production during a zymosan treatment in rats. J Nutr Sci Vitaminol 48:305–310

Tsuda T, Horio F, Uchida K, Aoki H, Osawa T (2003) Dietary cyanidin 3-O-β-D-glucoside-rich purple corn colour prevents obesity and ameliorates hyperglycemia. J Nutr 133:2125–2130

Van Dam RM, Hu FB, Rosenberg L, Krishnan S, Palmer JR (2006) Dietary calcium and magnesium, major food sources, and risk of type 2 diabetes in U.S. Black women. Diabetes Care 29:2238–2243

Willett W, Manson J, Liu S (2002) Glycemic index, glycemic load, and risk of type 2 diabetes. Am J Clin Nutr 76:274S–280S

Xia M, Ling WH, Ma J, Kitts DD, Zawistowski J (2003) Supplementation of diets with the black rice pigment fraction attenuates atherosclerotic plaque formation in apolipoprotein deficient mice. J Nutr 133:744–751

Yang Y, Andrews MC, Hu Y, Wang D, Qin Y, Zhu Y (2011) Anthocyanin extract from black rice significantly ameliorates platelet hyperactivity and hypertriglyceridemia in dyslipidemic rats induced by high fat diets. J Agric Food Chem 59:6759–6764

Yawadio R, Tanimori S, Morita N (2007) Identification of phenolic compounds isolated from pigmented rices and their aldose reductase inhibitory activities. Food Chem 101:1616–1625

Yoon HH, Paik YS, Kim JB, Hahn TR (1995) Identification of anthocyanidins from Korean pigmented rice. J Agric Chem Biotechnol 38:581–583

Yoshida H, Tomiyama Y, Mizushina Y (2010) Lipid components, fatty acids and triacylglycerol molecular species of black and red rices. J Food Chem 123:210–215

Zhao C, Giusti MM, Malik M, Moyer MP, Magnuson BA (2004) Effects of commercial anthocyanin-rich extracts on colonic cancer and nontumorigenic colonic cell growth. J Agric Food Chem 52:6122–6128

Chapter 8
Glycaemic Properties of Brown Rice

S. Shobana, M. Jayanthan, V. Sudha, R. Unnikrishnan,
R.M. Anjana, and V. Mohan

Introduction

Carbohydrates are the most important source of food energy and contribute to almost 40–80% of total food energy intake in different populations (FAO 1998); half of total calories are derived from carbohydrates (particularly cereals such as rice) in Asian Indian diets (Radhika et al. 2009a). In ancient times, cereals were consumed after minimal processing. Today, refined cereals such as white rice (WR) are predominantly consumed, and thus there has been a decrease in the whole grain consumption (Cleveland et al. 2000; Harnack et al. 2003). Several studies have shown a positive association of WR consumption and an inverse association between the consumption of whole grains such as brown rice (BR) and the risk of type 2 diabetes (Mohan et al. 2009; Sun et al. 2010). BR is a whole grain with intact bran and germ constituents and has higher fibre, vitamins, minerals and other health-beneficial compounds compared to WR. BR can be a wholesome, nutritious replacement for WR. Therefore, it is essential to understand the functional health benefits of BR in addition to its nutritional benefits. In this chapter, we discuss the glycaemic properties of BR and the various factors that affect it.

S. Shobana (✉) • M. Jayanthan • V. Sudha
Department of Foods, Nutrition & Dietetics Research, Madras Diabetes Research Foundation,
Chennai, Tamil Nadu, India
e-mail: shobanashanmugam@mdrf.in; jayanthan@mdrf.in; s2r_7@mdrf.in

R. Unnikrishnan • R. Anjana • V. Mohan
Department of Diabetology, Madras Diabetes Research Foundation,
Chennai, Tamil Nadu, India
e-mail: unnikrishnan_ranjit@yahoo.com; dranjana@drmohans.com;
drmohans@diabetes.ind.in

© Springer International Publishing AG 2017
A. Manickavasagan et al. (eds.), *Brown Rice*,
DOI 10.1007/978-3-319-59011-0_8

The Concept of Glycaemic Response (GR), Glycaemic Index (GI) and Glycaemic Load (GL)

Glycaemic response (GR) is the postprandial blood glucose response elicited when a food or meal containing carbohydrate is ingested. The GI is the standardized GR elicited by a portion of food containing 50 g (or in some cases 25 g) of available carbohydrate and is expressed as a percentage of the GR elicited by 50 g (or 25 g) of the reference carbohydrate (i.e. either a glucose solution or white wheat bread). Foods having carbohydrate that is digested, absorbed and metabolized quickly are high-GI foods (GI >70 on the glucose scale), whereas those that are digested, absorbed and metabolized slowly are low-GI foods (GI <55 on the glucose scale). The GL is the product of GI and the total available carbohydrate content in a given amount of food (Augustin et al. 2015).

Low-GI foods are digested slowly with slower release of glucose for absorption and cause a gradual rise in blood glucose with less demand on the pancreas to secrete insulin. Low-GI foods also help to control hunger and prevent excess calorie consumption and obesity (which predispose to diabetes and cardiovascular disease) (Esfahani et al. 2009). On the other hand, high-GI foods are rapidly digestible, release sugar faster for absorption, rapidly raise blood sugars and increase the demand for insulin, leading to β-cell exhaustion, obesity, insulin resistance and glucose intolerance. Kaur et al. (2016a) showed that the consumption of low-GI meal attenuated 24-h blood glucose profiles and promoted fat over carbohydrate oxidation compared to high-glycaemic-index meals in healthy Asians. This offers a potential dietary solution to reduce the burden of obesity-related chronic diseases.

Consumption of higher-GL diets (mainly from refined grains being high in GI in Indian diets) has been shown to increase the risk of type 2 diabetes in South Indian adults (Mohan et al. 2009; Radhika et al. 2009a, b).

Factors Affecting the Glycaemic Properties of Brown Rice

Varietal Variations

The glycemic properties of rice and rice based products has been reviewed recently (Kaur et al 2016b) Rice varieties differ in carbohydrate make-up (amylose content), kernel hardness, grain dimensions, etc. The amylose-to-amylopectin ratio in rice is an important determinant of glycaemic and insulinaemic responses (Miller et al. 1992; Trinidad et al. 2013). The GI of BR reported in literature is given in Table 8.1.

Miller et al. (1992) showed that the GI of brown variants of Doongara was of medium category as compared to Calrose and Pelde (high GI category) mainly due to their lower amylose content. This study result emphasizes screening of BR varieties for lower glycaemic properties. However, Panlasigui et al. (1991) indicated that amylose content alone cannot be a good predictor of glycaemic properties of rice and emphasized other physicochemical properties. Hence, screening of BR varieties for

Table 8.1 Glycaemic index of brown rice (BR)

Sample description	GI, classification	Subjects type, number	Reference food, duration	Remarks/findings	Reference
Brown rice	66 ± 5, medium	Healthy, 5–10	Glucose, 2 h	BR elicited lower GI compared to WR (GI = 72)	Jenkins et al. (1981)
Brown rice (Canada)	66 ± 5, medium	Healthy, 7	Glucose, 2 h	Wide variations exist in the GI of BR ranging from low to high GI	Atkinson et al. (2008)
Brown, steamed (USA)	50, low	Healthy, 8	Glucose, 3 h		
Brown rice, boiled in excess water for 25 mins (SunRice brand, Rice Growers Co-op, Australia)	72 ± 6, high	Healthy, 9	Glucose, 2 h		
Brown rice (China)	87 ± 2, high	Healthy, 10	Glucose, 2 h		
Medium-grain brown rice in 90 seconds, microwaved on high (SunRice brand, Rice Growers Co-op, Australia)	59 ± 8, high	Healthy, 10	Glucose, 2 h		
Brown rice parboiled, cooked 20 min, Uncle Ben's Natur-Reis® (Masterfoods, Belgium)	64 ± 7, medium	Healthy, 10	Glucose, 2 h		
Brown, high-amylose (IR42) rice, boiled 30 min (Philippines)	58, medium	Healthy, 10	Bread, 1 h	BR with higher amylose in spite of higher cooking time exhibited medium GI	
Calrose brown rice (Rice Growers Co-op, Australia)	87 ± 8, high	Healthy, 8	Glucose, 2 h	High-amylose Doongara rice variety (28% amylose) in spite of similar higher cooking time (30 min) as compared to normal-amylose rice varieties (Calrose and Pelde with 20% amylose) elicited a lower GI	Miller et al. (1992)
Doongara brown, high-amylose rice (Rice Growers Co-op, Australia)	66 ± 7, medium	Healthy, 8	Glucose, 2 h		
Pelde brown rice (Rice Growers Co-op, Australia)	76 ± 7, high	Healthy, 8	Glucose, 2 h		
Sunbrown Quick (Rice Growers Co-op, Australia)	80 ± 7, high	Healthy, 8	Glucose, 2 h		
Brown rice pasta	92 ± 8, high	Healthy, 6	Glucose, 2 h	BR pasta boiled for 19 min exhibited a very high GI	
Brown rice, short-grain rice (*Japonica*)	61.5 ± 4.7, medium	Healthy, 19	Glucose, 2 h	Lower GI for BR compared to WR (75.9)	Ito et al. (2005)

(continued)

Table 8.1 (continued)

Sample description	GI, classification	Subjects type, number	Reference food, duration	Remarks/findings	Reference
Pregerminated brown rice (PGBR) – short-grain rice (*Japonica*)	54.4 ± 5.1, medium	Healthy, 13	Glucose, 2 h	PGBR elicits lower GI compared to BR and WR	
1/3 PGBR – short-grain rice (*Japonica*) (PGBR and white rice blended in the ratio of 1:2)	67.4 ± 2.9, medium	Healthy, 13	Glucose, 2 h	The higher the ratio of PGBR to WR, the lower was the GI PGBR could be a healthy replacement for WR to control post-meal glycaemic response without increasing insulin concentration	
2/3 PGBR – short-grain rice (*Japonica*) PGBR and white rice blended in the ratio of 2:1	63.7 ± 5.3, medium	Healthy, 13	Glucose, 2 h		
Tai Ken, brown rice, (Union Rice Company; Taipei, Taiwan)	82 ± 0.22, high	Healthy, 10	White bread, 2 h	Soaking and cooking of BR would have enhanced gelatinization and GI	Lin et al. (2010)
Brown basmati rice	75 ± 7.8, high	Healthy, 10	Glucose, 2 h	Brown basmati rice elicited a higher glycaemic response due to higher cooking time (25 min)	Ranawana et al., (2009)
White and brown basmati rice (mixture 60% white basmati, 40% brown basmati)	59 ± 9, medium	Healthy, 10	Glucose, 2 h	A mixture of white and brown basmati rice elicited a medium GI even though the cooking time was similar as that of brown basmati rice (25 min)	
Thai red rice (unpolished)	76 ± 8, high	Healthy, 10	Glucose, 2 h	Higher GI may be due to higher cooking time (25 min)	
Commercial Malaysian brown rice, high-fibre rice B	60 ± 5.8, medium	Healthy, 10	Glucose, 2 h	GI does not solely depend on dietary fibre but also on the amylose content and extent of gelatinization; BR exhibited medium GI	Yusof et al. (2005)

Brown rice and milled rice with different amylose contents (AC) grown in the Philippines		Healthy, 9–10	Glucose, 2 h	Decrease in GI from milled rice to BR was greatest for glutinous (waxy), low-apparent AC rice with high GI Intermediate AC IR64 with medium GI showed lesser drop GI was highest for waxy rice and lowest for high-AC rice for both rice forms (brown and milled), but the decrease in GI with increasing AC was lower for brown rice than for milled rice	Trinidad et al. (2013)
IR64 (brown, AC 22.0%)	51 ± 1, low				
IR64 (milled, AC 22.9%)	57 ± 3, medium				
Sinandomeng (brown, AC 12.1%)	55 ± 2, medium				
Sinandomeng (milled, AC 12.6%)	75 ± 4, high				
Brown rice prepared by cross-breeding	51 ± 8, low	Healthy, 10	Glucose, 3 h	BR elicited the lowest glycaemic response compared to polished BR and WR	Karuppiah et al. (2011)
SunRice, low-GI brown rice (Doongara variety)	54	–	–	Low GI	http://www.gisymbol.com/sunrice-low-gi-brown-rice/

amylograph viscosity properties, gelatinization temperature and volume expansion after cooking in addition to the amylose content could help in the prediction of its glycaemic properties. Pathiraje et al. (2010) showed a lower GI for WR prepared from traditional rice varieties compared to "improved" varieties. Such studies are not available for BR. Deepa et al. (2010) reported the in vitro GI and starch digestibility of pigmented Indian brown rice varieties such as *Njavara* and *Jyothi* (74.8 for *Njavara*, 73.1 for *Jyothi*); BR prepared from pigmented varieties may have a higher proportion of phytochemicals (polyphenols) which may be beneficial for health. However, several red- and black-pigmented rice varieties are believed to be suitable for people with diabetes; valid GI evaluation is required before appropriate recommendations can be made. Ranawana et al. (2009) reported a higher GI for brown basmati rice (75) and attributed this to longer cooking time. Glutinous rice varieties are known to have higher amylopectin content, and hence, higher GI values have been reported for glutinous WR (Ranawana et al. 2009). Such reports are not available for glutinous BR varieties.

Cooking Methods, Extent of Gelatinization, Expansion upon Cooking

Cooking and cooling affects starch digestibility through the degree of gelatinization and retrogradation of rice starch. Amylose retrogradation leads to type 3 resistant starch (RS3) formation which is resistant to digestion as it is thermostable and melts above 120 °C. In contrast, retrograded amylopectin has low melting points (46–65 °C) and therefore melts upon reheating (Boers et al. 2015) and may lose its property. Subjecting high-amylose BR to retrogradation would be beneficial.

The extent of hydration of starch granules, the method and degree of gelatinization of starch and the volume of expansion upon cooking affect the GR (glycaemic response) of food. Panlasigui et al. (1991) showed that WR from IR42 rice variety with the maximum volume of expansion upon cooking exhibited a higher GI. This is due to the increased surface area and higher susceptibility of starch to amylolytic attack. Hence, rice (either brown or polished) should be cooked optimally without much disruption to the grain matrix to achieve moderate GR. Wolever et al. (1986) mentioned that boiling WR for a shorter duration (5 min) produced less swollen and intact grains eliciting a lower GR compared to rice boiled for a longer duration (15 min). This should apply for BR also. Duration of cooking largely influences the GI (Ranawana et al. 2009). Daomukda et al. (2011) showed that the degree of gelatinization of cooked BR decreased with decreasing ratios of water and the texture of cooked BR was harder with lower water ratios. They also reported a lesser degree of gelatinization for steaming. Studies in this direction would help devise healthy cooking practices for BR. It is also very important to determine the gelatinization temperature of rice varieties before taking them for hydrothermal processing (parboiling) so as to maintain optimal gelatinization of starch to modulate the GR.

BR takes a longer time to cook due to the presence of bran which is rich in fat and forms a physical barrier between the cooking medium and starchy endosperm, and this delays the rate of hydration of kernel during the process of cooking. Prolonged cooking may rupture the bran layer and leach out the endosperm constituents leading to higher volume expansion upon cooking. Careful selection of high-amylose variety and optimal cooking with appropriate rice-to-water ratio may be helpful in achieving favourable glycaemic properties while cooking BR.

Processing

Different processing methods cause different impact on anatomical as well as chemical constituents of BR. Food structure matrix and integrity of the anatomical constituents determine its cooking characteristics and GR. Any disruption to these may increase the starch bioavailability. There are many studies, which report the effect of processing on GI of WR. Puffed WR, due to its porous structure, exhibited higher GI, and mixed reports exist for parboiled WR, with some studies showing a reduction in GI compared to unparboiled (Casiraghi et al. 1993; Wolever et al. 1986; Pathiraje et al. 2010; Larsen et al. 2000) in contrast to others that do not (Larsen et al. 1996). Such studies do not exist for BR. We (Shobana et al. 2016) reported a GI of 57.6 for BPT parboiled BR; however, we did not evaluate the GI of raw BR from the same variety. Selection of appropriate paddy variety and standardization of optimal parboiling conditions to achieve lower GI for BR are required. Parboiling imparts hardness to the kernel. Rice (either WR or BR) produced with higher retrogradation may be more chewy, and the higher degree of mastication required may in turn lead to higher glycaemic responses as reported in the case of WR (Ranawana et al. 2010; Ranawana et al. 2014). Sun et al. (2015) reported a lower GR for WR eaten with chopsticks as compared to a spoon. However, such kind of studies do not exist for BR, and it should be borne in mind that BR in general is chewy compared to WR and the texture of cooked rice further depends on the variety and starch make-up. Parboiled high-amylose rice varieties, being harder, may require more chewing, and the cooked BR texture can be an important determinant of GR.

Instantiation has been shown to increase digestibility even in high-amylose WR (Brand et al. 1985; Holt and Brand Miller 1995). However, Ranawana et al. (2009) reported a lower GI for easy-cook long grain and a higher GI for easy-cook basmati WR. One of our in-house studies showed higher GI for instant BR (in-house data, unpublished). For the preparation of quick-cooking rice, WR/BR is hydrothermally processed and dried. This is further cooked before consumption, and these processes may disrupt the grain matrix largely and may lead to higher GR. Hence, with instant BR varieties, one can expect a higher GI.

Similarly, for WR noodles and vermicelli, Juliano et al. (1989) reported a lower GI. Such information is not available for BR noodles, and it is clear that the selection of a high-amylose variety for this purpose would be advantageous. Factors influencing the glycaemic properties of BR are given in Fig. 8.1.

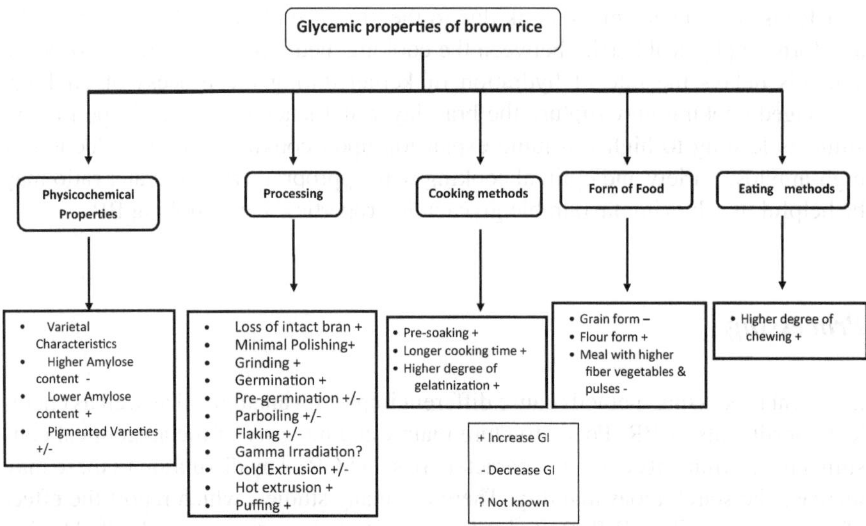

Fig. 8.1 Possible factors influencing the glycaemic properties of brown rice

The bran and endosperm cell walls of the BR are made of non-starchy polysaccharides, and this acts as a barrier protecting the starchy component. Any processing that disrupts this barrier (e.g. polishing and pulverizing) increases the accessibility of digestive enzymes and thereby increases digestibility. The effect of polishing on GI of rice has been demonstrated by Karupaiah et al. (2011), who reported a lower glycaemic and insulinaemic index for new BR variant (GI = 51, II = 39) compared to polished BR (GI = 79, II = 63) and control WR of Cap Rambutan variety (GI = 86, II = 68). Boers et al. (2015) reported that polishing exerted a definite influence on GI. Our group (Shobana et al. 2016) has recently shown that Bapatla (BPT-5204, an Indian rice variety) BR elicited the lowest GR (GI = 57.6) compared to undermilled rice (UMR, GI = 73) and WR (GI = 79.6) (prepared from the same rice variety). We postulated that the differences in the degree of milling, distinct cooking characteristics attributed to the intactness of the bran content and the rate of hydration and gelatinization could be the possible reasons for higher GI of UMR and WR. Hand-pounded rice, which was consumed in the past, may contain higher levels of bran and germ constituents compared to WR; however, some amounts of bran would have been lost during hand pounding and winnowing process. Hence, compared to BR, one would expect a higher GI for hand-pounded rice owing to the loss of minimal amounts of bran during processing. Earlier studies from our centre have shown a lower GR for BR and pulse-based diets compared to WR-based diets in Asian Indians (Mohan et al. 2014).

O'Dea et al. (1981) showed lower digestibility for both BR and WR in the grain form, compared to the ground form. This may be due to the increased surface area. The method of milling also affects glycaemic responses. Different milling methods may produce starch degradation to different extents in rice.

Imam et al. (2012) reported that the antidiabetic properties of germinated BR are due to bioactive components like γ-aminobutyric acid (GABA), γ-oryzanol, dietary fibre, phenolics, vitamins, acylated steryl β-glucoside and minerals. However, germination increases the reducing sugar content and hydrolyses starch which may affect the glycaemic properties of germinated BR. However, the extent of hydrolysis depends on the duration of germination. Ito et al. (2005) reported that pregerminated BR elicits a lower GR compared to normal BR (Table 8.1).

Miller et al. (1992) reported very high GI for brown rice pasta.GR for pasta products may depend on the following factors such as type of rice used, amylose content, processing conditions, extent of retrogradation, RS formation, cooking method and several other factors. Careful consideration of the aforementioned factors and appropriate cooking methods are required to achieve lower GR for the same.

There are several studies on the post-meal glycaemic response of WR-based meals. Inclusion of lentils, cheese and barley to WR was found to lower the GI (Sugiyama et al. 2003; Sakuma et al. 2009; Hettiaratchi et al. 2011). However, such studies are not available for BR.

Conclusions

The GI of BR varies between low and high depending on the various factors outlined above. Studies are required to establish optimal cooking methods for BR to achieve lower GR. Research towards devising optimal parboiling methods to produce BR with lower glycaemic properties are needed. Polishing, grinding, germination, popping, puffing and instantiation processes may increase the GI of BR. However, such studies are not available. Utilization of functional ingredients such as soluble fibre in formulations for the preparation of BR-based products could be helpful in reducing the GR. There is an urgent need for GI databases on BR and its products especially for the high-carbohydrate-consuming populations like Indians. Food regulatory authorities need to insist on display of GI values on the food packs. Healthcare professionals and food scientists should work together to develop more scientific evidence-based BR products with favourable glycaemic and health properties.

References

Atkinson FS, Foster-Powell K, Brand-Miller JC (2008) International tables of glycemic index and glycemic load values: 2008. Diabetes Care 31(12):2281–2283

Augustin LS, Kendall CW, Jenkins DJ, Willett WC, Astrup A, Barclay AW, Björck I, Brand-Miller JC, Brighenti F, Buyken AE, Ceriello A (2015) Glycemic index, glycemic load and glycemic response: an International Scientific Consensus Summit from the International Carbohydrate Quality Consortium (ICQC). Nutr Metab Cardiovasc Dis 25(9):795–815

Boers HM, ten Hoorn JS, Mela DJ (2015) A systematic review of the influence of rice characteristics and processing methods on postprandial glycaemic and insulinaemic responses. Br J Nutr 114(07):1035–1045

Brand JC, Nicholson PL, Thorburn AW, Truswell AS (1985) Food processing and the glycemic index. Am J Clin Nutr 42(6):1192–1196

Casiraghi MC, Brighenti F, Pellegrini N, Leopardi E, Testolin G (1993) Effect of processing on rice starch digestibility evaluated by in vivo and in vitro methods. J Cereal Sci 17(2):147–156

Cleveland LE, Moshfegh AJ, Albertson AM, Goldman JD (2000) Dietary intake of whole grains. J Am Coll Nutr 19(sup3):331S–338S

Daomukda, N., Moongngarm, A., Payakapol, L. and Noisuwan, A., 2011. Effect of cooking methods on physicochemical properties of brown rice. In: 2nd international conference on environmental science and technology (IPCBEE), vol 6

Deepa G, Singh V, Naidu KA (2010) A comparative study on starch digestibility, glycemic index and resistant starch of pigmented ('Njavara'and 'Jyothi') and a non-pigmented ('IR 64') rice varieties. J Food Sci Technol 47(6):644–649

Esfahani A, Wong JM, Mirrahimi A, Srichaikul K, Jenkins DJ, Kendall CW (2009) The glycemic index: physiological significance. J Am Coll Nutr 28(sup4):439S–445S

FAO/WHO (1998) Carbohydrates in human nutrition: report of joint FAO/WHO expert consultation. FAO Food Nutr Pap 66:1–140

Harnack L, Walters SAH, Jacobs DR (2003) Dietary intake and food sources of whole grains among US children and adolescents: data from the 1994-1996 continuing survey of food intakes by individuals. J Am Diet Assoc 103(8):1015–1019

Hettiaratchi UPK, Ekanayake S, Welihinda J, Perera MSA (2011) Glycemic and insulinemic responses to breakfast and succeeding second meal in type 2 diabetics. Int J Diab Dev Count 31(4):199–206

Holt SH, Miller JB (1995) Increased insulin responses to ingested foods are associated with lessened satiety. Appetite 24(1):43–54

Imam MU, Azmi NH, Bhanger MI, Ismail N, Ismail M (2012) Antidiabetic properties of germinated brown rice: a systematic review. Evid Based Complement Alternat Med 2012

Ito Y, Mizukuchi A, Kise M, Aoto H, Yamamoto S, Yoshihara R, Yokoyama J (2005) Postprandial blood glucose and insulin responses to pre-germinated brown rice in healthy subjects. J Med Investig 52(3, 4):159–164

Jenkins DJ, Wolever TM, Taylor RH, Barker H, Fielden H, Baldwin JM, Bowling AC, Newman HC, Jenkins AL, Goff DV (1981) Glycemic index of foods: a physiological basis for carbohydrate exchange. Am J Clin Nutr 34(3):362–366

Juliano BO, Perez CM, Komindr S, Banphotkasem S (1989) Properties of Thai cooked rice and noodles differing in glycemic index in noninsulin-dependent diabetics. Plant Foods Hum Nutr 39(4):369–374

Karupaiah T, Aik CK, Heen TC, Subramaniam S, Bhuiyan AR, Fasahat P, Zain AM, Ratnam W (2011) A transgressive brown rice mediates favourable glycaemic and insulin responses. J Sci Food Agric 91(11):1951–1956

Kaur B, Chin RQY, Camps S, Henry CJ (2016a) The impact of a low glycaemic index (GI) diet on simultaneous measurements of blood glucose and fat oxidation: a whole body calorimetric study. J Clin Transl Endocrinol 4:45–52

Kaur B, Ranawana V, Henry J (2016b) The glycemic index of rice and rice products: a review, and table of GI values. Crit Rev Food Sci Nutr 56(2):215–236

Larsen HN, Christensen C, Rasmussen OW, Tetens IH, Choudhury NH, Thilsted SH, Hermansen K (1996) Influence of parboiling and physico-chemical characteristics of rice on the glycaemic index in non-insulin-dependent diabetic subjects. Eur J Clin Nutr 50(1):22–27

Larsen HN, Rasmussen OW, Rasmussen PH, Alstrup KK (2000) Glycaemic index of parboiled rice depends on the severity of processing: study in type 2 diabetic subjects. Eur J Clin Nutr 54(5):380

Lin MHA, Wu MC, Lu S, Lin J (2010) Glycemic index, glycemic load and insulinemic index of Chinese starchy foods. World J Gastroenterol 16(39):4973–4979

Miller JB, Pang E, Bramall L (1992) Rice: a high or low glycemic index food? Am J Clin Nutr 56(6):1034–1036

Mohan V, Radhika G, Sathya RM, Tamil SR, Ganesan A, Sudha V (2009) Dietary carbohydrates, glycaemic load, food groups and newly detected type 2 diabetes among urban Asian Indian population in Chennai, India (Chennai urban rural epidemiology study 59). Br J Nutr 102(10):1498–1506

Mohan V, Spiegelman D, Sudha V, Gayathri R, Hong B, Praseena K, Anjana RM, Wedick NM, Arumugam K, Malik V, Ramachandran S (2014) Effect of brown rice, white rice, and brown rice with legumes on blood glucose and insulin responses in overweight Asian Indians: a randomized controlled trial. Diabetes Technol Ther 16(5):317–325

O'dea K, Snow P, Nestel P (1981) Rate of starch hydrolysis in vitro as a predictor of metabolic responses to complex carbohydrate in vivo. Am J Clin Nutr 34(10):1991–1993

Panlasigui LN, Thompson LU, Juliano BO, Perez CM, Yiu SH, Greenberg GR (1991) Rice varieties with similar amylose content differ in starch digestibility and glycemic response in humans. Am J Clin Nutr 54(5):871–877

Pathiraje PMHD, Madhujith WMT, Chandrasekara A, Nissanka SP (2010) The effect of rice variety and parboiling on in vivo glycemic response. Trop Agric Res 22(1):26–33

Radhika G, Van Dam RM, Sudha V, Ganesan A, Mohan V (2009a) Refined grain consumption and the metabolic syndrome in urban Asian Indians (Chennai urban rural epidemiology study 57). Metabolism 58(5):675–681

Radhika G, Ganesan A, Sathya RM, Sudha V, Mohan V (2009b) Dietary carbohydrates, glycemic load and serum high-density lipoprotein cholesterol concentrations among South Indian adults. Eur J Clin Nutr 63(3):413–420

Ranawana DV, Henry CJK, Lightowler HJ, Wang D (2009) Glycaemic index of some commercially available rice and rice products in great Britain. Int J Food Sci Nutr 60(sup4):99–110

Ranawana V, Leow MK, Henry CJK (2014) Mastication effects on the glycaemic index: impact on variability and practical implications. Eur J Clin Nutr 68(1):137–139

Ranawana V, Monro JA, Mishra S, Henry CJK (2010) Degree of particle size breakdown during mastication may be a possible cause of inter individual glycemic variability. Nutr Res 30(4):246–254

Sakuma M, Yamanaka-Okumura H, Naniwa Y, Matsumoto D, Tsunematsu M, Yamamoto H, Taketani Y, Takeda E (2009) Dose-dependent effects of barley cooked with white rice on postprandial glucose and desacyl ghrelin levels. J Clin Biochem Nutr 44(2):151–159

Shobana S, Lakshmipriya N, Ramya Bai M, Gayathri R, Ruchi Vaidya, Sudha ., Malleshi N.G, Krishnaswamy K, Henry C.J.K., Anjana R.M., Unnikrishnan R, Mohan V (2016) Even minimal polishing of an Indian parboiled brown rice variety leads to increased glycemic responses. Asia Pac J Clin Nutr (In press)

Sugiyama M, Tang AC, Wakaki Y, Koyama W (2003) Glycemic index of single and mixed meal foods among common Japanese foods with white rice as a reference food. Eur J Clin Nutr 57(6):743–752

Sun Q, Spiegelman D, van Dam RM, Holmes MD, Malik VS, Willett WC, Hu FB (2010) White rice, brown rice, and risk of type 2 diabetes in US men and women. Arch Intern Med 170(11):961–969

Sun L, Ranawana DV, Tan WJK, Quek YCR, Henry CJ (2015) The impact of eating methods on eating rate and glycemic response in healthy adults. Physiol Behav 139:505–510

Trinidad TP, Mallillin AC, Encabo RR, Sagum RS, Felix AD, Juliano BO (2013) The effect of apparent amylose content and dietary fibre on the glycemic response of different varieties of cooked milled and brown rice. Int J Food Sci Nutr 64(1):89–93

Wolever TMS, Jenkins DJA, Kalmusky J, Jenkins A, Janet DMK, Giordano C, Giudici S, Josse RG, Wong GS (1986) Comparison of regular and parboiled rice: explanation of discrepancies between reported glycemic responses to rice. Nutr Res 6(4):349–357

Yusof BNM, Talib RA, Karim NA (2005) Glycaemic index of eight types of commercial rice in Malaysia. Malaysian J Nutr 11:151–163

Chapter 9
Nutritional and Health Benefits of Rice Bran Oil

Amanat Ali and Sankar Devarajan

Rice bran oil (RBO) is gaining popularity among other traditionally used cooking oils because of its better cooking quality, prolonged shelf life and well-balanced fatty acid composition as well as the presence of many antioxidant components. RBO has lower viscosity and relatively high smoke point, which make it as healthy cooking oil. RBO is rich in vitamin E (both tocopherols and tocotrienols) and bioactive phytonutrients, which include phytosterols, γ-oryzanol, squalene and triterpene alcohols. All of these compounds exhibit high antioxidant, anti-inflammatory, hypocholesterolaemic, antidiabetic and anticancer activities. The dietary intake of RBO has been reported to lower the levels of blood cholesterol, blood pressure and blood glucose and can help to reduce inflammation and symptoms of metabolic syndrome. RBO helps to boost the immune system and prevent the process of premature ageing and age-related neurodegenerative diseases. Because of its cardiac-friendly phytochemicals and antioxidant potentials, RBO has been categorized as healthy edible oil for human consumption and has attained the status of "heart-healthy oil". As per scientific evidences, it is suggested that a daily intake of 50 g of RBO besides dietary and lifestyle modifications may be considered enough to attain its beneficial effects in reducing the risk of chronic diseases, in particular the cardiovascular diseases. This chapter discusses the nutritional and phytochemical components of RBO, their mechanism of action as well as health benefits in the prevention and management of chronic diseases.

A. Ali (✉)
Department of Food Science and Nutrition, College of Agricultural and Marine Sciences, Sultan Qaboos University, POB 34, Al-Khoud 123, Muscat, Sultanate of Oman
e-mail: amanat@squ.edu.om

S. Devarajan
Nutrition & Dietetics Program, Department of Human Sciences, University of Arkansas at Pine Bluff, Pine Bluff 71601, Arkansas, USA
e-mail: devsankara@gmail.com

© Springer International Publishing AG 2017
A. Manickavasagan et al. (eds.), *Brown Rice*,
DOI 10.1007/978-3-319-59011-0_9

Introduction

Rice (*Oryza sativa*) is one of the most widely available and popularly consumed cereal grains, which is regarded as the staple food for more than 50% of world's population. More than 90% of world's rice is consumed in Asia, the highest amount in China. The global per capita consumption of rice has increased over the past years from 50 to 65 kg per annum. The worldwide rice consumption during the year 2015–2016 has been estimated to be 478.441 million metric tonnes, whereas the United States Department of Agriculture estimates that the world's rice production during the year 2016–2017 will be about 481.5 million metric tons (Statistica 2016). Out of top 11 rice producing countries in the world, 10 are from Asia. China and India rank the top of the list and contribute over 25% and 23%, respectively, of the world's rice production. The milling of paddy can yield about 70% of rice (endosperm). The outer layer of rice grain (pericarp) is called the bran that constitutes about 10% of the rough rice grain. Depending on the rice variety and type of extraction, the yield of RBO may be between 18% and 22% (Sayre and Saunders 1990).

The crude RBO is mainly obtained through the solvent extraction process. To produce the edible grade vegetable oil, it is then refined and processed further either chemically or physically to meet the standards of specifications. The quality of RBO is, however, affected by the processing steps that are applied during the refining of RBO, which can affect the retention/availability of oryzanol and various other bioactive components in the commercial refined RBO. The process of refining may consists of acid degumming, centrifugation, clarification, bleaching, deodorization and winterization (Rajam et al. 2005). Chemical refining of crude RBO yields better product in terms of colour, cloud point and other physical characteristics (Danielski et al. 2005; Rajam et al. 2005). Although the chemical refining is preferred over physical refining, it can lead to significant losses in some minor bioactive components in the refined oil (Patel and Naik 2004; van Hoed et al. 2006; Prasad et al. 2011). The oryzanol content of RBO extracted from the bran of 18 different Indian paddy cultivars ranged from 1.63% to 2.72% (Krishna et al. 2001). The presence of higher quantities of γ-oryzanol in the physically refined oil may help to improve its oxidative stability. The γ-oryzanol-rich RBO possesses strong antioxidant activity that helps to protect the body cells from the damaging effects of very-low-density lipoproteins (Xu et al. 2001). Based on its high antioxidant potential, RBO has been categorized as valuable edible oil for human consumption (Bopitiya and Madhujith 2014).

Physiochemical Characteristics and Cooking Properties of RBO

RBO is the most available and very-well-studied rice product. RBO is a pale yellow, translucent and odourless, having pleasant mild nutty flavour with lightly sweet neutral taste. The vegetable oils are good sources of unsaturated fatty acids. RBO contains 38.4% oleic acid, 34.4% linoleic acid and 2.2% α-linolenic acid. The saturated fatty acids present in RBO are 2.9% stearic acid and 21.5% palmitic acid (Sayre and Saunders 1990). RBO is free from trans fats. Although the RBO contains only small amounts of α-linolenic acid, it is sufficient enough for the de novo synthesis of other omega-3 polyunsaturated fatty acids such as eicosapentaenoic acid (EPA) and docosahexaenoic acid (DHA) in tissue phospholipids as compared to other vegetable oils. The proportionate amount of fatty acids in RBO, however, may vary with the extraction process (cold or hot extraction).

Because of the presence of high levels of unsaturated fatty acids and many bioactive components, vegetable oils differ in their fatty acid compositions and therefore behave differently when heated. The typical fatty acid composition, lower viscosity and relatively high smoke point (~254 °C), makes it versatile for different types of cooking. RBO can be used for sautéing, grilling and marinades and can also be of great value in salad dressings. RBO is considered as perfect cooking oil for stir-frying or deep-frying (Sayre and Saunders 1990) because it takes less time to prepare foods and can save the energy as well. The foods cooked at high temperatures appear to absorb less oil, almost 15% less during frying (Mishra and Sharma 2014). Foods cooked in RBO have better taste and flavour and likely to be less oily during eating. It has a high storage stability, as its various bioactive components and vitamin E contents act as antioxidant and protect it not only from oxidation and rancidity but also are responsible for its higher thermal stability (Bergman and Xu 2003; Fang et al. 2003; Mezouri and Eichner 2007). The presence of γ-oryzanol and γ-tocotrienol in RBO may play a protective role on the availability of α-tocopherol in deep-frying (Hamid et al. 2014).

Bioactive Components of RBO

RBO presents several advantages over other cooking oils due to the presence of many bioactive antioxidant components such as tocopherols, γ-oryzanol and tocotrienols, which are responsible for its oxidative stability and health benefits (Kim and Godber 2001; Wilson et al. 2000). The crude RBO encompasses a rich unsaponifiable fraction (~5%), which consists of sterols (43%), triterpene alcohols (28%), 4-methyl-sterols (10%) and other less polar components (Sayre and Saunders 1990; Mezouri and Eichner 2007; Macchar et al. 2012). The total phenolic content (TPC) of RBO can vary depending upon the rice varieties and oil extraction process. The extracts obtained from two Sri Lankan rice varieties (BG 400 white, and LD 365 red) exhibited dose-dependent free radical scavenging activity. No pro-oxidant activity was observed in the RBO extracts when tested even at the highest level (Bopitiya and Madhujith 2014).

The phytosterols in RBO include β-sitosterol, campesterol, stigmasterol, squalene and γ-oryzanol. γ-Oryzanol is often recognized as the most active component of RBO and consists of a mixture of ferulic acid esters of triterpene alcohols (Metwally et al. 1974; Norton 1995; Akihisa et al. 2000; Lloyd et al. 2000; Fang et al. 2003; Patel and Naik 2004). The amount of γ-oryzanol in crude RBO can vary between 1% and 2% depending on the extraction method. During the chemical refining process, it is neutralized and can be transferred to soap stock. The use of physical refining process under light conditions may however be able to preserve most of γ-oryzanol (Krishna et al. 2001). Other vegetable oils do not contain the cardioprotective γ-oryzanol, and therefore RBO is regarded as the heart-friendly oil. Rice brain oil is also a rich source of vitamin E (both tocopherols and tocotrienols). It contains variable quantities of tocotrienols, especially β- and γ-tocotrienols, but it is naturally very rich in tocopherols (Rukmini and Raghuram 1991; Rogers et al. 1993). Vitamin E not only helps to boost immunity but also has anti-mutagenic properties.

γ-Oryzanol has similar functions as vitamin E for growth promotion, capillary functions in the skin, improved blood circulation and stimulation of hormonal secretions (Luh et al. 1991; Bergman and Xu 2003; Fang et al. 2003). γ-Oryzanol has structural similarities to cholesterol and may compete with it for the binding sites and may increase the faecal excretion of cholesterol and its metabolites (Mäkynen et al. 2012; Kota et al. 2013). γ-Oryzanol has also been reported to help in the inhibition of gastric acid secretion and can decrease the postexercise muscle fatigue (Szcześniak et al. 2016). An ideal edible oil should contain saturated, mono-unsaturated and polyunsaturated fatty acids in proportions of 1:1.5:1 ratio to meet the recommended intake of fatty acids. However, this is not the case in practical terms as all the edible oils differ in their fatty acid composition. The RBO has almost similar ratio of fatty acids as recommended by the WHO and AHA for lowering the blood cholesterol levels (Lai et al. 2012; Friedman 2013). The polyunsaturated fatty acids (PUFA) in RBO exert greater hypolipidaemic activities as compared to other vegetable oils containing linoleic acid and therefore may help to lower the cardiovascular risk (Friedman 2013). Studies, however, suggest that cholesterol-lowering properties of RBO could mainly be due to its unsaponifiable fraction of bioactive components rather than because of its fatty acid composition (Abumweis et al. 2008; Liang et al. 2014).

Significance of RBO in Human Health

RBO is gaining popularity as compared to other traditionally used cooking oils (such as corn oil, sunflower oil, safflower oil, canola oil, olive oil, etc.) because of its better cooking characteristics, prolonged shelf life and well-balanced fatty acid composition as well as the presence of a number of bioactive substances. RBO is commonly used in many Asian cultures, where it is regarded as "premium edible oil". In Japan, RBO is commonly known as a "heart oil", whereas in Western countries, it has attained the status of a "healthy food" (CAC 2003). It is also now becoming popular in the USA and other parts of the world because of its relatively low price and many health benefits (Liang et al. 2014).

The rate of mortality due to cardiovascular events in the Asian and Far Eastern Asian countries is much lower than the Europeans and North Americans, which may be attributed to their dietary patterns. The diets in the Asian and Far Eastern Asian countries are generally low in saturated fatty acids, are poor in cholesterol and are rich in rice and legume-based vegetable proteins. The hypocholesterolaemic properties of vegetable oils are associated with their unsaturated fatty acids contents, mainly oleic acid, linoleic acid and α-linolenic. In general, RBO has been shown to have the potential to lower cholesterol, blood pressure and blood glucose level and can help to reduce inflammation and symptoms of metabolic syndrome. It may help in weight loss and therefore in controlling obesity. RBO has been shown to be effective in the prevention and management of cardiovascular disease (CVD) risk factors if consumed as part of a healthy diet (Zavoshy et al. 2012). It may also help to boost the immune system and prevent diabetes, cardiovascular diseases, cancer and premature ageing (Manosroi et al. 2012a, b). RBO has therefore a great potential in the development of pharmaceutical and cosmetic products (Ammar et al. 2012). RBO relieves the menopausal symptoms, increases the cognitive function and may lower the incidence of allergic reactions (Mehdi et al. 2015).

Health Benefits of RBO

Anti-hyperlipidaemic and Hypocholesterolaemic Effects

The data from numerous studies have shown that the intake of RBO reduced the plasma total cholesterol (TC), triglycerides (TG) and low-density lipoprotein cholesterol (LDL-C) and increased the high-density lipoprotein cholesterol (HDL-C) levels in rodents, rabbits, non-human primates and humans (Cicero and Derosa 2005; Lai et al. 2012; Macchar et al. 2012; Shakib et al. 2014; Devarajan et al. 2016a). The mechanism of action of RBO on lipid metabolism is however not yet conclusive. The presence of appreciable amounts of unsaponifiable fractions (triterpene alcohols, phytosterols, γ- oryzanol and tocotrienols) in RBO have been shown to have beneficial effects on lipid metabolism in terms of their antioxidant, hypolipidaemic and anti-atherogenic properties (Lee et al. 2005; Tabassum et al. 2005; Macchar et al. 2012; Hota et al. 2013; Dhavamani et al. 2014). The specific bioactive components in RBO were responsible for its anti-hyperlipidaemic properties, whereas the particular fatty acids (mono- and polyunsaturated) seem to have some impact (Nicolosi et al. 1991; Rong et al. 1992). The phytosterols, in particular the ss-sitosterol and 4-desmethylsterols and not the 4,4′-dimethylsterols in RBO, have been shown to reduce the plasma TC and LDL-C levels. These phytosterols may either affect the absorption of dietary cholesterol from the gut or may enhance the attachment of cholesterol to bile acids, which are then excreted in the faeces (Vissers et al. 2000).

The data from earlier studies on rats indicated that rats fed diets containing RBO at 10% level for 8 weeks showed lower plasma TC, LDL-C and VLDL-C and increased HDL-C levels, while no changes in TG were observed on cholesterol-containing or cholesterol-free diets (Sharma and Rukmini 1986). Feeding RBO also reduced the liver cholesterol and TG levels. An increased excretion of neutral sterols and bile acids in the faeces was also observed (Sharma and Rukmini 1986, 1987). RBO showed better results for liver lipids as compared to groundnut oil. A further decrease in serum TC levels was observed when RBO was supplemented with γ-oryzanol at 0.5% level in the diet (Seetharamaiah and Chandrasekhara 1988, 1989). Seetharamaiah and Chandrasekhara (1990) examined the effects of γ-oryzanol on the biliary secretion and faecal excretion of cholesterol, phospholipids and bile acids in male albino rats. They didn't observe any change in bile flow and its composition when rats were fed control diet supplemented with 0.5% γ-oryzanol. However, supplementation of high-cholesterol diet with γ-oryzanol indicated increased bile flow and total bile acid excretion with simultaneous 20% decrease in cholesterol absorption. These results suggest that γ-oryzanol and some other components in the unsaponifiable fraction of RBO such as tocotrienols and tocopherols can increase the faecal excretion of bile acids and neutral sterols (Sharma and Rukmini 1986; Seetharamaiah and Chandrasekhara 1989).

Adding phytosterols to hypercholesterolaemic rat diets, in particular the cyclo-artenol, significantly reduced both the plasma cholesterol and triglycerides (Rukmini and Raghuram 1991). Supplementation of RBO with γ-oryzanol appeared to be strongly associated with alleviating the cardiovascular disease risk factors, especially when the rats were fed a high-fat diet (Edwards and Radcliffe 1994; Radcliffe et al. 1997). The rats fed γ-oryzanol supplemented diet also showed 25% reduction in cholesterol absorption as compared to control. The rats fed diet containing 10% refined RBO showed significantly lower serum total, free esterified and (LDL + VLDL) cholesterol values as compared to those fed 10% groundnut oil diet. RBO also exhibited an increase in HDL-C levels. Purushothama et al. (1995) studied the impact of long-term feeding of RBO on lipids and lipoprotein metabolism in rats. The rats fed RBO showed lower levels of plasma TC, LDL-C and VLDL-C, TG and phospholipids as compared to those fed on peanut oil. However, only the rats receiving 20% RBO in their diet, showed a 20% increase in high-density lipoprotein cholesterol (HDL-C) level, as compared to rats fed on peanut oil (Purushothama et al. 1995).

The hypolipidaemic response of RBO has also been studied in non-human primates (Nicolosi et al. 1991). The use of RBO or its blends at 20–25% of total energy intake as dietary fat showed significant reduction in the serum TC, LDL-C and apolipoprotein-B level. Ausman et al. (2005) reported that the lipid-lowering properties of physically refined RBO (PRBO) may be attributed to decreased cholesterol absorption and not to the hepatic cholesterol synthesis. They suggested that a reduction in fatty streak formation, the early signs of atherosclerosis with PRBO, may be due to its non-triglyceride fraction. Wilson and his colleagues (2007) not only compared the cholesterol-lowering potential of various vegetable oils but also compared the impact of various individual bioactive components of

RBO such as trans-ferulic acid and γ-oryzanol as compared to RBO alone in hypo-cholesterolaemic hamsters. They fed high-cholesterol diet (HCD) to hamsters as control group and compared the effect of feeding HCD with 10% RBO, HCD plus 0.5% trans-ferulic acid and HCD with 0.5% γ-oryzanol. The serum LDL + VLDL and total plasma cholesterol levels after 10 weeks reduced considerably in experimental groups fed on diet with 10% RBO, diet containing 0.5% trans-ferulic acid and diet containing 0.5% γ-oryzanol as compared to control group (Wilson et al. 2007). The animals fed on diets containing γ-oryzanol and RBO showed significant reduction in plasma lipid hydroperoxides and triglycerides. The results indicated that γ-oryzanol might have potentiated the lowering of plasma LDL and VLDL levels and may also help to raise the HDL cholesterol level as compared to trans-ferulic acid.

The cholesterol-lowering efficacy of RBO is much superior to that is apparently judged based on its fatty acid composition. This may be associated with the presence of other bioactive constituents, mainly γ-oryzanol in RBO (Moldenhauer et al. 2003). The naturally occurring γ-oryzanol and vitamin E synergistically work to scavenge the free radicals and thereby protect the cells from oxidative stress (Kennedy and Burlingame 2003). Different mechanisms have been suggested about the anti-atherogenic action of RBO. γ-Oryzanol is considered as the possible fundamental component of RBO due to its anti-atherosclerotic action. It inhibits the intestinal absorption of cholesterol, increases the bile flow and accelerates the excretion of cholesterol in the faeces (Kanbara et al. 1992; Cicero and Gaddi 2001). Tsuji et al. (2003) studied the effects of hypocholesterolaemic diets containing RBO and different concentrations of γ-oryzanol on serum cholesterol levels in rats. They observed that the reduced TC level in rats fed on RBO is attributed to the antioxidant properties of γ-oryzanol. The data from the animal model studies confirm the anti-hyperlipidaemic properties of γ-oryzanol and its overall impact to lower the CVD risk.

RBO in Clinical Trials on Humans

The early studies of Suzuki and Oshima (1970) reported the anti-hyperlipidaemic properties of RBO in healthy young Japanese women. The study showed that the daily use of 60 g of a blend of RBO and safflower oil (70:30) was more effective in lowering the plasma TC levels; even only after 7 days of treatment, the RBO and sunflower oils, when given either alone or in different proportionate combinations, were not that effective. The data from various studies in which RBO was given for 4–14 weeks period indicated that the RBO at a dose level up to 50 g/day was effective in reducing the TC, LDL-C, TG and apolipoprotein levels with a simultaneous increase in HDL-C concentrations (Suzuki and Oshima 1970; Raghuram et al. 1989; Lichtenstein et al. 1994; Qureshi et al. 1997). In young female volunteers, who daily consumed five eggs for seven consecutive days, the blended oil exerted the hypocholesterolaemic effects. In addition to this, a significant association was observed with increased levels in plasma HDL-C (Tsuji et al. 1989). Ishihara and his colleagues (1982) evaluated the impact of γ-oryzanol supplementation on 40 women with postmenopausal syndrome. Treatment with 300 mg of γ-oryzanol/day for 4–8 weeks showed a significant decrease in plasma TC, LDL-C and TG levels with a simultaneous increase in HDL-C concentration in hyperlipoproteinaemic subjects. The plasma lipid peroxide was also lower in subjects who had previously elevated levels. No particular changes in liver and renal functions or no other side effects were observed.

These results were subsequently confirmed in another study, in which 12 moderately nonobese hyperlipoproteinaemic subjects were asked to substitute their usual cooking oils with RBO. It was observed that the patients who received RBO showed 16% and 25% decrease in plasma TC and 32% and 35% reduction in plasma TG after 15 and 30 days of treatment, respectively, as compared to control group (Ishihara 1984). Raghuram et al. (1989) also observed that the subject with higher baseline levels of TC and TG showed faster and greater decrease in lipid levels on RBO. Similar results were observed in both hypercholesterolaemic and hypertriglyceridaemic patients, when they were given 300 mg/day of γ-oryzanol for 3 months without any side effects (Yoshino et al. 1989). Lichtenstein et al. (1994) conducted a comparative double-blind Latin-square design study on elderly people for a period of 32 days to evaluate the effects of various edible oils (rice bran, canola, corn or olive oil) on their plasma lipid profile. They, however, did not observe any statistically significant differences in the plasma TC and LDL-C concentrations in subjects, who consumed the RBO-, canola oil- or corn oil-enriched diets.

The results of a double-blind 12-week-long clinical trial on hypercholesterolaemic human subjects, who were given a supplement of tocotrienol-rich fraction that was obtained from specially processed RBO, together with a standard National Cholesterol Education Program (NCEP) Step-1 diet, indicated a significant decrease in plasma TC and LDL-C levels as compared to control. The serum apolipoprotein B, lipoprotein (a) (Lp(a)), platelet factor 4 and thromboxane B2 also decreased significantly as compared to baseline levels (Qureshi et al. 1997). The combined treatment of NCEP Step-1 diet and tocotrienol-rich fraction of RBO resulted in 25% reduction in plasma LDL-C levels (Qureshi et al. 2002). The tocotrienol treatment decreased the Lp(a) plasma levels, whereas neither the NCEP Step-1 diet nor any anti-hypercholesterolaemic drugs showed such impact. It appears that the RBO and its bioactive components may be able to safely improve the plasma lipid profile in hypercholesterolaemic patients. The data from various clinical trials clearly indicates that the consumption of RBO-rich diets can significantly improve the levels of HDL-C in hypercholesterolaemic human subjects (Berger et al. 2004).

Rajnarayana and his colleagues (2001) gave 75 ml of RBO thrice daily as the cooking medium with breakfast, lunch and dinner to nine healthy human volunteers, aged between 42 and 57 years for a period of 50 days. They observed that the volunteers who consumed RBO showed significantly lower levels of lipid peroxides, triglycerides, LDL, VLDL and TC. They suggested that the bioactive components of RBO have antioxidant and lipid-lowering activities. Most and his colleagues (2005) evaluated the effects of defatted rice bran and RBO in an average American diet on the blood lipid profile of moderately hypercholesterolaemic persons. They observed that the consumption of RBO-containing diet significantly decreased the LDL-C level by 7%, whereas HDL-C level remained unchanged. They concluded that RBO and not the fibre in diet lowered cholesterol in healthy, moderately hypercholesterolaemic adults that may be associated with the bioactive unsaponifiable components of RBO (Most et al. 2005). Kuriyan et al. (2005) in a crossover study design assessed the consumption of RBO and refined sunflower oil in hyperlipidaemic human subjects for a period of 3 months. They observed that RBO as the main cooking oil significantly reduced serum cholesterol and triglyceride levels as compared to sunflower oil in hyperlipidaemic human subjects.

The most potent and commonly used class of drugs to prevent dyslipidaemia are 3-hydroxy-3-methylglutaryl coenzyme A (HMG-CoA) reductase inhibitors (statins). Statins are effective in decreasing the rate of mortality from coronary artery disease, the incidence of myocardial infarction, stroke and peripheral vascular diseases (Shah and Goldfine 2012). However, a number of adverse effects have been reported with the use of statins including asymptomatic increases in liver transaminases and myopathy (Björnsson et al. 2012; Maji et al. 2013; Castilla-Guerra et al. 2016; Gurwitz et al. 2016). Shakib et al. (2014) compared the effects of RBO versus statins on blood glucose, glycosylated haemoglobin (HbA1C) and serum lipid profiles in patients with type 2 diabetes. The RBO group was given a low-calorie diet, and the patients consumed 30 g/day RBO as salad dressing. They also used RBO as the main cooking oil for 6 months. The patients in atorvastatin group received a low-calorie diet together with 40 mg/day of atorvastatin drug for 6 months. The diabetic and moderately hyperlipidaemic patients showed significant increases in the fasting and postprandial blood glucose, HbA1C and liver transaminase (alanine transaminase ALT and aspartate transaminase AST) levels on atorvastatin, whereas there was a reduction in all these parameters in RBO group. Significant reductions were however observed after 6 months in lipid profile levels, blood urea, serum uric acid and erythrocyte sedimentation rate (ESR) in patients in both the RBO and atorvastatin groups. They concluded that the use of RBO together with dietary modifications may be effective in lowering the fasting and postprandial blood glucose, HbA1c and serum lipid levels, reduce the TC/HDL-C ratio and therefore may reduce the risk of cardiovascular diseases. They also observed that RBO has anti-inflammatory properties and may exert a hypouricaemic action. Based on their findings, Shakib et al. (2014) suggested that RBO may be used as a safe alternative natural hypolipidaemic agent in place of atorvastatin. Atorvastatin may induce side effects in some patients who show intolerance to statins. Chithra et al. (2015) evaluated the anti-atherogenic effects of Njavara RBO (NjRBO) on atherosclerosis by modulating the enzymes and genes involved in lipid metabolism in rats fed a high-cholesterol diet (HCD). They hypothesized that NjRBO possesses anti-atherogenic properties that may modulate the lipid metabolism by up-regulating the genes involved in reverse cholesterol transport and antioxidative defence mechanisms through the induction of gene and protein expression of paraoxonase 1 (PON1).

The quality of most of the published studies is generally poor because of relatively low number of study participants and shorter durations, which were not sufficient enough to draw statistically valid conclusions. It has therefore been suggested that more randomized clinical trials with large number of study subjects and longer experimental time periods should be warranted to validate these findings and to confirm whether the RBO can be regarded as safe and efficacious in long-term treatments for mild to moderate hyperlipoproteinaemias. The consumption of a blend of RBO and safflower oil (70:30) together with other lifestyle changes has been shown to help in lowering blood lipid profile and inflammatory biomarkers such as oxidized LDL and high sensitivity C-reactive proteins (hs-CRP) in hyperlipidaemic patients (Upadya et al. 2015). They concluded that this strategy may in turn help to prevent lifestyle diseases. The results of a recent meta-analysis indicated that consumption of RBO reduced the LDL-C and TC concentrations and had favourable effects on HDL-C concentrations in men. No considerable changes were however observed related to other lipid profile components. They concluded that the consumption of RBO may help in prevention and control of CVD (Jolfaie et al. 2016).

Hypoglycaemic and Antidiabetic Effects

The tocotrienol-rich fraction (TRF) of RBO has been shown to act as an antioxidant to effectively decrease the glycosylated haemoglobin (HbA1C) in diabetic rats (Wan Nazaimoon and Khalid 2002). Supplementation of RBO helped to improve the glycaemic control and lipid profile in streptozotocin (STZ)-induced diabetic rats (Chen and Cheng 2006). It has been proposed that γ-oryzanol may play an effective role in the prevention and management of type 2 diabetes (Ohara et al. 2009). It has been shown that γ-oryzanol can regulate the secretion of insulin and blood glucose levels by normalizing the liver enzyme activities and therefore may lower the risk of hyperglycaemia induced by high-fat diets (Son et al. 2011). Ghatak and Panchal (2012a) studied the hypoglycaemic potential of γ-oryzanol in streptozotocin-induced diabetic rats having an elevated serum glucose level of 340–400 mg/dL. They observed a decline in the serum glucose levels of rats within 2–4 h after the administration of oryzanol at a dose of 50 and 100 mg/kg BW.

The data from the animal models as well as from the clinical trials on human diabetic patients indicated the blood glucose-lowering potential of tocotrienol-rich fraction (TRF) of RBO (Siddiqui et al. 2010). Tocotrienols in RBO are considered to lower blood TC concentrations by inhibiting the HMG-CoA reductase activity in the biosynthetic pathways of cholesterol metabolism (Houston et al. 2009). Tocotrienols have cardioprotective properties by improving the postischaemic ventricular functions and reducing the myocardial infarction (Vasanthi et al. 2012). The bioactive components and antioxidants present in RBO as well as its oleic acid and conjugated linoleic acid (CLA) contents may help to boost the metabolic rate, regulate the blood glucose and lipid profile, reduce inflammation, lose weight and control obesity (Ros 2003; Zhao et al. 2004). Shakib and his colleagues (2014) concluded from their study that the use of RBO together with dietary modifications may be effective in lowering the fasting and postprandial blood glucose and glycosylated haemoglobin (HbA1c) levels.

The dietary components can moderately help the dyslipidaemic patients to reduce the risk of cardiac diseases. In order to reduce or to maintain an adequate cholesterolaemia, the Adult Treatment Panel III (ATP III) of the National Cholesterol Educational Program (NCEP) suggests to introduce four portions of soy proteins (25 g/day), 2 g/day of phytosterols and 10–25 g/day of vegetable soluble fibres (like psyllium, guar gum, pectin, oats) in the daily diet (Kris-Etherton et al. 2002). Cicero and Derosa (2005) reviewed the available data on the pharmacology and toxicology of rice bran and its main components including RBO, specifically to its potential efficacy in reducing the CVD risk. The use of RBO can be considered as an appropriate dietary strategy that may help to reduce the liver lipid contents and therefore may be useful in treating non-alcoholic steatohepatitis, which appear to be associated with metabolic syndrome and CVD (Marchesini et al. 2003). The pigmented rice berry bran oil (RBBO) may have beneficial effects on diabetes by reducing the oxidative stress. It was suggested that the bioactive components of RBO can play a role in the prevention and management of diabetes mellitus (Posuwan et al. 2013). Ghatak and Panchal (2012b) observed that γ-oryzanol content in RBO was effective in ameliorating the neuropathic pain in diabetic patients. They suggested that due to its γ-oryzanol content, the RBO can favourably affect diabetic neuropathy.

RBO has been shown to inhibit high insulin response due to its polyphenols (Chou et al. 2009). In addition to this, its high MUFA content may also help to indirectly decrease the hyperinsulinemia (Li et al. 2005). The American Diabetes Association recommends that in order to improve their hyperlipidaemia and to prevent heart-related diseases, the diabetic patients should consume those vegetable oils, which contain high amounts of oleic acid (Lai et al. 2012). The polyphenols, tocotrienols and y-oryzanol fractions of RBO may therefore help to alleviate the endothelial dysfunction and subsequently reduce the insulin resistance (Manila et al. 2014). Insulin resistance leads to abnormal lipid metabolism increasing the risk of CVD in diabetic patients. RBO, because of its optimal fatty acid composition and constituent bioactive components, which have high absorption capacity in the gastrointestinal tract, can not only inhibit the intestinal absorption of cholesterol and block the synthesis of cholesterol analogues but can also increase the excretion of its metabolites from the body and thus may reduce the incidence of cardiovascular diseases. One of the mechanisms of diabetes aetiology is the increased apoptosis of insulin-secreting cells. RBO has been shown to indirectly inhibit the caspase inactivation and thereby may lead to inhibition of β-cell apoptosis, which can reduce the chances of diabetes. Caspases are a family of endoproteases (cysteine-aspartic proteases, cysteine aspartases or cysteine-dependent aspartate-directed proteases), which play essential roles in the programmed cell death and inflammation (McIlwain et al. 2015). RBO can suppress the progression of diabetes and therefore can be helpful in developing the dietary strategies in the prevention and management of diabetes. RBO can also help to promote the blood circulation, regulate the endocrine and autonomic functions and support the growth and development in humans and animals. Substitution of RBO or canola oil (CO) for sunflower oil was shown to attenuate the lipid disorders in postmenopausal type 2 diabetic women. They observed that RBO was more effective in improving the lipid profile as compared to canola oil (Salar et al. 2016). It is evident that consuming RBO can improve the plasma lipid profile; however its mechanism of action on diabetic hyperlipidaemia and the development of diabetes are still not clear. It should therefore be further explored for its potential health benefits in the management and control of hyperglycaemia and diabetes.

Effects of RBO on Oxidative Stress and Cancer Risk

Mitochondrial dysfunction can lead to excessive production of reactive oxygen species (ROS) and free radicals, which are produced as a result of certain metabolic abnormalities. They cause cellular damages through the oxidation of proteins lipids and DNA and therefore result in oxidative stress and progression of various chronic diseases (Giacco and Brownlee 2010; Kaneto et al. 2010; Waly et al. 2010; Ju and Zullaikah 2013). Polyphenols play an important role in modulating the differentially regulated pathways in endothelial cells and thus can help in maintaining the vascular homeostasis. The published data underlines the significance of phytochemicals in inhibiting the pathways that activate the nuclear transcription factor-kappa B (NF-κB) that is linked to a variety of inflammatory diseases (Surh et al. 2001; Bellik et al. 2012). Polyphenols protect the endothelial cells against various stimuli by downregulating the tumour necrosis factor alpha (TNF-α) (Suganya et al. 2016). The scientific data suggests that certain food ingredients and phytochemical antioxidants can prevent digestive disease processes, may improve the mitochondrial functions and may prevent or slow down the progression and development of age-related neurodegenerative diseases (Ellis et al. 2016; Serafini and Peluso 2016).

The distinctive properties of RBO and its high antioxidant potential can better help to prevent cellular lipid and protein oxidation (Iqbal et al. 2004; Rajnarayana et al. 2001; Hsieh et al. 2005). Hagl et al. (2016) concluded that rice bran extracts including RBO have great nutraceutical potential in the prevention of mitochondrial dysfunctions and may attenuate the oxidative stress in neurodegenerative diseases. It has been shown that that tocotrienols exhibit stronger antioxidant activity than tocopherols, which is attributed to their high capacity to donate phenolic hydrogen to various free radicals. The γ-tocotrienol (γ-T3) component of RBO can induce the expression of TNF-related apoptosis-inducing ligand (TRAIL) in human cancer cells and can promote the tumour cell apoptosis via cascade reactions (Kannappan et al. 2010a). The y-T3 has also been shown to affect the cell signalling pathways through the induction of protein tyrosine phosphatase SHP-1 and can sensitize the tumour cells to chemotherapeutic agents for apoptosis (Kannappan et al. 2010b). The y-T3 has also been reported to induce mitochondria-mediated apoptosis in human gastric adenocarcinoma SGC-7901 cells (Sun et al. 2009). The palmitic and linoleic acids have also been shown to induce the endoplasmic reticulum (ER) stress and apoptosis in hepatoma cells (Zhang et al. 2012).

Shih et al. (2011) reported that RBO showed preventive effects in delaying the colon carcinogenesis. They observed higher hepatic antioxidant status including the glutathione (GSH) and thiobarbituric acid reactive substance levels as well as the superoxide dismutase and catalase activities, in RBO-fed rats. They concluded that this higher antioxidant status in RBO-fed rats might be responsible in delaying the carcinogenesis. The inclusion of RBO in rat diets can improve their antioxygenic potential and may protect against oxidative stress (Rana et al. 2004). MUFA and conjugated linoleic acid (CLA) present in RBO can also exert antitumour effects. CLA has been shown to ameliorate the inflammation-induced colorectal cancer in mice through the activation of peroxisome proliferator-activated receptor (PPAR-y) (Evans et al. 2010). RBO has therefore the potential to play an important role as antitumour food on apoptosis through three different pathways: (1) by inducing death of receptors activating caspase cascade reactions, (2) by blocking JAK-STAT signal pathways by inducing SHP-1 and (3) by blocking mtDNA mutation from oxidative stress by ROS. All these pathways ultimately lead to tumour cell apoptosis (Liang et al. 2014). The oral intake of RBO in rats indicated beneficial effects on stress response and on learning and memory functions. A decrease in the stress-induced behavioural and neurochemical changes was also observed (Mehdi et al. 2015). It is suggested that the various bioactive components of RBO may exert synergistic effect in combating the reactive oxygen species (ROS) and may therefore help in the prevention of cellular oxidative damage, which needs to be studied further.

Health Benefits of RBO Blends

The blends of RBO with other less expensive vegetable oils are now gaining a greater popularity as cooking media because of their cost-benefit ratio and health benefits. Blending RBO with other vegetable oils such as olive oil, groundnut oil, sunflower oil and sesame oil has been shown to improve the quality of blends in terms of their physicochemical properties, fatty acid composition, antioxidant potential and in vivo antioxidant status (Choudhary et al. 2015; Umesha and Naidu 2015; Devarajan et al. 2016b). Rats fed on high-cholesterol and cholesterol-free diets showed significantly ($p < 0.05$) lower levels of TC, TG and LDL-C and increased level of HDL-C in animals when they were given RBO blends (containing either safflower oil or sunflower oil in 70:30 ratio). The RBO blends also showed a reduction in liver cholesterol and TG concentrations and increased the excretion of neutral sterols and bile acids in faeces (Sunitha et al. 1997). The higher contents of tocopherols and tocotrienols in RBO also improved the oxidative stability of these oil blends. Thus blending of RBO with other oils can not only improve the plasma lipid profile but may also result in economic advantages (Sunitha et al. 1997).

Koba (1997) studied the cholesterol-lowering ability of different blends of RBO and safflower oil in rats. No significant differences were observed in the serum and liver cholesterol levels among rats when fed different oil blends. However, the HDL-C level of rats fed the RBO-containing diets (especially in rats fed higher proportions of RBO) was higher than that of rats fed only safflower oil. The HDL-C-to-TC ratio, a desirable outcome for CVD risk factor, also improved (Koba et al. 2000). The additional improvement in lipid metabolism by RBO-safflower oil blends cannot be explained based on their fatty acids or plant sterols composition because the blending of RBO with sunflower oil did not exert the same anti-hypercholesterolaemic properties (Sugano and Tsuji 1997). The cholesterol-lowering abilities of RBO diet was greater than that anticipated from its constituent fatty acids. Accinni et al. (2006) evaluated the supplementation effects of γ-oryzanol, tocotrienols, niacin and omega-3 polyunsaturated fatty acids on oxidative stability, lipid profile and inflammatory responses in volunteers with abnormal blood lipid levels. During a 4-month trial period, all groups given different dietary supplements showed improvement in their blood lipid profile, and the best profile was in patients with γ-oryzanol-supplemented diet. Feeding blended oils to rats containing RBO, sesame oil and coconut oil with balanced fatty acid composition helped to lower their serum and liver lipids (Reena and Lokesh 2007).

The blends of RBO with soybean oil, in particular with palm oil, have also been shown to further reduce the risk of atherosclerosis in hypercholesterolaemic women (Utarwuthipong et al. 2009). In a double-blind, controlled, randomized parallel group study, Malve and his colleagues studied the LDL-cholesterol-lowering activity of a blend of RBO and safflower oil (8:2) in patients with hyperlipidaemia (Malve et al. 2010). The control group included the patients who continued to use the same oil, which they were using before. At the end of a 3-month trial, 82% of the patients from the group who consumed the blend of RBO and safflower oil (8:2) had LDL levels <150 mg/dL as against 57% in the control group. They concluded that the substitution of usual cooking oil with RBO and safflower oil (8:2) blend was helpful in reducing the LDL-C levels and shifting the patients to lower lipid risk category. This may also be due to their improved fatty acid composition and bioactive components, as they showed better antioxidant and anti-inflammatory effects (Choudhary et al. 2013). The incorporation of alpha-linolenic acid (ALA) and eicosapentaenoic acid (EPA) and docosahexaenoic acid (DHA) into RBO through lipase-catalysed inter-esterification has been shown to offer health benefits (Chopra and Sambaiah 2009). Because of their proven cholesterol-lowering potential, the plant sterols are added to different food products to enhance their potential to decrease the blood LDL-cholesterol levels (Scoggan et al. 2008). The consumption of a plant sterol-based spread derived from RBO as a part of normal diet proved effective in reducing plasma lipid levels in mildly hypercholesterolaemic individuals (Eady et al. 2011). Daily consumption of RBO-modified milk (containing 18 g RBO for 5 weeks) significantly decreased TC level and tended to decrease LDL-C level in patients with type 2 diabetes. However, no significant influence on insulin resistance was observed (Lai et al. 2012).

Both sesame oil and RBO are known for their optimum unsaturated fatty acids and antioxidants contents. In a randomized dietary approach study, a blend of 20% unrefined cold-pressed lignans-rich sesame oil and 80% physically refined γ-oryzanol-rich RBO as cooking oil in mild-to-moderate hypertensive patients was studied on 300 hypertensive patients. Sesame oil-RBO blend was supplied to hypertensive patients, and they were asked to consume it as the only cooking oil. Significant reduction in blood pressure was observed in hypertensives treated with the blend RBO and sesame oil. TC, LDL-C, TG and non-HDL-C levels reduced, while HDL-C levels increased significantly in these patients after 60 days use of sesame oil-RBO blend. It was suggested that consuming a blend of sesame oil and RBO had a significant antihypertensive and lipid-lowering effect (Devarajan et al. 2016a).

In another study, Devarajan and his colleagues (2016b) determined the antihyperglycaemic potential of the blend of sesame oil and physically refined RBO (20:80) in type 2 diabetes mellitus (T2DM). Sesame oil-RBO blend was supplied to the T2DM patients, and they were asked to consume it as the regular cooking oil in place of any other edible oils for 8 weeks. At wk. 4 and wk. 8, the T2DM patients treated with sesame oil-RBO blend showed significant reduction in fasting and postprandial glucose ($p < 0.001$). HbA1c, TC, TG and LDL-C levels were also significantly reduced, while HDL-C level significantly increased at wk. 8 in T2DM patients treated with the sesame oil-RBO blend. It was concluded that the use of sesame oil-RBO blend lowers hyperglycaemia and improves lipid profile in patients with T2DM.

Uses of RBO in Baking, Pharmaceutical and Cosmetic Industries

The recent trends in the baking industry indicate a reduction in the use of fats and oils as well as to replace the plastic fats with liquid vegetable oils (Chung and Pomeraz 1983; Kamran et al. 2005). It has been recommended that not only the total amount of fat be lowered in high-fat baked products but also the animal fats should be replaced with polyunsaturated vegetable oil products (Salz 1982). The RBO can maintain its nutritive quality even at high temperatures and can be used to make margarine and shortening to be used in baking industry. Refined RBO has been shown to replace the bakery shortening in bread preparation. Kaur et al. (2012) concluded that bakery shortening can successfully be replaced with refined RBO (up to 50%) in bread making with improved baking qualities.

The phytonutrients in RBO can have the potential application in the context of their utility as functional ingredients for the development of nutraceuticals and nutritional supplements to fight against many disease conditions (Jariwalla 2001). Squalene, a bioactive compound in RBO, is easily absorbed by the skin and keeps it soft, supple and smooth. RBO has anti-inflammatory properties and has been shown to reduce the effects of menopause like hot flashes. The nano-emulsions based on the bioactive components of RBO have been shown to improve the skin health (Daniela et al. 2011; Wuttikul and Boonme 2016). Rigo et al. (2015) demonstrated that the nano-encapsulation of RBO showed protective properties to prevent and repair the skin damages caused by excessive exposure to UV-B radiation.

Safety and Toxicological Aspects of RBO

Since the RBO is categorized as healthy cooking oil, its chemical and nutrient composition, nutrient quality and toxicological safety are required to be assessed appropriately. RBO did not show any mutagenicity when tested by bacterial reverse mutation in Ames mutagenicity assays (Polasa and Rukmini 1987). The bioactive components of RBO did not reveal any toxicity and carcinogenicity in in vivo assays in mice and rats (Tamagawa et al. 1992). On the other hand, they markedly inhibited the inflammation and tumour-promoting effects of 12-O-tetradecanoylphobol-13-acetate (TPA) in mice (Yasukawa et al. 1998). Nutritional and toxicological studies did not show any abnormalities in animals fed with either RBO or groundnut oil. No side effects were also observed in adults and children even at high doses of phytosterols from the RBO as they are poorly absorbed and can effectively be excreted via biliary route (Becker et al. 1993; Weststrate and Meijer 1998).

The Cosmetic Ingredient Review (CIR) Expert Panel however showed its concerns about the presence of contaminants such as pesticides residues in RBO used for cooking and recommends that the level of these contaminants should not exceed the currently allowed safe limits. The CIR Expert Panel has concluded that the rice-derived ingredients are safe as cosmetic ingredients in the practices being used and the concentrations as described in their safety assessment (Anonymous 2006). RBO is used in cosmetics as a skin-conditioning and surfactant-cleansing agent. RBO was not found to be a sensitizer and was negative in ocular toxicity assays and Ames assay. Its component such as γ-oryzanol was also found to be negative in bacterial and mammalian mutagenicity assays. Oral carcinogenicity studies done on components of rice bran (phytic acid and γ-oryzanol) were negative. The phytochemical bioactive components of RBO can be a new source of cosmetic raw materials. The cosmetic formulations, for example, gels and creams, which contain the rice bran bioactive compounds such as ferulic acid, γ-oryzanol and phytic acid, showed better clinical anti-ageing activities (Manosroi et al. 2012a, b). Oluremi et al. (2013) observed that the crude RBO may contain some heavy metals, and therefore it should be refined to reduce the Fe and Cu overloads, as they may appear in higher quantities than recommended by the CODEX range. Araghi et al. (2016) evaluated the toxicity and safety aspects of RBO in chicken embryo model. They demonstrated that RBO showed no especial toxicity in chicken embryo model, and therefore it might be regarded as safe for human consumption. In view of its safety, hypolipidaemic and hypoglycaemic activities, the RBO is considered as good alternative and valuable source of edible oil.

Conclusion

The rate of diabetes with associated risk of CVD is continuously on the increase worldwide (IDF 2015). Type 2 diabetes mellitus (T2DM) is a complex multifactorial condition that is caused by inappropriate dietary and lifestyle patterns and inheritance factors (Waly et al. 2010; Tuomilehto and Schwarz 2016). T2DM is characterized by insulin resistance and is often accompanied with cardiovascular disease risk factors, including obesity, dyslipidaemia and hypertension (Hanley et al. 2002; Semple 2016). Dietary strategies are considered as the first line of defence in the prevention and management of diabetes, cardiovascular diseases and cancers. RBO with its excellent fatty acid composition and bioactive antioxidants has demonstrated beneficial effects to improve the plasma lipid profile in rodents, rabbits, non-human primates and humans. Consumption of RBO has been shown to have a direct relationship with its antihypertensive, antidiabetic, lipid-lowering and anti-carcinogenic properties (Wilson et al. 2000 2007; Most et al. 2005; Salar et al. 2016; Dhavamani et al. 2014; Devarajan et al. 2016a, b; Szcześniak et al. 2016). Because of its well-balanced and richness of unsaturated fat, bioactive components and versatile cooking properties, RBO has gained popularity as healthy cooking oil. Vast majority of scientific data greatly augments the importance of RBO and its significant physiological action in health and diseases. Although RBO has unique physiological and biological properties, the clear cut mechanisms of RBO and its bioactive components on health and diseases still need to be elucidated. As evidenced from several observational and animal studies, it is well documented that the RBO has an imperative role in the prevention, management and control of chronic diseases, and therefore RBO would certainly be a valuable dietary addition as functional food in everyday diet.

References

Abumweis SS, Barake R, Jones PJ (2008) Plant sterols/stanols as cholesterol lowering agents: a meta-analysis of randomized controlled trial. Food Nutr Res 52:1–35

Accinni R, Rosina M, Bamonti F, Della Noce C, Tonini A, Bernacchi F (2006) Effects of combined dietary supplementation on oxidative and inflammatory status in dyslipidemic subjects. Nutr Metab Cardiovasc Dis 16:121–127

Akihisa T, Yasukawa K, Yamaura M, Ukiya M, Kimura Y, Shimizu N, Arai K (2000) Triterpene alcohol and sterol ferulates from rice bran and their anti-inflammatory effects. J Agric Food Chem 48(6):2313–2319

Ammar HO, Al-Okbi SY, Mostafa DM, Helal AM (2012) Rice bran oil: preparation and evaluation of novel liquisolid and semisolid formulations. Int J Pharm Compd 16(6):516–523

Anonymous (2006) Amended final report on the safety assessment of *Oryza sativa* (rice) bran oil, *Oryza sativa (rice)* germ oil, rice bran acid, *Oryza sativa* (rice) bran wax, hydrogenated rice bran wax, *Oryza sativa* (rice) bran extract, *Oryza sativa* (rice) extract, *Oryza sativa* (rice) germ powder, *Oryza sativa* (rice) starch, *Oryza sativa* (Rice) bran, hydrolyzed rice bran extract, hydrolyzed rice bran protein, hydrolyzed rice extract, and hydrolyzed rice protein. Int J Toxicol 25(Suppl 2):91–120

Araghi A, Seifi S, Sayrafi R, Sadighara P (2016) Safety assessment of rice bran oil in a chicken embryo model. Avicenna J Phytomed 6(3):351–356

Ausman L, Rong N, Nicolosi R (2005) Hypocholesterolemic effect of physically refined rice bran oil: studies of cholesterol metabolism and early atherosclerosis in hypercholesterolemic hamsters. J Nutr Biochem 16(9):521–529

Becker M, Staab D, von Bergmann K (1993) Treatment of severe familial hypercholesterolemia in childhood with sitosterol and sitostanol. J Pediatr 122:292–296

Bellik Y, Boukraâ L, Alzahrani HA, Bakhotmah BA, Abdellah F, Hammoudi SM, Iguer-Ouada M (2012) Molecular mechanism underlying anti-inflammatory and anti-allergic activities of phytochemicals: an update. Molecules 18(1):322–353

Berger A, Rein D, Schafer A, Monnard I, Gremaud G, Lambelet P (2004) Similar cholesterol-lowering properties of rice bran oil, with varied c-oryzanol, in mildly hypercholesterolemic men. Eur J Nutr 44(3):163–173

Bergman CJ, Xu Z (2003) Genotype and environment effects on tocopherols, tocotrienols and gamma-oryzanol contents of Southern US rice. Cereal Chem 80(4):446–449

Björnsson E, Jacobsen EI, Kalaitzakis E (2012) Hepatotoxicity associated with statins: reports of idiosyncratic liver injury postmarketing. J Hepatol 56(2):374–380

Bopitiya D, Madhujith T (2014) Antioxidant potential of rice bran oil prepared from red and white rice. Trop Agric Res 26(1):1–11

CAC (Codex Alimentarius Commission) (2003) The need for inclusion of rice bran oil in the standards for named vegetable oils. In: Joint FAO/WHO food standards programme, Codex Committee on fats and oils, 18th session, 3–7 February 2003, London

Castilla-Guerra L, Del Carmen Fernandez-Moreno M, Colmenero-Camacho MA (2016) Statins in stroke prevention: present and future. Curr Pharm Des 22(30):4638–4644

Chen CW, Cheng HH (2006) A rice bran oil diet increases LDL receptor and HMG-CoA reductase mRNA expressions and insulin sensitivity in rats with streptozotocin/nicotinamide-induced type 2 diabetes. J Nutr 136(6):1472–1476

Chithra PK, Sindhu G, Shalini V, Parvathy R, Jayalekshmy A, Helen A (2015) Dietary Njavara rice bran oil reduces experimentally induced hypercholesterolemia by regulating genes involved in lipid metabolism. Br J Nutr 113(8):1207–1219

Chopra R, Sambaiah K (2009) Effects of rice bran oil enriched with n-3 PUFA on liver and serum lipids in rats. Lipids 44(1):37–46

Chou TW, Ma CY, Cheng HH (2009) A rice bran oil diet improves lipid abnormalities and suppress hyperinsulinemic responses in rats with streptozotocin/nicotinamide-induced type 2 diabetes. J Clin Biochem Nutr 45:29–36

Choudhary M, Grover K, Sangha J (2013) Effect of blended rice bran and olive oil on cardiovascular risk factors in hyperlipidemic patients. Food Nutr Sci 4:1084–1093

Choudhary M, Grover K, Kaur G (2015) Development of rice bran oil blends for quality improvement. Food Chem 173:770–777

Chung OK, Pomeraz Y (1983) Recent trends in usage of fats and oils as functional ingredients in the baking industry. J Am Oil Chem Soc 60:1848–1851

Cicero AFG, Derosa G (2005) Rice bran and its main components: potential role in the management of coronary risk factors. Curr Top Nutraceu Res 3(1):29–46

Cicero AFG, Gaddi A (2001) Rice bran oil and gamma-oryzanol in the treatment of hyperlipoproteinaemias and other conditions. Phytother Res 15:277–289

Daniela S, Tatiana A, Naira R, Josiane B, Gisely SV, Gustavo CO, Pedro A (2011) Formation and stability of oil-in-water nanoemulsion containing rice bran oil: in vitro and in vivo assessments. J Nanobiotechnol 9(44):1–9

Danielski L, Zetzl C, Hense H, Brunner G (2005) A process line for the production of raffinated rice oil from rice bran. J Supercrit Fluids 34:133–141

Devarajan S, Chatterjee B, Urata H, Zhang B, Ali A, Singh R, Ganapathy S (2016a) A blend of sesame and rice bran oils lowers hyperglycemia and improves the lipids. Am J Med 129(7):731–739

Devarajan S, Singh R, Chatterjee B, Zhang B, Ali A (2016b) A blend of sesame oil and rice bran oil lowers blood pressure and improves the lipid profile in mild-to-moderate hypertensive patients. J Clin Lipidol 10(2):339–349

Dhavamani S, Rao YPC, Lokesh BR (2014) Total antioxidant activity of selected vegetable oils and their influence on total antioxidant values in vivo: a photochemiluminescence based analysis. Food Chem 164:551–555

Eady S, Wallace A, Willis J, Scott R, Frampton C (2011) Consumption of a plant sterol-based spread derived from rice bran oil is effective at reducing plasma lipid levels in mildly hypercholesterolaemic individuals. Br J Nutr 15:1–12

Edwards MS, Radcliffe JD (1994) A comparison of the effect of rice bran oil and corn oil on lipid status in the rat. Biochem Arch 10:87–94

Ellis R, Seal ML, Adamson C, Beare R, Simmons JG, Whittle S, Allen NB (2016) Brain connectivity networks and longitudinal trajectories of depression symptoms in adolescence. Psychiatry Res 21(260):62–69

Evans NP, Misyak SA, Schmelz EM (2010) Conjugated linoleic acid ameliorates inflammation-induced colorectal cancer in mice through activation of PPARγ. J Nutr 140(3):515–521

Fang NB, Yu SG, Badger TM (2003) Characterization of triterpene alcoholand sterol ferulates in rice bran using LC–MS/MS. J Agric Food Chem 51:3260–3267

Friedman M (2013) Rice brans, rice bran oils, and rice hulls: composition, food and industrial uses, and bioactivities in humans, animals, and cells. J Agric Food Chem 61(45):10626–10641

Ghatak SB, Panchal SS (2012a) Anti-diabetic activity of oryzanol and its relationship with the antioxidant property. Int J Diab Dev Ctries 32:185–192

Ghatak SB, Panchal SS (2012b) Protective effect of oryzanol isolated from crude rice bran oil in experimental model of diabetic neuropathy. Revista Brasileira de Farmacognosia, Brazilian. J Pharmacogn 22(5):1092–1103

Giacco F, Brownlee M (2010) Oxidative stress and diabetic complications. Circ Res 107(9):1058–1070

Gurwitz JH, Go AS, Fortmann SP (2016) Statins for primary prevention in older adults: uncertainty and the need for more evidence. JAMA 316(19):1971–1972

Hagl S, Berressem D, Grewal R, Sus N, Frank J, Eckert GP (2016) Rice bran extract improves mitochondrial dysfunction in brains of aged NMRI mice. Nutr Neurosci 19(1):1–10

Hamid AA, Dek MS, Tan CP, Zainudin MA, Fang EK (2014) Changes of major antioxidant compounds and radical scavenging activity of palm oil and rice bran oil during deep-frying. Antioxidants 3:502–515

Hanley AJ, Williams K, Stern MP, Haffner SM (2002) Homeostasis model assessment of insulin resistance in relation to the incidence of cardiovascular disease: the San Antonio Heart Study. Diabetes Care 25:1177–1184

Hota D, Chakrabarti A, Dutta P, Singh I (2013) A comparative evaluation of anti-hyperlipidemic efficacy of rice bran oil, olive oil and groundnut oil. Clinical study report. Postgraduate Institute of Medical education and Research, Chandigarh, pp. 1–84. www.ricela.com/images/Final-Report.pdf

Houston MC, Fazio S, Chilton FH, Wise DE, Jones KB, Barringer TA, Bramlet DA (2009) Nonpharmacologic treatment of dyslipidemia. Prog Cardiovasc Dis 52:61–94

Hsieh RH, Lien LM, Lin SH, Chen CW, Cheng HJ, Cheng HH (2005) Alleviation of oxidative damage in multiple tissues in rats with streptozotocin-induced diabetes by rice bran oil supplementation. Ann N Y Acad Sci 1042:365–371

International Diabetes Federation (IDF) (2015) IDF Diabetes Atlas, Seventh Edition. International Diabetes Federation (IDF), Chaussée de La Hulpe 166, B-1170 Brussels, Belgium. Online version of IDF Diabetes Atlas: www.diabetesatlas.org

Iqbal J, Minhajuddin M, Beg ZH (2004) Suppression of diethylnitrosamine and 2-acetylaminofluorene-induced hepatocarcinogenesis in rats by tocotrienol-rich fraction isolated from rice bran oil. Eur J Cancer Prev 13(6):515–520

Ishihara M (1984) Effect of gamma-oryzanol on serum lipid peroxide level and clinical symptoms of patients with climacteric disturbances. Asia Oceania J Obstet Gynaecol 10:317–323

Ishihara M, Ito Y, Nakakita T, Maehama T, Hieda S, Yamamoto K, Ueno N (1982) Clinical effect of gamma-oryzanol on climacteric disturbance on serum lipid peroxides. Nihon Sanka Fujinka Gakkai Zasshi 34(2):243–251

Jariwalla RJ (2001) Rice-bran products: phytonutrients with potential applications in preventive and clinical medicine. Drugs Exp Clin Res 27:17–26

Jolfaie NR, Rouhani MH, Surkan PJ, Siassi F, Azadbakht L (2016) Rice bran oil decreases total and LDL cholesterol in humans: a systematic review and meta-analysis of randomized controlled clinical trials. Horm Metab Res 48(7):417–426

Ju YH, Zullaikah S (2013) Effect of acid-catalyzed methanolysis on the bioactive components of rice bran oil. J Taiwan Inst Chem Eng 44:924–928

Kamran S, Masood SB, Faqir MA, Muhammad N (2005) Improved quality of baked products by rice bran oil. Int J Food Safety 5:1–8

Kanbara R, Fukuo Y, Hada K, Hasegawa T, Terashi A (1992) The influence of sonic stress on lipid metabolism and the progress of atherosclerosis in rabbits with hypercholesterolemia – studies on antiaterosclerotic effect of γgammaoryzanol in sonic stress. Jpn J Atheros 20:159–163

Kaneto H, Katakami N, Matsuhisa M, Matsuoka T (2010) Role of reactive oxygen species in the progression of type 2 diabetes and atherosclerosis. Mediat Inflamm:1–11

Kannappan R, Yadav VR, Aggarwal BB (2010a) γ-Tocotrienol but not γ-tocopherol blocks STAT3 cell signaling pathway through induction of protein-tyrosine phosphatase SHP-1 and sensitizes tumor cells to chemotherapeutic agents. J Biol Chem 285(43):33520–33528

Kannappan R, Ravindran J, Prasad S (2010b) Gamma-tocotrienol promotes TRAIL-induced apoptosis through reactive oxygen species/extracellular signal-regulated kinase/p53-mediated upregulation of death receptors. Mol Cancer Ther 9(8):2196–2207

Kaur A, Jassal V, Thind SS, Aggarwal P (2012) Rice bran oil an alternate bakery shortening. J Food Sci Technol 49(1):110–114

Kennedy G, Burlingame B (2003) Analytical, nutritional and clinical methods analysis of food composition data on rice from a plant genetic resources perspective. Food Chem 80:589–596

Kim JS, Godber JS (2001) Oxidative stability and vitamin E levels increased in restructured beef roast with added rice bran oil. J Food Qual 24:17–26

Koba K (1997) Effect of safflower oil/rice bran oil mixture on serum cholesterol concentration in rats. J Nagasaki Prefect Women's Jr Coll 45:33–37

Koba K, Liu JW, Bobik E, Sugano M, Huang YS (2000) Cholesterol supplementation attenuates the hypocholesterolemic effect of rice bran oil in rats. J Nutr Sci Vitaminol 46(2):58–64

Kota SK, Jammula S, Kota SK et al (2013) Nutraceuticals in dyslipidemia management. J Med Nutr Nutraceut 2(1):26–40

Kris-Etherton PM, Hecker KD, Bonanome A, Coval SM, Binkoski AE, Hilpert KF, Griel AE, Etherton TD (2002) Bioactive compounds in foods: their role in the prevention of cardiovascular disease and cancer. Am J Med 113(Suppl 9B):71S–88S

Krishna AGG, Khatoon S, Sheila PM, Sarmandal CV, Indira TN, Mishra A (2001) Effect of refining of crude rice bran oil on the retention of oryzanol in the refined oil. Am Oil Chem Soc 78:127–131

Kuriyan R, Gopinath N, Vaz M, Kurpad VA (2005) Use of rice bran oil in patients with hyperlipidemia. Natl Med J India 18:292–296

Lai MH, Chen YT, Chen YY, Chang JH, Cheng HH (2012) Effects of rice bran oil on the blood lipids profiles and insulin resistance in type 2 diabetes patients. J Clin Biochem Nutr 51(1):15–18

Lee JW, Lee SW, Kim MK, Rhee C, Kim IH, Lee KW (2005) Beneficial effect of the unsaponifiable matter from rice bran oil on oxidative stress in vitro compared with α-tocopherol. J Sci Food Agric 85:493–498

Li Y, Hou MJ, Ma J, Tang ZH, Zhu HL, Ling WH (2005) Dietary fatty acids regulate cholesterol induction of liver CYP7alpha1 expression and bile acid production. Lipids 40:455–462

Liang Y, Gao Y, Lin Q, Luo F, Wu W, Lu Q, Liu Y (2014) A review of research progress on the bioactive ingredients and physiological activities of rice bran oil. Eur Food Res Technol 238:169–176

Lichtenstein AH, Ausman LM, Carrasco W, Gualtieri LJ, Jenner JL, Ordovas JM, Nicolosi RJ, Goldin BR, Schaefer EJ (1994) Rice bran oil consumption and plasma lipid levels in moderately hypercholesterolemic humans. Arterioscler Thromb 14:549–556

Lloyd BJ, Siebenmorgen TJ, Beers KW (2000) Effects of commercial processing on antioxidants in rice bran. Cereal Chem 77:551–555

Luh BS, Barber S, Benedito de Barber C (1991) Rice bran: chemistry and technology. In: Luh BS (ed) Rice production and utilization. Van Nostrand Reinhold, New York, p 313

Macchar D, Kansara U, Ghatak SB, Bhadada SV, Panchal SS (2012) Antihyperlipidemic activity of various combinations of rice bran oil and safflower oil on Triton WR-1339 induced hyperlipidemia in rats. JPB Sci 1(1):16–26

Maji D, Shaikh S, Solanki D, Gaurav K (2013) Safety of statins. Indian J Endocrinol Metab 17(4):636–646

Mäkynen K, Chitchumroonchokchai C, Adisakwattana S, Failla M, Ariyapitipun T (2012) Effect of gamma-oryzanol on the bioaccessibility and synthesis of cholesterol. Eur Rev Med Pharmacol Sci 16(1):49–56

Malve H, Kerkar P, Mishra N, Loke S, Rege NN, Marwaha-Jaspal A, Jainani KJ (2010) LDL-cholesterol lowering activity of a blend of rice bran oil and safflower oil (8:2) in patients with hyperlipidaemia: a proof of concept, double blind, controlled, randomised parallel group study. J Indian Med Assoc 108(11):785–788

Manila C, Maria LJ, Angelica C, Rosalia RR, Marla DH (2014) Rice bran enzymatic extract supplemented diets modulate adipose tissue inflammation markers in Zucker rats. Nutrition 30(4):466–472

Manosroi A, Chutoprapat R, Abe M, Manosroi W, Manosroi J (2012a) Anti-aging efficacy of topical formulations containing niosomes entrapped with rice bran bioactive compounds. Pharm Biol 50(2):208–224

Manosroi A, Ruksiriwanich W, Abe M, Sakai H, Aburai K, Manosroi W, Manosroi J (2012b) Physico-chemical properties of cationic niosomes loaded with fraction of rice (*Oryza sativa*) bran extract. J Nanosci Nanotechnol 12(9):7339–7345

Marchesini G, Bugianesi E, Forlani G, Cerrelli F, Lenzi M, Manini R, Natale S, Vanni E, Villanova N, Melchionda N, Pizzetto M (2003) Nonalcoholic fatty liver, steatohepatitis, and the metabolic syndrome. Hepatology 37:917–923

McIlwain DR, Berger T, Mak TW (2015) Caspase functions in cell death and disease. Cold Spring Harb Perspect Biol 7(4):pii: a026716. doi:10.1101/cshperspect.a026716

Mehdi BJ, Tabassum S, Haider S, Perveen T, Nawaz A, Haleem DJ (2015) Nootropic and anti-stress effects of rice bran oil in male rats. J Food Sci Technol 52(7):4544–4550

Metwally AM, Habib AM, Khafagy SM (1974) Sterols and triterpene alcohols from rice bran oil. Planta Med 25:68–72

Mezouri S, Eichner K (2007) Comparative study on the stability of crude and refined rice bran oil during long-term storage at room temperature. Eur J Lipid Sci Technol 109:98–205

Mishra R, Sharma HK (2014) Effect of frying conditions on the physico-chemical properties of rice bran oil and its blended oil. J Food Sci Technol 51(6):1076–1084

Moldenhauer KA, Champagne ET, McCaskill DR, Guraya H (2003) Functional products from rice. In: Mazza G (ed) Functional foods. Technomic Publishing Co. Inc., Lancaster, pp 71–89

Most MM, Tulley R, Morales S, Lefevre M (2005) Rice bran oil, not fiber, lowers cholesterol in humans. Am J Clin Nutr 81:64–68

Nicolosi RJ, Ausman LM, Hegsted DM (1991) Rice bran oil lowers serum total and low density lipoprotein cholesterol and apo B levels in nonhuman primates. Atherosclerosis 88(2–3):133–142

Norton RA (1995) Quantitation of steryl ferulate and p-coumarate esters from corn and rice. Lipids 30(3):269–274

Ohara K, Uchida A, Nagasaka R, Ushio H, Ohshima T (2009) The effects of hydroxycinnamic acid derivatives on adiponectin secretion. Phytomed 16:130–137

Oluremi OI, Solomon AO, Saheed AA (2013) Fatty acids, metal composition and physico-chemical parameters of Igbemo Ekiti rice bran oil. J Environ Chem Ecotoxicol 5(3):39–46

Patel M, Naik SN (2004) Gamma-oryzanol from rice bran oil-a review. J Sci Ind Res 63:569–578

Polasa K, Rukmini C (1987) Mutagenicity tests of cashew nut shell liquid, rice-bran oil and other vegetable oils using the Salmonella typhimurium/microsome system. Food Chem Toxicol 25:763–766

Posuwan J, Prangthip P, Leardkamolkarn V, Yamborisut U, Surasiang R, Charoensiri R, Kongkachuichai R (2013) Long-term supplementation of high pigmented rice bran oil (*Oryza sativa* L.) on amelioration of oxidative stress and histological changes in streptozotocin-induced diabetic rats fed a high fat diet; Riceberry bran oil. Food Chem 138(1):501–508

Prasad NMN, Sanjay KR, Khatokar SM, Vismaya MN, Swamy NS (2011) Health benefits of rice bran – a review. J Nutr Food Sci 1:108. doi:10.4172/2155-9600.1000108

Purushothama S, Raina PL, Hariharan K (1995) Effect of long term feeding of rice bran oil upon lipids and lipoproteins in rats. Mol Cell Biochem 146:63–69

Qureshi AA, Bradlow BA, Salser WA, Brace LD (1997) Novel tocotrienols of rice bran modulate cardiovascular disease risk parameters of hypercholesterolemic humans. J Nutr Biochem 8:290–298

Qureshi AA, Sami SA, Khan FA (2002) Effects of stabilized rice bran, its soluble and fiber fractions on blood glucose levels and serum lipid parameters in humans with diabetes mellitus Types I and II. J Nutr Biochem 13:175–187

Radcliffe JD, Imrhan VA, Hsueh AM (1997) Serum lipids in rats fed diets containing rice bran oil or high-linoleic acid safflower oil. Biochem Arch 13:87–95

Raghuram T, Brahmaji G, Rukmini C (1989) Studies on hypolipidemic effects of dietary rice bran oil in human subjects. Nutr Rep Int 35:889–895

Rajam L, Kumar D, Sundaresan A, Arumughan CA (2005) Novel process for physically refining rice bran oil through simultaneous degumming and dewaxing. J Am Oil Chem Soc 82:213–220

Rajnarayana K, Prabhakar MC, Krishna DR (2001) Influence of rice bran oil on serum lipid peroxides and lipids in human subjects. Indian J Physiol Pharmacol 45(4):442–444

Rana P, Vadhera S, Soni G (2004) In vivo antioxidant potential of rice bran oil (RBO) in albino rats. Indian J Physiol Pharmacol 48(4):428–436

Reena MB, Lokesh BR (2007) Hypolipidemic effect of oils with balanced amounts of fatty acids obtained by blending and interesterification of coconut oil with rice bran oil or sesame oil. J Agric Food Chem 55(25):10461–10469

Rigo LA, da Silva CR, de Oliveira SM, Cabreira TN, Cabreira TN, de Bona da Silva C, Ferreira J, Beck RC (2015) Nanoencapsulation of rice bran oil increases its protective effects against UVB radiation-induced skin injury in mice. Eur J Pharm Biopharm 93:11–17

Rogers EJ, Rice SM, Nicolosi RJ, Carpenter DR, Muclelland CA, Romanczyc LR Jr (1993) Identification and quantitation of gamma-oryzanol components and simultaneous assessment of tocopherols in rice bran oil. J Am Oil Chem Soc 70:301–307

Rong N, Ausman LM, Nicolosi RJ (1992) Oryzanol decreases cholesterol absorption and aortic fatty streaks in hamsters. Lipids 32:303–309

Ros E (2003) Dietary cis-monounsaturated fatty acids and metabolic control in type 2 diabetes. Am J Clin Nutr 78(3):617S–625S

Rukmini C, Raghuram TC (1991) Nutritional and biochemical aspects of the hypolipidemic action of rice bran oil, a review. J Am Coll Nutr 10:593–601

Salar A, Faghih S, Pishdad GR (2016) Rice bran oil and canola oil improve blood lipids compared to sunflower oil in women with type 2 diabetes: a randomized, single-blind, controlled trial. J Clin Lipidol 10(2):299–305

Salz KM (1982) Fat and cholesterol intakes of white adults in Columbia, Maryland. J Am Diet Assoc 81(5):541–546

Sayre B, Saunders R (1990) Rice bran and rice bran oil. Lipid Techol 2:72–76

Scoggan KA, Gruber H, Chen Q, Plouffe LJ, Lefebvre JM, Wang B (2008) Increased incorporation of dietary plant sterols and cholesterol correlates with decreased expression of hepatic and intestinal ABCG5 and ABCG8 in diabetic BB rats. J Nutr Biochem 15:32–38

Seetharamaiah GS, Chandrasekhara N (1988) Effect of oryzanol on fructose induced hypertriglyceridaemia in rats. Indian J Med Res 88:278–281

Seetharamaiah GS, Chandrasekhara N (1989) Studies on hypocholesterolemic activity of rice bran oil. Atherosclerosis 78:219–223

Seetharamaiah GS, Chandrasekhara N (1990) Effect of oryzanol on cholesterol absorption and biliary and fecal bile acid in rats. Indian J Med Res 92:471–475

Semple RK (2016) EJE PRIZE 2015: how does insulin resistance arise, and how does it cause disease? Human genetic lessons. Eur J Endocrinol 174(5):R209–R223

Serafini M, Peluso I (2016) Functional foods for health: the interrelated antioxidant and anti-inflammatory role of fruits, vegetables, herbs, spices and cocoa in humans. Curr Pharm Des 22(44):6701–6715. [Epub ahead of print]

Shah RV, Goldfine AB (2012) Statins and risk of new-onset diabetes mellitus. Circulation 126:e282–e284

Shakib MC, Gabrial S, Gabrial G (2014) Rice bran oil compared to atorvastatin for treatment of dyslipidemia in patients with type 2 diabetes. Maced J Med Sci 7(1):95–102

Sharma RD, Rukmini C (1986) Rice bran oil and hypocholesterolemia in rats. Lipids 21:715–717

Sharma RD, Rukmini C (1987) Hypocholesterolemic activity of unsaponifiable matter of rice bran oil. Indian J Med Res 85:278–281

Shih CK, Ho CJ, Li SC, Yang SH, Hou WC, Cheng HH (2011) Preventive effects of rice bran oil on 1, 2-dimethylhydrazine/dextran sodium sulphate-induced colon carcinogenesis in rats. Food Chem 126:562–567

Siddiqui S, Khan MR, Siddiqui WA (2010) Comparative hypoglycemia and nephroprotective effects of tocotrienol rich fraction (TRF) from palm oil and rice bran oil against hyperglycemia induced nephropathy in type 1 diabetic rats. Chem Biol Interact 188(3):651–658

Son MJ, Rico CW, Nam SH, Kang MY (2011) Effect of oryzanol and ferulic acid on the glucose metabolism of mice fed with a high-fat diet. J Food Sci 76:7–10

Statistica (2016) https://www.statista.com/topics/1443/rice/. Accessed Dec 2016 and https://www.worldriceproduction.com/

Sugano M, Tsuji E (1997) Rice bran oil and cholesterol metabolism. J Nutr 127:521–524

Suganya KS, Govindaraju K, Kumar VG, Karthick V, Parthasarathy K (2016) Pectin mediated gold nanoparticles induces apoptosis in mammary adenocarcinoma cell lines. Int J Biol Macromol 93(Pt A):1030–1040

Sun W, Xu W, Liu H (2009) Gamma-tocotrienol induces mitochondria-mediated apoptosis in human gastric adenocarcinoma SGC-7901 cells. J Nutr Biochem 20(4):276–284

Sunitha T, Manorama R, Rukmini C (1997) Lipid profile of rats fed blends of rice bran oil in combination with sunflower and safflower oil. Plant Foods Hum Nutr 51:219–224

Surh YJ, Chun KS, Cha HH, Han SS, Keum YS, Park KK, Lee SS (2001) Molecular mechanisms underlying chemopreventive activities of anti-inflammatory phytochemicals: downregulation of COX-2 and iNOS through suppression of NF-kappaB activation. Mutat Res 480–481:243–268

Suzuki S, Oshima S (1970) Influence of blending of edible fats and oils on human serum cholesterol level (part 2). Jpn J Nutr 28:3–9

Szcześniak KA, Ostaszewski P, Ciecierska A, Sadkowski T (2016) Investigation of nutriactive phytochemical-gamma-oryzanol in experimental animal models. J Anim Physiol Anim Nutr (Berl) 100(4):601–617

Tabassum S, Aggarwal S, Ali SM, Beg ZH, Khan AS, Afzal K (2005) Effect of rice bran oil on the lipid profile of steroid responsive nephrotic syndrome. Indian J Nephrol 15:10–13

Tamagawa M, Otaki Y, Takahashi T, Otaka T, Kimura S, Miwa T (1992) Carcinogenity study of gamma-oryzanol in B6C3F1 mice. Food Chem Toxicol 30:49–56

Tsuji E, Itoh H, Itakura H (1989) Comparison of effects of dietary saturated and polyunsaturated fats on the serum lipids levels. Clin Ther Cardiovasc Dis 8:149–151

Tsuji E, Takahashi M, Kinoshita S, Tanaka M, Tsuji K (.2003) Effects of different contents of a-oryzanol in rice bran oil on serum cholesterol levels. In: XIIIth international symposium on atherosclerosis. Atherosclerosis Supp. 4:278

Tuomilehto J, Schwarz PE (2016) Preventing diabetes: early versus late preventive interventions. Diabetes Care 39(Suppl 2):S115–S120

Umesha SS, Naidu KA (2015) Antioxidants and antioxidant enzymes status of rats fed on n-3 PUFA rich Garden cress (*Lepidium sativum* L) seed oil and its blended oils. J Food Sci Technol 52(4):1993–2002

Upadya H, Devaraju CJ, Joshi SR (2015) Anti-inflammatory properties of blended edible oil with synergistic antioxidants. Indian J Endocrinol Metab 19(4):511–519

Utarwuthipong T, Komindr S, Pakpeankitvatana V, Songchitsomboon S, Thongmuang N (2009) Small dense low-density lipoprotein concentration and oxidative susceptibility changes after consumption of soybean oil, rice bran oil, palm oil and mixed rice bran/palm oil in hypercholesterolaemic women. J Int Med Res 37(1):96–104

Van Hoed V, Depaemelaere G, Villa Ayala J, Santiwattana P, Verhé R, De Grey W (2006) Influence of chemical refining on the major & minor components of rice bran oil. JAOCS 83:315–321

Vasanthi HR, Parameswari RP, Das DK (2012) Multifaceted role of tocotrienols in cardioprotection supports their structure: function relation. Genes Nutr 7(1):19–28

Vissers MN, Zock PL, Meijer GW, Katan MB (2000) Effect of plant sterols from rice bran oil and triterpene alcohols from sheanut oil on serum lipoprotein concentrations in humans. Am J Clin Nutr 72:1510–1515

Waly MI, Essa MM, Ali A, Al-Shuhaibi YM, Al-Farsi YM (2010) The global burden of type 2 diabetes: a review. Int J Biol Med Res 1:326–329

Wan Nazaimoon WM, Khalid BA (2002) Tocotrienols-rich diet decreases advanced glycosylation end-products in non-diabetic rats and improves glycemic control in streptozotocin-induced diabetic rats. Malays J Pathol 24(2):77–82

Weststrate JA, Meijer GW (1998) Plant sterol-enriched margarines and reduction of plasma total- and LDL-cholesterol concentrations in normocholesterolaemic and mildly hypercholesterolaemic subjects. Eur J Clin Nutr 52(5):334–343

Wilson T, Ausman L, Lawton C, Hegsted D, Nicolosi R (2000) Comparative cholesterol lowering properties of vegetable oils: beyond fatty acids. J Am Coll Nutr 19(5):601–607

Wilson TA, Nicolosi RJ, Woolfrey B, Kritchevsky D (2007) Rice bran oil and oryzanol reduce plasma lipid and lipoprotein cholesterol concentrations and aortic cholesterol ester accumulation to a greater extent than ferulic acid in hypercholesterolemic hamsters. J Nutr Biochem 18:105–112

Wuttikul K, Boonme P (2016) Formation of microemulsions for using as cosmeceutical delivery systems: effects of various components and characteristics of some formulations. Drug Deliv Transl Res 6(3):254–262

Xu Z, Hua N, Godber JS (2001) Antioxidant activity of tocopherols, tocotrienols & γ-oryzanol components from rice bran against cholesterol oxidation accelerated by 2,2'-Azo-bis (2-methylpropionamidin) Dihydrochloride. J Agric Food Chem 49: 2077–2081.

Yasukawa K, Akihisa T, Kimura Y, Tamura T, Takido M (1998) Inhibitory effect of cycloartenol ferulate, a component of rice bran, on tumor promotion in two-stage carcino-genesis in mouse skin. Biol Pharm Bull 21(10):1072–1076

Yoshino G, Kazumi T, Amano M, Takeiwa M, Yamasaki T, Takashima S, Iwai M, Hatanaka H, Baba S (1989) Effects of gamma-oryzanol on hyperlipidemic subjects. Curr Ther Res 45:543–552

Zavoshy R, Noroozi M, Jahanihashemi H (2012) Effect of low calorie diet with rice bran oil on cardiovascular risk factors in hyperlipidemic patients. J Res Med Sci 17(7):626–631

Zhang Y, Xue R, Zhang ZN (2012) Palmitic and linoleic acids induce ER stress and apoptosis in hepatoma cells. Lipids Health Dis 11:1–8

Zhao G, Etherton TD, Martin KR et al (2004) Dietary alpha-linolenic acid reduces inflammatory and lipid cardiovascular risk factors in hypercholesterolemic men and women. J Nutr 134(11):2991–2997

Chapter 10
Microbial Association in Brown Rice and Their Influence on Human Health

P. Panneerselvam, Asish K. Binodh, Upendra Kumar,
T. Sugitha, and A. Anandan

Introduction

Rice (*Oryza sativa* L.) is an important staple food in most of the countries in this globe. It has been marked as a globally viable crop due to its flexibility at various environmental and soil conditions (FAO 2013). Rice provides 20% of the world's dietary energy supply, which is a higher portion than those of wheat and maize (19% and 5%, respectively). Rice is not only a good source of protein and carbohydrate, but it also contains bioactive substances beneficial for human health, which is mainly located in the germ and bran layers of the grain (Juliano 1992). The global rice production was 731.2 million tons, and the milled rice was up to the tune of 487.5 million tons (FAO 2013). Asian countries contributed 662.9 million tons of rice in 2012, which accounts 91% of the global production. According to FAO 2013, rice production in 2012 was 27% higher than 2002 and 5.7% higher than that in 2010. This clearly shows that rice production is increasing steadily.

India and China jointly produces 50% of world's rice from 50% of the world's total rice area, whereas Indonesia, Bangladesh, Vietnam, and Thailand reports 9%, 6%, 5%, and 4% of world's rice, respectively. The only major rice producing country other than Asia is the United States, accounting for 1.5% of the total world production. The European Union contributes only 0.5% to the global rice production from Italy, Spain, France, Portugal, and Greece. Most of the rice produced in these regions is utilized domestically, and less than 5% of world rice production is traded internationally (Fairhurst and Dobermann 2002). Some of the major exporters are

P. Panneerselvam (✉) • U. Kumar • A. Anandan
ICAR- National Rice Research Institute, Cuttack, Odisha, India
e-mail: panneerccri@rediffmail.com

A.K. Binodh • T. Sugitha
Agricultural College and Research Institute, Killikulam, Tuticorin, Tamil Nadu, India

© Springer International Publishing AG 2017
A. Manickavasagan et al. (eds.), *Brown Rice*,
DOI 10.1007/978-3-319-59011-0_10

Thailand, the United States, and Vietnam, which export about five, three, and two million tons a year, respectively.

There are many different types of rice with many different qualities to suit different consumer preferences, which include aroma, flavor, grain length, color, stickiness, and texture (Binodh et al. 2006). Nutritional content may also vary between different types of rice. Over 8000 varieties of rice have been reported so far. Rice can be categorized by its size as being short, medium, and long grain. Short grain, which has the highest starch content, makes the rice stickier, while long grain is lighter and tends to remain separate when cooked. In addition, there are a number of "speciality" rices available, including colored rice, aromatic rice, and wild rice. Varieties differing from the *indica* or *japonica* with respect to quality and characterized by some special quality features may be defined as speciality rice. Some of the speciality rice, viz., arborio (starchy white rice), basmati (aromatic rice), sweet rice, jasmine rice (soft-textured long grain aromatic rice), Bhutanese red rice (red-colored rice grown in the Himalayas has an earthy and nutty taste), forbidden rice (turning purple upon cooking from black color), and some wild rice (nutty flavor and high protein content), have been identified by the researchers as well as farmers (Chaudhary 2003). Rice is also classified according to the degree of milling as brown rice and white rice. All varieties of rice can be processed post-harvest as either brown or white rice affecting aroma, texture, and nutritive value (Zareiforoush et al. 2016). At present, brown rice is gaining popularity among people because of their aroma, taste, nutritive value, and nutrigenomic properties.

Brown Rice Vs White Rice

Rough rice can be separated into husk and brown rice (BR) through hulling process, and brown rice is further differentiated into starchy endosperm (92%), embryo (2%), and bran (6%). Brown rice is often referred to as whole rice in which the inedible fraction, i.e., the outer hull, got stripped off. White rice (WR), normally termed as polished rice, is obtained by removing the bran and germ along with all the incredible nutrients that reside within these important layers. This caused nutritional imbalance when white rice is consumed alone as staple food despite its composition (about 90% starch in dry solids). "Beriberi," a disease caused due to vitamin B1 deficiency, is one of the representative illnesses related to the nutritional imbalance. Despite the significant food value losses resulting from milling and polishing, most rice is preferred in its polished state (Cho and Lim 2016).

Brown rice (BR), which contains bran and embryo, is a treasure pot for numerous nutritional and bioactive components like functional lipids, vitamins, amino acids, dietary fiber, phenolic compounds, phytosterols, gamma-aminobutyric acid (GABA), and minerals. Although brown rice is cherished with many health-related merits, it rarely finds consumer acceptance. White rice is considered tastier and easier to cook, whereas the oil-rich aleurone layer of brown rice will easily turn rancid during storage and exhibit rough eating texture (Ohtsubo et al. 2005).

To increase the feasibility and consumer acceptance of brown rice, germinated brown rice has been gaining momentum nowadays in terms of enhanced potential health promoting functions apart from its eating quality. Germination increases the rate of water absorption and softens the cooked brown rice kernels, thereby improving its eating quality. The residual enzymes are activated during the germination process, thus inducing the formation of various metabolic components having bioactive functions. The germinated brown rice (GBR) and its syrup exert positive effect in case of type II diabetics, hypercholesterolemics, and hypertensive people due to its bioactive compounds. Germinated brown rice (GBR) is used as a nutritive ingredient for many food preparations, including noodles, bread, cookies, rice ball, tea, milk, and breakfast cereals.

Microbial Profiling of Brown Rice

Brown rice (BR) is more nutritious than white rice (WR) as the bran and embryo of brown rice contain rich in fibers and vitamins (Cui et al. 2010). Hence brown rice may harbor more microbial association as compared to white rice. In some countries particularly in Korea, brown rice is soaked in water and supplied as GBR to the consumers at an industrial level. During the time of germination, it will improve the quality of rice by hydrolyzing the high molecular weight polymers, which in turn reduces the molecular size and produces many bio-functional substances. The production of bio-functional substances will enhance the following substances, viz., γ-aminobutyric acid (GABA), free amino acids, dietary fiber, magnesium, potassium, zinc, inositols, tocotrienols, ferulic acid, γ-oryzanol, and prolylendopeptidase inhibitor, and reduce phytic acid (Kayahara and Tsukahara 2000; Ohisa et al. 2003; Ohtsubo et al. 2005). The germination of brown rice normally occurs in a humid and warm environment, and this suitable condition will encourage the multiplication of microorganisms (Lu et al. 2010). As brown rice consumption by the people is getting popular, the nutrient profiling has been documented extensively; however, there was not much information on profiling of microbial association. Some research findings indicated that the microbial population was observed to be higher in brown rice as compared to white rice; however, the germinated brown rice had higher microbial population than brown rice (Martinez et al. 2013).

Microbial Association in Germinated Brown Rice (GBR)

To unlock the bio-nutrients in brown rice, and make it available, germination of a brown rice seems to be a smart technique. Germination is referred as a biological process which is induced by the activation of residual enzymes (Lu et al. 2010). The compositional changes during germination of brown rice can be optimized by understanding its metabolic pathway. Germination is usually initiated by soaking the seeds in water, uptake of water by dry seeds, which eventually undergoes sprouting. Soaking of brown rice in water for certain duration of time invites the native microbial flora around the environment to act upon it. In other words, it involves different steps of fermentation process. Each biological process is activated by various groups of microflora and in turn produces different biomolecules which are available in the medium. Hence, optimizing the germination conditions for brown rice for the commercial production of GBR with consistent physical properties and quality is important (Puri et al. 2014).

The factors influencing the germination of brown rice include intrinsic parameters such as nature of cultivar, milling and storage conditions, as well as the other factors including humidity, temperature, aeration and relative oxygen content, light, and pH. Moisture is essential for germination process, and it can be provided by two means: (i) soaking and temporary immersion of the seeds in water and (ii) atmospheric germination. The latter process (atmospheric germination process) exhibited higher degree of germination with longer root and sprout than the simple water soaking (Lu et al. 2010). The optimum germination conditions for BR include soaking at 25°C for 24 h, followed by atmospheric germination at 30°C for 3 days (Capanzana and Buckle 1997). Importantly, extended period of germination induces more microbial growth and causes too much growth of root and sprout, often rendering GBR unsuitable as food material. Sometimes it may cause cross contamination and even accumulate with microbial toxins. Regular washing, soaking, and atmospheric germination are reported to suppress microbial growth and off-flavor formation. In order to control microbial growth, chitosan solution or electrolyzed water may be used with rinsing solutions (Lu et al. 2010; Oh and Choi 2000). Besides germination rate and microbial infection, the metabolic changes in biofunctional constituents should be considered at the time production of GBR.

There are different types of microbial association that were reported in brown rice. Some of these are beneficial for the consumers, whereas some are even harmful and they are listed in Tables 10.1 and 10.2. In GBR, the lactic acid bacteria particularly *Weissella confusa*, *Pediococcus pentosaceus*, and *Lactobacillus fermentum* were commonly recorded (Kim et al. 2012). The mesophilic bacterial association in brown rice is gradually decreasing during the time of further polishing or dehulling process. Park et al. 2003 recorded that the initial mesophilic bacterial count was 3.2 × 10^4 CFU/g in brown rice, where as it was 3.5 × 10^3 CFU/g in dehulled rice; this information clearly indicates that the removal of every layer from brown rice will reduce the microbial population. The other findings showed that the microbial population was increased during the time of germination (Imam and Ismail, 2013). Lu et al. (2010) reported that the aerobic bacterial population was increased from 3.91 to 6.52 log CFU/g after 36 h of germination.

In an investigation by Ankolekar et al. (2009), out of 178 rice samples collected from US retail food stores had *Bacillus* sp. (3.3 × 10^1 CFU/g) in 94 samples. Sarrias et al. (2002) observed less *B. cereus* population (10^2 CFU/g) in husked and white rice samples than unhusked rice (10^3 CFU/g). Similarly, 70 of 100 Turkish white rice samples had mold contamination in the range of 1.0 × 10^1 to 1.5 × 10^4 CFU/g, and this mold growth was effectively inhibited in properly dried (13 to 15%

Table 10.1 Beneficial microbes associated with rice

Name	Group	Benefits	References
Cellulomonas flavigena	Gram-positive bacteria	Cellulose degradation	Cottyn et al. 2001
Brevibacillus brevis, Brevibacillus laterosporus, Brevibacterium spp.	Gram-positive bacteria	Produces L. amino acids in the gut	Cottyn et al. 2001
Paenibacillus macerans, Paenibacillus polymyxa, Paenibacillus spp.	Gram-positive bacteria	2,3 Butanediol and ethanol production	Adalakha et al. 2015
Bacillus licheniformis, Bacillus subtilis group, *Bacillus megaterium, Bacillus pumilus*	Gram-positive bacteria	Exhibits antagonistic property by producing chitosan, conversion of glutamate to GABA	Kim et al. 2014
Lactic acid bacteria, *Leuconostoc mesenteroides*	Gram-positive bacteria	Improves nutritional value of food and control of intestinal infections, improves digestion of lactose, controls different types of cancer, and controls serum cholesterol levels	Gilliland 1990
Monascus purpureus	Fungi	Pigment production (red)	Adams 1998; Chavan and Kadam 1989; Harlander 1992; Sankaran 1998; Soni and Sandhu 1990
Hansenula anomala, Saccharomyces sp., *Geotrichum candidum, Trichosporon pullulans, Torulopsis etchellsii*	Yeast	Increase of all essential amino acids and in the reduction of antinutrients (such as phytic acid), enzyme inhibitors, and flatus sugars Production of alpha-amylase	Chavan and Kadam 1989; Shortt 1998. Steinkraus et al. 1993
Mucorrouxianus, Mucor indicus, Rhizopus oryzae, R. chinensis	Fungi	Alcohol production	Adams 1998; Chavan and Kadam 1989; Harlander 1992; Sankaran 1998; Soni and Sandhu 1990
Endomycopsis fibuligera	Yeast	Production of alpha-amylase	Sandhu et al. 1987

moisture) samples. Some findings have recorded the presence of *E. coli* in white rice grains from retailed market samples; however the average count of *E. coli* was below the detection limit, and this study also indicated that the numbers of coliforms increased markedly, i.e., 4.2×10^3 and 4.3×10^3 CFU/g rice in brown rice and germinated brown rice grains, respectively (Lee et al. 2007). The coliform counts

Table 10.2 Microbes causing adverse effects associated with rice

Name	Group	Adverse effect	References
Clavibacter michiganensis	Gram-positive bacteria	Red stripe in rice	Cottyn et al. 2001
Microbacterium saperdae, Microbacterium barkeri, Microbacterium liquefaciens, Microbacterium arborescens	Gram-positive bacteria	Bacteremia in children, interstitial pulmonary inflammation	Laffineur et al. 2003; Giammanco et al. 2006
Bacillus cereus and Bacillus thuringiensis	Gram-positive bacteria	Enterotoxin production	Kim et al. 2014
Corynebacterium aquaticum	Gram-positive bacteria	Lipolytic activity hampers long-term storage of brown rice	Delucca et al. 1978
Staphylococcus saprophyticus	Gram-positive bacteria	Food-borne pathogen	Cottyn et al. 2001
Pseudomonas avenae subsp. *avenae* *Pseudomonas setariae*	Gram-negative bacteria	Bacterial brown stripe	Shakya and Chung 1983
Pseudomonas glumae	Gram-negative bacteria	Grain rot, bacterial seedling rot	Uematsu et al. 1976
Pseudomonas plantarii	Gram-negative bacteria	Seedling blight	Azegami et al. 1987
Erwinia chrysanthemi	Gram-negative bacteria	Foot rot	Ou 1985
Pseudomonas syringae pv. *oryzae*	Gram-negative bacteria	Bacterial sheath rot	
Xanthomonas campestris pv. *oryzae*	Gram-negative bacteria	Bacterial blight	
Xanthomonas campestris	Gram-negative bacteria	Bacterial leaf streak	
Aspergillus flavus, Fusarium fujikuroi, Candida	Fungi	Toxin production	Tanaka et al. 2007
Pyricularia oryzae	Fungi	Blast	Mew and Gonzales 2002
Fusarium moniliforme	Fungi	Bakanae, foot rot	
Cercospora oryzae	Fungi	Narrow brown leaf spot	
Acrocylindriumoryzae, Sarocladium attenuatum	Fungi	Sheath rot, black kernel	
Helminthosporium oryzae	Fungi	Brown spot	
Gibberella saubinetii, Botryosphaeria saubinetii	Fungi	Scab	
Rhizoctonia solani	Fungi	Sheath blight	
Phyllosticta glumarum	Fungi	Glume blight	
Tilletia horrida, Neovossia horrida	Fungi	Kernel smut, False smut	

(3.10 log CFU/g rice) were also noticed in raw rice. The above information cautioning consumer, in view of consumer safety, proper documentation of harmful microbial association in brown rice and its products is also required.

Profile of Amylolytic Enzymes in Germinated Brown Rice

The dry seeds require sufficient energy for their metabolic changes, which is mostly from the degradation of residual starch that constitutes 90% of brown rice (BR). Alpha-amylase, which randomly breaks the α-1,4 bonds in starch, is considered as the principal enzyme for starch degradation. Since the α-amylase is active in the late stage, glucose produced from sucrose and maltose is the main source of energy in the early stage of germination (Palmiano and Juliano 1972). The α-amylase was observed to be produced during the protein synthesis in the late stage rather than pre-existing in seeds. The dry seeds were also found to contain a small amount of α-amylase enzyme (Moongngarm and Saetung 2010). Conversely, the higher amount of β-amylase is present in BR endosperm and becomes activated with the start of imbibitions, which sequentially breaks down the starch to maltose units (Palmiano and Juliano 1972). Other than these two main enzymes, the α-1,6 bond of starch is digested by a debranching enzyme during germination. As a result of these amylolytic actions, the decrease in the amount of amylose in brown rice (BR) was noticed during germination. This is one of the key factors in determining the quality and texture of cooked rice.

Jiamyangyuen and Ooraikul (2008) studied the influence of germination on sensory properties and the eating quality of brown rice (BR) and reported that cohesiveness and softness in cooked germinated brown rice (GBR) increased by germination, whereas the stickiness and blandness decreased. Moongngarm (2010) observed that the BR pasting viscosity of flour severely decreased as germination time increased. The smooth and densely packed starch granules are reported to change their shape to rough and eroded during germination. This phenomenon happened possibly from the degradations of starch with its granule-associated proteins by the amylases and proteases activated during germination or from associated microbes. The role of protease and amylase producing microbes in germinating brown rice still remains unexplored.

List of Beneficial and Harmful Microbes in Brown Rice

The rice seed provides a habitat for a rich diversity of microorganisms consisting of bacteria, fungi, microscopic algae, as well as members of the microfauna such as plant nematodes. The majority of bacteria carried by rice are located on the surface or slightly more deeply seated under the hull. Because there is little known on the identities of total bacterial populations associated with rice seeds, plant pathologists often group bacteria by the effect they may have on the plant into categories of neutral, beneficial, or deleterious.

Bioactive Molecules in Germinated Brown Rice and the Microbial Influence

The germinated brown rice is rich in bioactive molecules which have significant nutrigenomic effect in many metabolic disorders such as blood pressure regulation, type II diabetes, and reduction in obesity and blood cholesterol level (Chao and Lim 2016). Evidences based on the studies with rats on GBR diet increased the levels of neutral sterols and fecal excretion of bile acid (Miura et al. 2006) and produced a hypotensive effect and reduction in blood triglyceride concentration and total cholesterol, decrease in weight blood sugar, and increase in high-density lipoprotein–cholesterol. The brown rice (BR) has substantial amount of dietary fiber, which accounts approximately 2% of total BR solids. Gamma-oryzanol, a lipophilic fraction, is the major constituent in BR. Other bio-functional compounds in BR include minerals, vitamin E, phytosterols, phenolic compounds, and phytic acid, which are mostly present in the bran and germ of BR. The germinated brown rice (GBR) contains some of these bio-functional substances at higher levels than raw BR, indicating that germination may enhance the bioactivity of BR. As in GBR, the gamma-aminobutyric acid (GABA), a unique bio-functional compound in GBR, is well known to be produced during germination. Brown rice favors the growth of fungi and yeast as compared to rough or white rice at 85% RH. Prolonged duration of germination causes contamination and thereby affects the quality of GABA. Among the bacteria, quantitative microbiological risk assessment studies of *B. cereus* on brown rice showed that combined effects of ultrasonication and slight acidic electrolyzed water improved the microbial safety of brown rice (Tango et al. 2014).

Gamma-Aminobutyric Acid (GABA)

GABA is a nonprotein amino acid produced primarily from the decarboxylation of glutamic acid by the enzyme glutamate decarboxylase which is highly active during germination of rice (Liu et al. 2005). It acts as a main inhibitory neurotransmitter in the mammalian cortex. During water absorption, glutamate decarboxylase (GAD) in brown rice is activated and converts glutamic acid into gamma-aminobutyric acid (GABA). According to Komatsuzaki et al. (2007), glutamic acid is an amino acid, a form of stored protein in brown rice which is changed into an amide. The GABA quantity can be enhanced by soaking brown rice in warm water at 20–40°C (Shinmura et al. 2007). Some of the microbes, viz., lactic acid bacteria, *Bacillus licheniformis Bacillus subtilis* group, *Bacillus megaterium, and Bacillus pumilus*, present on the surface of brown rice also mediate the stimulation of GABA synthesis (Kim et al. 2014). The knowledge on the microbes and process involved in GABA synthesis is still in the dark.

The synthesis and bio-functions of gamma-aminobutyric acid (GABA) have been extensively studied so far. Moreover, the GABA content in brown rice is influenced by the genotype differences. Roohinejad et al. (2012) investigated the changes in GABA content in 18 rice varieties with 72 h immersion in distilled water (30°C) and found that all rice samples tested demonstrated increases in GABA content, but the increase in GABA content widely differed among rice varieties.

Oh and Choi (2000) observed that BR germinated for 72 h with soaking in chitosan solution at the concentration of 100 ppm recorded 7.6 times more GABA than that in native BR and 1.3 times more than that in the control germinated in distilled water. On the other hand, according to Lu et al. (2010), though electrolyzed oxidizing water was very effective in suppressing microbial growth while germination of brown rice, it resulted shortened radicals and sprouts. Activation of protease enzyme in brown rice (BR) at the time of germination increases the concentration of free amino acids in BR correspondingly (Komatsuzaki et al. 2007; Moongngarm and Saetung 2010; Veluppillai et al. 2009). By contrast, it was also observed that the protein content in BR slightly increased by germination (Lee et al. 2007; Moongngarm and Saetung 2010). This result might be attributed to the loss of decomposed starch and sugars which are soluble components during germination. The protease activity in brown rice might be due to protease producing microbes associated with brown rice which needs further exploration (Lee et al. 2007).

Dietary Fibers

Nondigestible dietary fibers enhance the beneficial microbes for human health (Gibson and Roberfroid 1995). Dietary fibers (DF) are edible carbohydrate polymers, which are normally not digested by the intestine enzymes of human beings. It is grouped into soluble and insoluble dietary fibers based on their water solubility. Soluble dietary fibers, viz., pectin, mucilage, and some hemicelluloses, play a major role in raising intestinal viscosity and lowering blood glucose as well as cholesterol levels. The insoluble dietary fibers include cellulose and some cellulosic substances, which help to maintain colon health by expanding fecal volume, assisting bowl movements, and removal of toxic metabolites. Cheng (1993) recorded that rice bran constitutes large quantity (24–34% of total bran solids) of dietary fiber, which corresponds to 1.4–3.3% in total BR. So, the consumption of BR is highly recommended nowadays to benefit from the herculean effects of dietary fiber. This insoluble dietary fiber is the sole responsible compound in germinated brown rice (GBR) for the antidiabetic effect (Seki et al. 2005). The destarched and defatted pre-germinated brown rice bran is free from oil and soluble compounds, and it is observed to reduce the postprandial glucose level. Although dietary fiber (DF) is found in many foods, consuming BR and GBR may ensure regular and sufficient uptake of this health promoting substance (Martinez et al. 2013).

During germination of brown rice, the composition of DF exhibited remarkable changes depending on germination method (Kim et al. 2001; Mohan et al. 2010; Ohtsubo et al. 2005). Many workers have observed that the changes in dietary fiber levels during germination are very well correlated with variations among different *Oryza sativa* at species level. For instance, Koshihikari variety displayed a 1.5 times increase in dietary fiber after 72 h of soaking (Ohtsubo et al. 2005). Moreover, Mohan et al. (2010) found that contents of total dietary fiber were increased by 28.2% in *indica* and 40.0% in *japonica*. The soluble dietary fiber contents were also increased by 61.0% and 79.0%, respectively.

On the contrary, the atmospheric germination after soaking is found to decrease in dietary fiber under long-term processing conditions (Jayadeep and Malleshi 2011; Kim et al. 2001). According to Jayadeep and Malleshi (2011), insoluble dietary fiber contents in BR decreased from 41.5 to 36.8 g/kg in variety IR 64 and 37.0 to 31.7 g/kg in variety BPT by atmospheric germination with simple soaking of 16 h. Kim et al. (2001) noticed that dietary fiber content was drastically reduced after 6 days of atmospheric germination. This may be due to microbial growth and off-flavor formation, and it happened due to the extensive soaking and germination. The overall process should be carried out within a period less than 6 days for safe and contamination-free GBR. The soluble dietary fiber in brown rice (BR) comprises of glucose, xylose, and mannose, whereas insoluble dietary fiber consists of mainly glucose, arabinose, and xylose (Rao and Muralikrishna, 2004). Although the most predominant sugar, i.e., glucose (58.9%), did not change significantly, germination induced decrease in the proportion of xylose (16.3% to 9.32%) and mannose (13.1% to 0.00%) in soluble dietary fiber. The arabinose percentage was increased nearly 6.5-fold (3.4% to 22.1%) compared to that in BR during germination. In case of insoluble dietary fiber, germination resulted in increases of rhamnose, arabinose, and xylose but decreases in galactose and glucose (Rao and Muralikrishna 2004).

Microbial Association and Biochemical Changes During Fermentation of Brown Rice

Brown rice is one of the richest sources of bio-functional compounds like dietary fibers, proteins, carbohydrates, vitamins, and minerals as discussed earlier. But the organoleptic properties of brown rice and their products are inferior because of its lower protein content, deficiency of some essential amino acids like lysine, low starch, the presence of some antinutrients, viz., phytic acid, tannins, and polyphenols, and the coarse nature of the grains (Chavan and Kadam 1989). Many techniques like genetic improvement and amino acid supplementation with protein concentrates have been employed for solving the problem of the odd nature of brown rice. In addition, some processing methods which include cooking, fermentation, milling and sprouting is being adopted to improve the nutritional properties of brown rice (Mattila-Sandholm 1998). Natural fermentation of cereals leads to a decrease in the level of carbohydrates, nondigestible poly- and oligosaccharides, and synthesis of amino acids and improves the availability of B group vitamins.

Some of the common fermenting bacteria associated with brown rice are species of *Bacillus, Lactobacillus, Leuconostoc, Micrococcus, Pediococcus*, and *Streptococcus*. In some rice-based products, fungal genera like *Aspergillus, Cladosporium, Fusarium, Paecilomyces, Penicillium*, and *Trichothecium* are most frequently observed. Among the fermenting yeast species, *Saccharomyces* is common, which results in alcoholic fermentation (Steinkraus 1998). During the course of fermentation, some microorganisms may involve in parallel, while the other microbes may act in a sequential manner with a changing dominant flora.

The nature of microbial flora developed in each fermented food depends on certain extrinsic factors like pH, temperature, moisture content, salt concentration, and the composition of the food matrix. Most fermented foods that are common in the Western world, as well as many of those from Asia, are dependent on lactic acid bacteria (LAB) to intervene the fermentation process (Conway 1996). Fermentation by lactic acid bacteria contributes toward the safety, enhanced nutritional value, shelf life, and acceptability of a different kind of cereal-based foods (Oyewole 1997).

Fermentation provides optimum pH conditions for enzymatic degradation of phytate, and the reduction in phytate may increase the amount of soluble calcium, iron, and zinc many folds (Chavan and Kadam 1989; Gillooly et al. 1984; Haard et al. 1999; Khetarpaul and Chauhan 1990; Nout and Motarjemi 1997; Stewart and Getachew 1962). The effect of fermentation on the protein and amino acids is a bit of a controversy. During the fermentation of corn meal, the level of available methionine, lysine, and tryptophan is found to increase (Nanson and Field 1984). In a similar manner, fermentation is found to significantly enhance the protein quality as well as the level of lysine in maize, millet, sorghum, and other cereals (Hamad and Fields 1979). It appears that the effect of fermentation on the nutritive value of foods is variable.

The fermentation process improves shelf life, texture, taste, and aroma of the final product. Several volatile compounds are formed during the time of brown rice fermentations, which contribute to a complex blend of flavors (Chavan and Kadam 1989). The presence of aromas representative of diacetylacetic acid and butyric acid make fermented products more appetizing. The microbiology of many of these products is quite complex and not known.

Lactic Acid Bacteria (LAB) in Brown Rice

Cleaned brown rice is soaked in water for a few days while the series of naturally occurring microbes will multiply, and this population is dominated by LAB. The endogenous grain amylases will produce the fermentable sugars, and it serves as a source of energy for the lactic acid bacteria during the process of fermentation. The term LAB is a broad group of gram-positive, catalase-negative, non-sporing rods and cocci, which utilize carbohydrates during fermentation and produce lactic acid as the major end product (Aguirre and Collins 1993). According to the pathways by which hexoses are metabolized, they are divided into homofermentative and heterofermentative.

In homofermentative process, some important group of bacteria, viz., *Pediococcus*, *Streptococcus*, *Lactococcus*, and *Lactobacillus*, will produce lactic acid as major end product from glucose fermentation. Whereas in heterofermentative process, the bacteria genera like *Leuconostoc, Lactobacillus,* and *Weissella* will produce lactate, CO_2, and ethanol from glucose (Aguirre and Collins 1993; Tamime and O'Connor 1995).The preservative role of lactic fermentation technology has been confirmed in many cereal products. The antibiosis characteristic of LAB might be due to the production of acids, hydrogen peroxide, antibiotics, and low molecular weight bacteriocins. The production of organic acids reduces the pH to below 4.0, and it will arrest the survival of some spoilage organisms present in cereal products (Daly 1991; Oyewole 1997).

LAB also produce hydrogen peroxide through the oxidation of reduced NADH by flavin nucleotides, which reacts rapidly with oxygen. Since LAB do not possess catalase, it can't break down the hydrogen peroxide produced. So the accumulated hydrogen peroxide is inhibitory to some of the microorganisms (Caplice and Fitzgerald 1999). Apart from this, the level of tannin may be decreased as a result of lactic acid fermentation. It helps to increase the absorption of iron, except in some high tannin brown rice, where no significant improvement in iron availability has been observed (Nout and Motarjemi 1997).

Another advantage of lactic acid fermentation is that the LAB-involved fermented products have viricidal (Esser et al. 1983) and antitumor effects (Oberman and Libudzisz 1996; Seo et al. 1996). In Africa and Asia, a number of indigenous products utilize rice in combination with legumes, thus improving the protein quality in the fermented product. Lactic acid bacteria also produce small ribosomal antimicrobial peptides called bacteriocins which can inhibit bacterial strains closely related or non-related to produced bacteria, and it can be utilized as one of the safe food additives (Yang et al. 2014).

Brown Rice as Probiotic Substrate

Probiotic foods are functional foods containing sufficient number of a certain live microorganism that modifies the gut microbiota thereby conferring health benefits to the host. Although most of the recently developed probiotics tend to be milk-based, new probiotic formulations with alternative substrates have been explored. Cereals are becoming one of the important alternatives to milk due to their inherent property to support the probiotic bacterial growth and also their protective resistance effect against bile (Charalampopoulos et al. 2002). Brown rice contains compounds like tocopherols, tocotrienols, anthocyanins, polyphenols, γ-oryzanol, enzymes, polyunsaturated fatty acids, and resistant starch (Hagiwara et al. 2004). All these have shown positive effects as modulators of blood pressure, glycemia, or serum cholesterol levels (Murata et al. 2007). It may also contain antioxidants that could help in the prevention of tumoral proliferation (Parrado et al. 2006) and even nutraceutical prebiotics that could reduce the colitis (Kataoka et al. 2007) through modification of the colonic microbiota.

Whole rice grains or fractions of the grain can also be modified for improving their nutritional or to enhance their functional properties. Traditionally, the two main reasons to ferment rice-based foods are (i) to arrest the activity of undesirable microorganisms and (ii) to create foods with better flavor, aroma, and texture

(Gelinas and McKinnon 2000). Despite the fermentability and functional potential of rice, only few studies attempted to use brown rice as a substrate for the production of probiotic foods (Saman et al. 2011). They concluded that brown rice proved to be a very good substrate for the growth of *Lactobacillus plantarum*.

Brown Rice as Prebiotics

Some nondigestible dietary fibers, such as inulin and fructooligosaccharides (FOS), have been shown to increase the levels of beneficial bacteria for general human health such as *Bifidobacterium* and *Lactobacillus*, resulting in reduced intestinal pH, improved intestinal environments, enhanced immunomodulation effect, and suppressed intestinal disorders (Macfarlane et al. 2006). Prebiotics are defined as the nondigestible food ingredients, and it brings beneficial effect to the host through stimulation of selected growth and activity of bacteria already inhabiting the colon. Brown rice fermented by *Aspergillus oryzae* (FBRA) is a kind of fermented processed food prepared from brown rice and rice bran. It has earlier been reported that FBRA has a defensive effect on azoxymethane-induced colon carcinogenesis and dextran sulfate sodium-induced acute colitis in rats (Kataoka et al. 2008). FBRA contains a large proportion of fiber, and dietary administration of 10% FBRA to rats increased the lactobacilli viable cell numbers in feces (Nemoto et al. 2011). Beta-glucan and arabinoxylan are the main constituents in the dietary fiber present in FBRA, which stimulates the growth of bifidobacteria and lactobacilli strains in vitro (Jaskari et al. 1998).

Nutrigenomics Effects of Brown Rice

Modulation of Blood Pressure

The nutrigenomic property of brown rice is conferred due to the bioactive molecule GABA, which has been elaborated in previous sections. GABA in brown rice exerts antihypertensive properties in spontaneously hypertensive rats (Hayakawa et al. 2002). The systolic blood pressure was drastically reduced within 8 h in hypertensive rats after single oral administration of GABA (0.5 mg/kg). The GABA-induced antihypertensive effect might be due to the result of reduced noradrenaline release through an action on presynaptic receptors for GABA (Hayakawa et al. 2002).

Brown Rice with Anticancer Property

GABA-enriched fractions extracted from GBR showed an inhibitory action on the proliferation of cancer cells and a stimulatory effect on immune responses (Oh and Oh 2003). GABA-enriched extracts from GBR does not have any inhibitory effects on human cervical cancer HeLa cell proliferation. But these extracts have apoptosis activity apart from having inhibitory effect on leukemia cell (mouse leukemia L1210 cells, human acute lymphoblastic leukemia Molt4 cells) proliferation via nitric oxide production (Oh and Oh 2003). The L1210 leukemia cells treated with GBR extracts showed a lot of DNA disintegration as compared to the leukemia cells plus BR extract. Oh and Oh (2003) observed that GABA-enriched fractions from GBR have improved immunoregulatory activities.

Brown Rice for Type II Diabetes

GABA shows other physiological functions in human and other vertebrates like strengthening of blood vessels, modulation of insulin secretion, and suppression of blood cholesterol increase. Some findings indicated that GABA will help to relieve from emotional unrest and stroke, and it is also found to enhance liver and kidney functions in addition to protection from chronic alcohol-related illness (Kawabata et al. 1999; Oh et al. 2003). It downregulates gluconeogenesis genes and controls type II diabetes, but the mechanism involved has to be thoroughly understood (Imam and Ismail 2013).

Harmful Microbial Association Particularly Toxigenic Mold in Brown Rice

Mycotoxin Contamination

In general, mycotoxin contamination is reported to be as less in rice as compared to many other cereal crops; however the rice is a good substrate for fungal growth. Also this substrate is commonly used as culture medium to test the toxigenic potential of different fungal cultures (Bars and Bars 1992). There are number of toxins that have been reported in rice, of which aflatoxin B1, fumonisin B1, and ochratoxin A are the highly toxic to mammals and have teratogenic, hepatotoxic, and mutagenic properties. Many mycotoxicosis cases have been reported in humans and animals due to the consumption of mycotoxin-contaminated food and feed (Reddy and Raghavender 2007). Though different fungal associations have been reported in rice, the major mycotoxigenic fungi are *Aspergillus* sp., *Fusarium* sp., and *Penicillium* sp. In rice, the following mycotoxins, viz., aflatoxins, fumonisins, trichothecenes, ochratoxin A, cyclopiazonic acid, zearalenone, deoxynivalenol (DON), citrinin, gliotoxin, and sterigmatocystin, have been reported.

Reddy et al. (2009) documented rice and rice-based products from various countries containing different types of mycotoxin contamination, and the details are as follows: aflatoxins were reported in China (0.99–3.87 µg/kg), India (0.1–308 µg/

kg), Korea (1.8–7.3 ng/g), the Philippines (0.27–11 µg/kg), and Vietnam (3.31–29.8 ng/g); fumonisins were noticed in Canada (0–10 ng/g), India (0.01–65 mg/kg), Korea (48.2–66.6 ng/g), and the United States (2.2–5.2 mg/kg); ochratoxin A was observed in Canada (0.3–2.4 ng/g), Korea (0.2–1 ng/g), Morocco (0.15–47 ng/g), Nigeria (24–1164 µg/kg), and Vietnam (0.75–2.78 ng/g); deoxynivalenol was observed in Germany (0–0.058 mg/kg); and zearalenone was observed in Canada (0.1–1 ng/g), Korea (21.7–47 ng/g), and Nigeria (24–116 µg/kg).

Aflatoxin Contamination

Lai et al. (2014) have registered the mycotoxin, particularly aflatoxin, contaminations of rice from different countries like Bangladesh (Dawlatana et al. 2002), Brazil (Almeida et al. 2012), China (Liu et al. 2006), Canada (Bansal et al. 2011), India (Reddy et al. 2009), Korea (Ee OK et al. 2014), Turkey (Aydin et al. 2011), Pakistan (Majeed et al. 2013), Tunisia (Ghali et al. 2008) and Vietnam (Nguyen et al. 2007). Aflatoxins are termed as adverse secondary metabolites produced by *Aspergillus flavus* and *Aspergillus parasiticus* (Binder et al. 2007), and its contamination in food causes a lot of health hazards (Williams et al. 2004).

Aflatoxins are reported to be carcinogenic and hepatotoxic; hence its contamination in different food products is being strictly monitored by WHO including the European Commission and Food and Drug Administration; these authorities have set restriction limits for various foodstuffs, so the consumption beyond this limits is unsafe (Van and Jonker 2004). Out of 18 types of various aflatoxins, G1, G2, B1, B2, M1, and M2 types are the important ones; however type B1 is reported to be very harmful to the human being (Krishnamurthy and Shashikala 2009). In Iran around 83% of import quality rice samples were contaminated with aflatoxins (2.09 µg kg^{-1}) (Mazaheri 2009).

The standard regulatory range limit, i.e., 10 ppb of aflatoxins, has been set in brown rice samples by the European Commission. In an investigation by Nisa et al. (2016), out of 50 brown rice samples collected from local market in Pakistan, nearly 92.0% of samples were contaminated with aflatoxin, of which 56% of samples were contaminated above permissible limits. They also have analyzed aflatoxin contamination in import quality samples, in which only 48% of samples were contaminated with aflatoxin, of which only 4% of samples were contained above permissible limits. This finding indicated that import quality of brown rice samples is observed to be good as compared to locally available samples. So, the consumption of locally available brown rice samples can be rendered unsafe and deleterious for human health due to its adverse effects.

In 2014, Asghar and his coworker reported that, out of 262 brown rice samples, 250 samples had aflatoxin contamination with an average of 3.89 µg/kg; however, around 71% samples were recorded below the maximum tolerated level (MTL), which is recommended by the European Union (4 µg/kg). They also suggested that aflatoxin contamination can be managed in brown rice by adopting proper harvesting practices, storage, and transportation.

Ochratoxin A Contamination

Ochratoxin A is produced by the fungal species, viz., *Aspergillus ochraceus, A. niger, A. carbonarius,* and *Penicillium verrucosum,* which is a possible carcinogenic compound to humans (IARC 1993b). The European regulatory authority has fixed 5 and 3 µg/kg of OTA in cereals and cereal products, respectively, as maximum residue levels (European commission 2006). The occurrence of OTA in rice has been reported in many countries, viz., the United Kingdom, Brazil, Canada, Vietnam, Korea, Morocco, Malaysia, Turkey, Tunisia, and Pakistan (Lai et al. 2014)

Around 15.0% of rice samples collected from Yangtze Delta region of China was contaminated with OTA at an average level of 3.5 µg/kg (Li et al. 2014). Similarly, 26.0% of rice samples from Morocco had OTA contamination with an average level of 3.5 µg/kg (Juan et al. 2008). Zaied et al. (2009) observed that 28% of rice samples from Tunisia contained OTA with an average amount of 44 µg/kg. In Pakistan, 50% of rice samples were contaminated with OTA with an average level of 12.94 µg/kg (Majeed et al. 2013). Lai et al. 2014 have collected 370 rice samples from six provinces in China and analyzed for the presence of aflatoxins (AFs) and ochratoxin A (OTA), of which 63.5% and 4.9% rice samples had been positive in AF and OTA contamination, respectively. However, the average amount of aflatoxin B1, total AFs, and OTA were recorded as 0.60, 0.65 and 0.85 µg/kg, respectively. They also stated that only 1.4% and 0.3% of collected samples had above the EU limit of aflatoxin B1 and OTA, respectively, which indicates that rice commodities are generally very safe. Most of the findings indicated that mycotoxin contamination in rice occurred mainly by improper processing and inappropriate storage (Reddy et al. 2009; Saleemullah et al. 2006). In general, the good processed samples are totally free from mycotoxin contamination. Hence, implementation of strict measures can totally eliminate mycotoxin contamination in rice.

Sterigmatocystin Contamination

In general, fungi invade through the germ (Fennel et al. 1973). Particularly in brown rice, fungal invasion begins at the germ unless the grain is stored under 100% humidity. There are many research findings indicated that *Aspergillus versicolor* was commonly detected in the most of the grains including brown rice (Takahashi et al. 1984). The toxin sterigmatocystin produced by *A. versicolor* is a typical endo-type toxin; hence the major portion (98%) remains within mycelia even after 1 month (Takahashi et al. 1984). The above information also showed that the distribution of sterigmatocystin mainly depends on the invaded fungal mycelium in the grains.

Fumonisin Contamination

In the year 1988, the fumonisin contamination was noticed in South Africa (Gelderblom et al. 1988). This toxin is produced by the following fungal general, viz., *Fusarium verticillioides, F. proliferatum, F. oxysporum, F. globosum, Fusarium* sp., and *Alternaria alternata* f. sp. *lycopersici* (Scott 2012). The fumonisin contamination in foods causes esophageal cancer (Marasas 2001; Missmer et al. 2006) and was reported that it is also a possible carcinogen to humans (IARC 1993a). The rough rice samples collected from *Fusarium* sheath rot disease-infected plants showed 27% of samples had *F. proliferatum* infection, and 11% of samples contained fumonisin contamination with the level of 0.5–6.2 ppm. The level of fumonisin contamination is observed to be drastically reduced at the time of hulling and polishing of rice. For example, a sample of rough rice containing 5.13 ppm fumonisin yielded hulls with 16.8 ppm fumonisin, brown rice with 0.87 ppm, and bran with 3.53 ppm; interestingly no fumonisin was observed in polished rice.

Conclusion and Future Thrust

The above scientific information on the microbes associated with brown rice may open new vistas to exploit brown rice commercially. Germination and fermentation are the two biological processes involved in inducing the biomolecules in brown rice which improve the physical quality of BR, such as cooking properties and eating texture. The idea about beneficial microbes particularly the prebiotic and probiotic potential of lactic acid bacteria can be harnessed for developing more brown rice-based products. The role of these microbes during germination and fermentation needs to be explored. The shelf life of germinated brown rice may be enhanced with the bacteriocins produced by lactic acid bacteria. The fructooligosaccharides produced by LAB in brown rice can be used as a potential prebiotic. Mycotoxin problem in rice can be managed by adopting good agricultural/hygienic practices, proper storage, and different kinds of cooking methods (boiling, stewing, steaming and pressure cooking), etc. The brown rice-based products and brown rice can be rendered microbiologically safe by formulating certain microbial standards and satisfying HACCP standards. Moreover, a vivid knowledge on the impact of brown rice from different varieties and cultivars on the microbial biodiversity and its biomolecule constituents should be investigated to bring brown rice more feasible with fabulous consumer acceptance.

References

Adams MR (1998) Fermented weaning foods. In: Wood JB (ed) Microbiology of fermented foods. Blackie Academic, London, pp 790–811

Adlakha N, Pfau T, Ebenhoh O, Yazdani SS (2015) Insight into metabolic pathways of the potential biofuel producer *Paenibacillus polymyxa* ICGEB2008. Biotechnol Biofuels 58:159

Aguirre M, Collins MD (1993) Lactic acid bacteria and human clinical infection. J Appl Bacteriol 62:473–477

Almeida MI, Almeida NG, Carvalho KL, Gongalves GAA, Silva CN, Santos EA, Garcia JC, Vargas EA (2012) Co-occurrence of aflatoxins B1, B2, G1 and G2, ochratoxin A, zearalenone, deoxynivalenol, and citreo viridin in rice in Brazil. Food Addit Contam Part 259 A 29:694–703

Ankolekar C, Rahmati T, Labbé RG (2009) Detection of toxigenic *Bacillus cereus* and *Bacillus thuringiensis* spores in U.S. rice. Int J Food Microbiol 128:460–466

Asghar MA, Iqbal J, Ahmed A, Khan MA (2014) Occurrence of aflatoxins contamination in brown rice from Pakistan. Iran J Public Health 43(3):291

Aydin A, Aksu H, Gunsen U (2011) Mycotoxin levels and incidence of mould in Turkish rice. Environ Monit Assess 178:271–280

Azegami K, Nishiyama K, Watanabe Y, Kadota I, Ohuchi A, Fukazawa C (1987) *Pseudomonas plantarii* sp. nov., the causal agent of rice seedling blight. Int J Syst Bacteriol 37:144–152

Bansal J, Pantazopoulos P, Tam J, Cavlovic P, Kwong K, Turcotte AM, Lau BP, Scott PM (2011) Surveys of rice sold in Canada for aflatoxins, ochratoxin A and fumonisins. Food Addit Contam 28:767–774

Bars LJ, Bars LP (1992) Fungal contamination of aromatic herbs, aflatoxinogenesis and residues in infusions. Microbiol Alim Nutr 10:267–271

Binder EM, Tan LM, Chin LJ, Handl J, Richard J (2007) Worldwide occurrence of mycotoxins in commodities feeds and feed ingredients. Animal Feed Sci Technol J 137(3):265–282

Binodh AK, Kalaiyarasi R, Thiyagarajan K, Manonmani S (2006) Physicochemical and cooking quality characteristics of promising varieties and hybrids in rice. Indian J Genet Plant Breed 66(2):107–112

Capanzana MV, Buckle KA (1997) Optimisation of germination conditions by response surface methodology of a high amylose rice (*Oryza sativa*) cultivar. LWT Food Sci Technol 30:155–163

Caplice E, Fitzgerald GF (1999) Food fermentations: role of microorganisms in food production and preservation. Int J Food Microbiol 50:131–149

Charalampopoulos D, Wang R, Pandiella SS, Webb C (2002) Application of cereals and cereal components in functional foods: a review. Int J Food Microbiol 79:131–141

Chaudhary RC (2003) Speciality rices of the world. Effect of WTO and IPR on its production trend and marketing. Food Agric Environ 1(2):34–41

Chavan JK, Kadam SS (1989) Critical reviews in food science and nutrition. Food Sci 28:348–400

Cheng HH (1993) Total dietary fiber content of polished, brown and bran types of *japonica* and *indica* rice in Taiwan: resulting physiological effects of consumption. Nutr Res 13:93–101

Cho DH, Lim ST (2016) Germinated brown rice and its bio-functional compounds. Food Chem 196:259–271

Conway PL (1996) Selection criteria for probiotic microorganisms. Asia Pac J Clin Nutr 5:10–14

Cottyn B, Regalado E, Lanoot B, De Cleene M, Mew TW, Swings J (2001) Bacterial populations associated with rice seed in the tropical environment. Phytopathology 91:282–292

Cui L, Pan Z, Yue T, Atungulu GG, Berrios J (2010) Effect of ultrasonic treatment of brown rice at different temperatures on cooking properties and quality. Cereal Chem 87:403–408

Daly C (1991) Lactic acid bacteria and milk fermentations. J Chem Technol Biotechnol 51:544–548

Dawlatana M, Coker RD, Nagler MJ, Wild CP, Hassan MS, Blunden G (2002) The occurrence of mycotoxins in key commodities in Bangladesh: surveillance results from 1993 to 1995. J Nat Toxins 11:379–386

Delucca II, Anthony J, Stephen JP, Robert LO (1978) Isolation and identification of lipolytic microorganisms found on rough rice from two growing areas. J Food Prot 1:4–66

Ee OK H, Kim DM, Kim D, Chung SH, Chung MS, Park KH, Chun HS (2014) Mycobiota and natural occurrence of aflatoxin, deoxynivalenol, nivalenol and zearalenone in rice freshly harvested in South Korea. Food Control 37:284–291

Esser P, Lund C, Clemensen J (1983) Antileukemic effects in mice from fermentation products of *Lactobacillus bulgaricus*. Milchwissenschaft 38:257–260

European Commission Regulation (2006) Commission regulation (EC) no. 1881/2006 of 19 272 December 2006 setting maximum levels for certain contaminants in foodstuffs. Official 273. J Eur Union L364:5–18

Fairhurst T, Dobermann A (2002) Rice in the global food supply. WORLD 454–349

FAO (2013) FAO rice market monitor. http://www.fao.org/economic/est/publications/rice-publications/rice-market-monitor-rmm/en/

Fennel DI, Borthast RJ, Lillenhij EB, Paterson RE (1973) Bright greenish yellow fluorescence and associated fungi in corn naturally contaminated with aflatoxin. Cereal Chem 50:404–414

Gelderblom WC, Jaskiewicz K, Marasas WF, Thiel PG, Horak RM, Vleggaar R, Kriek NP (1988) Fumonisins –novel mycotoxins with cancer-promoting activity produced by *Fusarium moniliforme*. Appl Environ Microbiol 54:806–811

Gelinas P, McKinnon C (2000) Fermentation and microbiological processes in cereal foods. In: Kulp K, Ponte JG Jr (eds) Handbook of cereal science and technology. Marcel Dekker, New York

Ghali R, Hmaissia-Khlifa K, Ghorbel H, Maaroufi K, Hedili A (2008) Incidence of aflatoxins, ochratoxin A and zearalenone in Tunisian foods. Food Control 19:921–924

Giammanco GM, Pignato S, Grimont F, Santengelo C, Leonardi G, Giuffrida A, Legname V, Giammanco G (2006) Interstitial inflammation due to *Microbacterium* sp. after heart transplantation. J Med Microbiol 55:335–339

Gibson GR, Roberfroid MB (1995) Dietary modulation of the human colonic microbiota: introducing the concept of prebiotics. J Nutr 125(6):1401–1412

Gilliland SE (1990) Health and nutritional benefits from lactic acid bacteria. FEMS Microbiol Lett 87(1–2):175–188

Gillooly M, Bothwell TH, Charlton RW, Torrance JD, Bezwoda WR, Macphail AP, Deman DP, Novelli L, Morrau D, Mayet F (1984) Factors affecting the adsorption of iron from cereals. Br J Nutr 51:37–46

Haard NF, Odunfa SA, Lee CH, Quintero-Ramırez R, Lorence-Quinones A, Wacher-Radarte C (1999) Fermented cereals. A global perspective. FAO Agric Serv Bullet 138

Hagiwara H, Seki T, Ariga T (2004) The effect of pre-germinated brown rice intake on blood glucose and PAI-1 levels instreptozotocin-induced diabetic rats. Biosci Biotechnol Biochem 68:444–447

Hamad AM, Fields ML (1979) Evaluation of protein quality and available lysine of germinated and ungerminated cereals. J Food Sci 44:456–459

Harlander S (1992) Food biotechnology. In: Lederberg J (ed) Encyclopaedia of microbiology. Academic Press, New York, pp 191–207

Hayakawa K, Kimura M, Kamata K (2002) Mechanism underlying caminobutyric acid-induced antihypertensive effect in spontaneously hypertensive rats. Eur J Pharmacol 438:107–113

Imam M, Ismail M (2013) Nutrigenomic effects of germinated brown rice and its bioactives on hepatic gluconeogenic genes in type 2 diabetic rats and HEPG2 cells. Mol Nutr Food Res 00:1–11

International Agency for Research on Cancer (IARC) (1993b) Evaluation of carcinogenic risks of chemical to humans. Some naturally-occurring substances: food items and constituents. Heterocyclic aromatic amines and mycotoxins, vol 56. IARC Monographs, Lyon, pp 359–362

International Agency for Research on Cancer (IARC) (1993a) Monographs on the evaluation of carcinogenic risks to humans, No. 56. IARC Press, Lyon, pp 489–521

Jaskari J, Kontula P, Siitonen A, Jousimies-Somer H, Mattila-Sandholm T, Poutanen K (1998) Oat beta-glucan and xylan hydrolysates as selective substrates for *Bifidobacterium* and *Lactobacillus* strains. Appl Microbiol Biotechnol 49:175–181

Jayadeep A, Malleshi NG (2011) Nutrients, composition of tocotrienols, tocopherols, and c-oryzanol, and antioxidant activity in brown rice before and after biotransformation. CyTA – J Food 9:82–87

Jiamyangyuen S, Ooraikul B (2008) The physico-chemical, eating and sensorial properties of germinated brown rice. J Food Agric Environ 6:119–124

Juan C, Zinedine A, Idrissi L, Manes J (2008) Ochratoxin A in rice on the 281 Moroccan retail market. Int J Food Microbiol 126:83–85

Juliano BO (1992) Structure, chemistry and function of rice grain and its fraction. Cereal Foods World 37:772–774

Kataoka K, Kibe R, Kuwahara T, Hagiwara M, Arimochi H, Iwasaki T, Benno Y, Ohnishi Y (2007) Modifying effects of fermented brown rice on fecal microbiota in rats. Anaerobe 13:220–227

Kataoka K, Ogasa S, Kuwahara T, Bando Y, Hagiwara M, Arimochi H, Nakanishi S, Iwasaki T, Ohnishi Y (2008) Inhibitory effects of fermented brown rice on induction of acute colitis by dextran sulfate sodium in rats. Dig Dis Sci 53(6):1601–1608

Kawabata K, Tanaka T, Murakami T, Okada T, Murai H, Yamamoto T, Mori H (1999) Dietary prevention of azoxymethane-induced colon carcinogenesis with rice-germ in F344 rats. Carcinogenesis 20:2109–2115

Kayahara H, Tsukahara K(2000) Flavor, health, and nutritional quality of pre-germinated brown rice. In: International chemical congress of Pacific Basin Societies, December 14–19, Honolulu

Khetarpaul N, Chauhan BM (1990) Effect of fermentation by pure cultures of yeasts and lactobacilli on the available carbohydrate content of pearl millet. Trop Sci 31:131–139

Kim SL, Son YK, Son JL, Hur HS (2001) Effect of germination condition and drying methods on physicochemical properties of sprouted brown rice. Kor J Crop Sci 46:221–228

Kim KS, Kim BH, Kim MJ, Han JK, Kum JS, Lee HY (2012) Quantitative microbiological profiles of brown rice and germinated brown rice. Food Sci Biotechnol 21(6):1785–1788

Kim B, Bang J, Kim H, Kim Y, Kim BS, Beuchat LR, Ryu JH (2014) *Bacillus cereus* and *Bacillus thuringiensis* spores in Korean rice: prevalence and toxin production as affected by production area and degree of milling. Food Microbiol 42:89–94

Komatsuzaki N, Tsukahara K, Toyoshima H, Suzuki T, Shimizu N, Kimura T (2007) Effect of soaking and gaseous treatment on GABA content in germinated brown rice. J Food Eng 78(2):556–560

Krishnamurthy YL, Shashikala J (2009) Inhibition of aflatoxin B1 production of *Aspergillus flavus,* isolated from soybean seeds by certain natural plant products. Lett Appl Microbiol 43(5):469–474

Laffineur K, Avesani V, Cornu G, Charlier J, Janssens M, Wauters G, Delme M (2003) Bacteremia due to a novel Microbacterium species in a patient with Leukemia and description of *Microbacterium paraoxydans* sp. nov. J Clin Microbiol 41:2242–2246

Lai XW, Sun DL, Ruan CQ, Zhang H, Liu CL (2014) Rapid analysis of aflatoxins B1, B2, and ochratoxin A in rice samples using dispersive liquid-liquid microextraction combined with HPLC. J Sep Sci 37:92–98

Lee MJ, Park SY, Ha SD (2007) Reduction of coliforms in rice treated with sanitizers and disinfectants. Food Control 18:1093–1097

Li R, Wang X, Zhou T, Yangm DG, Wang Q, Zhou Y (2014) Occurrence of four mycotoxins in cereal and oil products in Yangtze Delta region of China and their food safety risks. Food Control 35:117–122

Liu LL, Zhai HQ, Wan JM (2005) Accumulation of g-aminobutyriac acidin giant-embryo rice grain in relation to glutamate decarboxylase activity and its gene expression during water soaking. Cereal Chem 82:191–196

Liu Z, Gao J, Yu J (2006) Aflatoxins in stored maize and rice grains in Liaoning Province, 292 China. J Stored Prod Res 42:468–479

Lu ZH, Zhang Y, Li LT, Curtis RB, Kong XL, Fulcher RG, Zhang G, Cao W (2010) Inhibition of microbial growth and enrichment of γ- aminobutyric acid during germination of brown rice by electrolyzed oxidizing water. J Food Protect 73:483–487

Macfarlane S, Macfarlane GT, Cummings JH (2006) Review article: prebiotics in the gastrointestinal tract. Aliment Pharmacol Ther 24:701–714

Majeed S, Iqbal M, Asi MR, Iqbal SZ (2013) Aflatoxins and ochratoxin A contamination in rice, corn and corn products from Punjab, Pakistan. J Cereal Sci 58:446–450

Marasas WFO (2001) Discovery and occurrence of the fumonisins: a historical perspective. Environ Health Perspect 109(Suppl. 2):239–243

Martínez I, Lattimer JM, Hubach KL, Case JA, Yang J, Weber CG, Louk JA, Rose DJ, Kyureghian G, Peterson DA, Haub MD (2013) Gut microbiome composition is linked to whole grain-induced immunological improvements. ISME J 7(2):269–280

Mattila-Sandholm T (1998) VTT on lactic acid bacteria. VTT Symp 156:1–10

Mazaheri M (2009) Determination of aflatoxins in imported rice to Iran. Food Chem Toxicol 47(8):2064–2066

Mew TW, Gonzales P (2002) A handbook of rice seed borne fungi. Science Publishers, Enfield

Missmer SA, Suarez L, Felkner M, Wang E, Merrill AH Jr, Rothman KJ, Hendricks KA (2006) Exposure to fumonisins and the occurrence of neural tube defects along the Texas–Mexico border. Environ Health Perspect 114:237–241

Miura D, Ito Y, Mizukuchi A, Kise M, Aoto H, Yagasaki K (2006) Hypocholesterolemic action of pre-germinated brown rice in hepatomabearing rats. Life Sci 79:259–264

Mohan BH, Malleshi NG, Koseki T (2010) Physico-chemical characteristics and non-starch polysaccharide contents of *indica* and *japonica* brown rice and their malts. LWT Food Sci Technol 42:784–791

Moongngarm A (2010) Influence of germination conditions on starch physicochemical properties, and microscopic structure of rice flour. Int Conf Biol Environ Chem 1:78–82

Moongngarm A, Saetung N (2010) Comparison of chemical compositions and bioactive compounds of germinated rough rice and brown rice. Food Chem 122:782–788

Murata K, Kamei T, Toriumi Y, Kobayashi Y, Iwata K, Fukumoto I, Yoshioka M (2007) Effect of processed rice with brown rice extracts on serum cholesterol level. Clin Exp Pharmacol Physiol 34((Suppl. 1)):87–89

Nanson NJ, Field ML (1984) Influence of temperature on the nutritive value of lactic acid fermented cornmeal. J Food Sci 49:958–959

Nemato H, Ikata K, Arimouchi H, Iwasaki T, Ohnishi Y, Kuwahara T, Kataoka K (2011) Effects of fermented brown rice on the intestinal environments in healthy adult. J Med Invest 58:235–245

Nguyen MT, Tozlovanu M, Tran TL, Pfohl-Leszkowicz A (2007) Occurrence of aflatoxin B1, citrinin and ochratoxin a in rice in five provinces of the central region of Vietnam. Food Chem 105:42–47

Nisa A, Zahra N, Yasha NB (2016) Comparative study of aflatoxins in brown rice samples of local and import quality. Int Food Res J 23:1

Nout MJR, Motarjemi Y (1997) Assessment of fermentation as a household technology for improving food safety: a joint FAO/WHO workshop. Food Control 8:221–226

Oberman H, Libudzisz Z (1996) Fermented milks. In: Woods JB (ed) Microbiology of fermented foods. Blackie Academic, London, pp 308–350

Oh SH, Choi WG (2000) Production of the quality germinated brown rice containing high c-aminobutyric acid by chitosan application. Kor Soc Biotechnol Bioeng J 15:615–620

Oh SH, Oh CH (2003) Brown rice extracts with enhanced levels of GABA stimulated immune cells. Food Sci Biotechnol 12:248–252

Ohisa N, Ohno T, Mori K (2003) Free amino acid and γ-aminobutyric acid contents of germinated rice. J Jpn Soc Food Sci Technol 50:316–318

Ohtsubo K, Suzuki K, Yasuiand Y, Kasumi K (2005) Bio-functional components in the processed pre-germinated brown rice by a twin-screw extruder. J Food Compos Anal 18:303–316

Ou SH (1985) Rice diseases, 2nd edn. Commonwealth Mycological Institute, Kew

Oyewole OB (1997) Lactic fermented foods in Africa and their benefits. Food Control 8:289–297

Palmiano EP, Juliano OJ (1972) Biochemical changes in the rice grain during germination. Plant Physiol 49:751–756

Park SK, Ko YD, Kwon SH, Shon MY, Lee SW (2003) Occurrence of off odor and distribution of thermophilic bacteria from rice and cooked rice stored at electric rice cooker. Kor J Food Preserv 10:70–74

Parrado J, Miramontes E, Jover M, Gutierrez JF, Collantesde Terán L, Bautista J (2006) Preparation of a rice bran enzymatic extract with potential use as functional food. Food Chem 98:742–748

Puri S, Dhillon B, Sodhi NS (2014) Effect of degree of milling on overall quality of rice- a review. Int J Adv Biotechnol Res 15(3):474–489

Rao RSP, Muralikrishna G (2004) Non-starch polysaccharide–phenolic acid complexes form native and germinated cereals and millet. Food Chem 84:527–531

Reddy BN, Raghavender CR (2007) Outbreaks of aflatoxicoses in India. Afr J Food Agric Nutr Dev 7(5):1–15

Reddy KRN, Reddy CS, Muralidharan K (2009) Detection of *Aspergillus* sp. and aflatoxin B1 in rice in India. Food Microbiol 26:27–31

Roohinejad S, Omidizadeh A, Mirhosseini H, Saari N, Mustafa S, MeorHussin AS, AbdManap MY (2012) Effect of pre-germination time on amino acid profile and gamma amino butyric acid (GABA) contents in different varieties of Malaysian brown rice. Int J Food Prop 14:1386–1399

Saleemullah AI, Kahlil IA, Shah H (2006) Aflatoxin contents of stored and artificially inoculated cereals and nuts. Food Chem 98:699–703

Saman P, Fucinos P, Vazquez JA, Pandiella SS (2011) Fermentability of rice and rice bran for the growth of human *Lactobacillus plantarum* NCIMB 8826. Food Technol Biotechnol 49(1):128–132

Sandhu DK, Vilkhu KS, Soni SK (1987) Production of α-amylase by *Saccharomycopsis fibuligera* (Syn. *Endomycopsis fibuligera*). J Ferment Technol 65:387–394

Sankaran R (1998) Fermented food of the Indian subcontinent. In: Wood JB (ed) Microbiology of fermented foods. Blackie Academic and Professional, London, pp 753–789

Sarrias JA, Valero M, Salmeron MC (2002) Enumeration, isolation, and characterization of Bacillus Cereus strains from Spanish raw rice. Food Microbiol 19:589–595

Scott PM (2012) Recent research on fumonisins: a review. Food Addit Contam Part A 29:242–248

Seki T, Nagase R, Torimitsu M, Yanagi M, Ito Y, Kise M, Ariga T (2005) Insoluble fiber is a major constituent responsible for lowering the post-prandialblood glucose concentration in the pre-germinated brown rice. Biol Pharm Bull 28:1539–1541

Seo J, Chang H, Ji W, Lee E, Choi M, Kim H, Kim J (1996) Aroma components of traditional Korean soy sauce and soy bean paste fermented with the same meju. J Microb Biotechnol 6:278–285

Shakya DD, Chung HS (1983) Detection of *Pseudomonas avenae* in rice seed. Seed Sci Technol 11:583–587

Shinmura H, Nakagawa K, Sasaki C, Aoto H, Onishi M (2007) Germinated brown rice. United State Patent, No. US 7935369 B2

Shortt C (1998) Living it up for dinner. Chem Ind 8:300–303

Soni SK, Sandhu DK (1990) Indian fermented foods: microbiological and biochemical aspects. Indian J Microbiol 30:135–157

Steinkraus KH (1998) Bio-enrichment: production of vitamins in fermented foods. In: Wood JB (ed) Microbiology of fermented foods. Blackie Academic and Professional, London, pp 603–619

Steinkraus KH, Ayres R, Olek A, Farr D (1993) Biochemistry of *Saccharomyces*. In: Steinkraus KH (ed) Handbook of indigenous fermented foods. Marcel Dekker, New York, pp 517–519

Stewart R, Getachew A (1962) Investigations of the nature of injera. Econ Bot 16:127–130

Takahashi H, Yasaki H, Nanayama U, Manabe M, Matsuura S (1984) Distribution of sterigmato-cystin and fungal mycelium in individual brown rice kernels naturally infected by Aspergillus versicolor (Mycotoxins). Cereal Chem (USA) 61(1):48–52

Tamime AY, O'Connor TP (1995) Kishk: a dried fermented milk/cereal mixture. Int Dairy J 5:109–128

Tanaka K, Sago Y, Zheng Y, Nakagawa H, Kushiro M (2007) Mycotoxins in rice. Int J Food Microbiol 119:59–66

Tango CN, Jun W, Hwan OD (2014) Modeling of *Bacillus cereus* growth in brown rice submitted to a combination of ultrasonication and slightly acidic electrolyzed water treatment. J Food Prot 12:2012–2218

Uematsu T, Yoshimura D, Nishiyama K, Ibaraki T, Fujii H (1976) Pathogenic bacterium causing seedling rot of rice. Ann Phytopathol Soc Jpn 42:464–471

Van E, Jonker MA (2004) Worldwide regulation on aflatoxins: the situation in 2002. J Toxicol Toxin Rev 23(2):273–293

Veluppillai S, Nithyanantharajah K, Vasantharuba S, Balakumar S, Arasaratnam V (2009) Biochemical changes associated with germinating rice grains and germination improvement. Rice Sci 16:240–242

Williams JH, Phillips TD, Jolly PE, Stiles JK, Jolly CM, Aggarwal D (2004) Human aflatoxicosis in developing countries: a review of toxicology, exposure, potential health consequences, and interventions. Am J Clin Nutr 80(5):1106–1022

Yang SC, Lin CH, Sung CT, Fang JY (2014) Antibacterial activities of bacteriocins: application in foods and pharmaceuticals. Front Microbiol 5(241):1–10

Zaied C, Abid S, Zorgui L, Bouaziz C, Chouchane S, Jomaa M, Bacha H (2009) Natural occurrence of ochratoxin A in Tunisian cereals. Food Control 20:218–222

Zareiforoush H, Minaei S, Alizadeh MR, Banakar A (2016) Qualitative classification of milled rice grains using computer vision and metaheuristic techniques. J Food Sci Technol 53(1):118–131

Part IV
Value Addition

Chapter 11
Germinated Brown Rice

Shabir Ahmad Mir, Manzoor Ahmad Shah, and Annamalai Manickavasagan

Introduction

Rice (*Oryza sativa* L.) is one of the important cereals produced in the world and the major staple food for almost half of the population worldwide (Lin et al. 2011). In recent years, much attention has been paid on the health benefits of brown rice. Brown rice is the dehusked paddy which retains the embryo and bran layers of the grain. Brown rice is rich in minerals, vitamins, and phytochemicals mostly present in the bran layer contributing to the biological activities (Mir et al. 2016). Principally owing to substantial bioactive components and functional properties, increasing attention has been given into biologically activated cereal grains in recent years, such as germinated brown rice (Wu et al. 2013). Germination is an effective and common process used to improve the nutritional quality of cereals consumed around the world. During germination, significant changes occur in biochemical, nutritional, and sensory characteristics. Germination induces the activation of hydrolytic enzymes which leads to the generation of biochemical activities.

Germinated brown rice has been suggested as an alternative approach to mitigate highly prevalent diseases providing nutrients and biologically active compounds. Germination process generally results in improved levels of vitamins, minerals, fibers, and phytochemicals such as ferulic acid, gamma-aminobutyric acid, γ-oryzanol,

S.A. Mir (✉)
Department of Food Technology, Islamic University of Science and Technology,
Awantipora, Jammu and Kashmir, India
e-mail: shabirahmir@gmail.com

M.A. Shah
Department of Food Science and Technology, Pondicherry University, Puducherry, India
e-mail: mashahft@gmail.com

A. Manickavasagan
School of Engineering, University of Guelph, Guelph, ON, Canada
e-mail: manicks33@hotmail.com

© Springer International Publishing AG 2017
A. Manickavasagan et al. (eds.), *Brown Rice*,
DOI 10.1007/978-3-319-59011-0_11

and antioxidant activity. Consumption of germinated brown rice is receiving increasing attention supported by scientific evidence on its beneficial health effects reducing the risk of diseases such as obesity, cardiovascular diseases, type-2 diabetes, and neurodegenerative diseases (Patil and Khan 2011; Chinma et al. 2015). Germinated brown rice has been identified as a natural and inexpensive substitute of conventional white rice to improve nutritive and health status of a large population.

Germination Conditions for Brown Rice

Brown rice is soaked in water for several hours and is incubated under certain temperature and humidity conditions until the germinated brown rice is harvested. The germination process affects significantly the physicochemical, texture, and nutritional properties of rice. This process modifies the internal structure of grains and nutrients, and thus the brown rice becomes more nutritious, easier to chew, and tastier (Zhang et al. 2014).

In brown rice, germination commonly refers to the process up to the point where the rice kernel has a radicle of 2–5 mm, and pre-germination is the stage with an expanded radicle exposed approximately 0.5–1 mm (Watanabe et al. 2004). Germinated brown rice is different from normal brown rice in that it has undergone the process of germination; more specifically, the rice embryo is sprouted under suitable environmental conditions. It is a biological process induced by activating the residual enzymes. Several researchers attempted to find out the optimum soaking condition and incubation condition to obtain germinated brown rice with high content of nutrients and bioactive compounds. Germination is usually initiated by water uptake of grains, which encounters the physicochemical conditions desirable for sprouting. Various factors influencing the germination process including intrinsic parameters, such as variety of brown rice and storage conditions, as well as external factors, such as temperature, humidity, presence of oxygen or air, light exposure, and pH for germination. Optimizing the germination conditions for brown rice is important for the production of germinated brown rice with consistent desired properties (Zhang et al. 2015).

Several studies have been carried out to optimize the germination conditions and maximize the beneficial attributes of germinated brown rice since the chemical composition of the grains change significantly during germination. Moisture is one of the essential factors for the germination of brown rice. The moisture is provided by two procedures either simple soaking of the seeds in water or temporary immersion followed by atmospheric germination. The atmospheric germination process leads to a higher degree of germination than the simple water soaking, thus resulting in the germinated brown rice having longer root and sprout (Lu et al. 2010). Generally, brown rice can be germinated by soaking it in warm water of 35–40 °C for about 10–12 h, draining water and keeping in moist condition for 20–24 h and, during soaking period, changing the water every 3–4 h to prevent fermentation (which usually produces undesirable odor) and to maintain consistent water temperature. The result yields a long sprout from the brown rice grain; at this stage, nutrient accumulation in the grain is maximum. Diverse schedules have been exploited by various researchers and industries for the production of germinated brown rice. Capanzana and Buckle (1997) reported that the optimum germination conditions for brown rice include soaking at 25 °C for 24 h, followed by atmospheric germination at 30 °C for 3 days. By contrast, Miyoshi et al. (1996) reported that the optimal germination period was only 18–24 h. The long grain brown rice soaked in distilled water at room temperature for 12 h and then germinated at 30 °C for 24 h with 65% relative humidity under biological oxygen demand was considered as optimum conditions for germination (Xu et al. 2012). The extended period of germination induce microbial growth and cause excessive growth of root and sprout, often rendering germinated brown rice unsuitable as food material (Cho and Lim 2016).

Gamma-aminobutyric acid is a nonprotein amino acid that acts as the main inhibitory neurotransmitter in the mammalian central nervous system. Some germination condition of brown rice was optimized mainly for gamma-aminobutyric acid production. The simple water immersion prior to germination may have the effect of raising the gamma-aminobutyric acid content. The soaking and germination duration, temperature, and pH in soaking water have been reported to enhance gamma-aminobutyric acid production (Thitinunsomboon et al. 2013). Saikusa et al. (1994) reported an optimal temperature of 40 °C for gamma-aminobutyric acid accumulation, while 30 °C was given by Ohtsubo et al. (2005). Thitinunsomboon et al. (2013) recommended combining repeated soaking in tap water at 35 °C for 3 h with incubation at 37 °C for 21 h. The optimal pH for gamma-aminobutyric acid accumulation ranged from 3.0 to 5.8 (Charoenthaikij et al. 2009). The germination of brown rice at 35 °C for 36 h with pH 7 in distilled soaking water optimized the highest content gamma-aminobutyric acid (Zhang et al. 2014). Germination carried out at 28 °C and 100% relative humidity for 12, 24, and 48 h; brown rice and soaked brown rice were also analyzed (Cornejo and Rosell 2015).

Roohinejad et al. (2011) observed changes in gamma-aminobutyric acid content in 18 rice varieties with 72-h immersion in distilled water (30 °C) and reported that all rice samples tested demonstrated increases in gamma-aminobutyric acid content, although the increase in gamma-aminobutyric acid content widely differed among rice varieties. Choi et al. (2004) examined the effects of pH and temperature for soaking on gamma-aminobutyric acid content in germinated brown rice. Comparing various temperatures (10, 25, 40, and 50 °C), it was found that rapid increases in gamma-aminobutyric acid content occurred when the temperature was 40 °C or higher. At 50 °C, however, the gamma-aminobutyric acid content decreased after 2 h of soaking. A pH near neutral (pH 6.0) yielded the highest gamma-aminobutyric acid content, but not statistically higher than those at pH 4.0 or 7.0. Hence, they reported that the optimum soaking temperature and pH for gamma-aminobutyric production were 40 °C and 6.0, respectively. The results indicate that gamma-aminobutyric acid production may be decreased when the kernel is exposed to a low temperature below 30 °C. Komatsuzaki et al. (2007) claimed that atmospheric germination after soaking yielded the higher gamma-aminobutyric levels than that from soaking alone. Compared with soaking process, atmospheric germination after soaking offers the higher degree of germination, but the kernels become more susceptible to microorganism adulteration. Lu et al. (2010) reported that electrolyzed oxidizing water resulted in shortened sprouts and radicles but increased gamma-aminobutyric content to nearly double the amount in the control germinated brown rice. In addition, the electrolyzed oxidizing water was highly effective in inhibiting microbial growth during germination.

Influence of Germination on Physicochemical Properties of Brown Rice

Modification of rice flour through the process of germination may be one of the means to prepare rice flour suitable for applying to food products. During the process of germination, a number of biochemical processes take place, leading to change in nutritional quality, chemical compositions, and activities of various enzymes. These changes would affect physicochemical properties of germinated rice flour, which directly influence the quality of rice and rice-based products.

The germinated product is achieved by soaking the whole kernel of brown rice in water until its embryo begins to bud. During the process of germination, the chemical compositions of the rice change drastically, because the biochemical activity produces essential compounds and energy, for the formation of the seedling. Hydrolytic enzymes are activated and these decompose large molecular substances, such as starch, non-starch polysaccharides, and proteins, to small molecular compounds. In the ungerminated rice sample, starch granules are characterized by a very smooth surface embedded in a continuous matrix. After 2–4 days of germination, visible changes occurred within granules. Starch granules lost their smooth surface, becoming rougher and slightly eroded (Moongngarm 2011; Cho and Lim 2016).

Morphological properties of flour from brown rice and germinated brown rice that has undergone different germination time were analyzed. For nongerminated brown rice, the flour was characterized by having a continuous structure with the starch granules surrounded by well-defined protein bodies and embedded in a cementing matrix. But this continuous structure was destroyed after germination, and the destruction was increased as germination proceeded. After 2 days of germination, the protein bodies started to disappear due to activated proteolytic enzyme action. Starch granules lost their smooth surface, becoming rougher and slightly eroded. After 4 days of germination, the continuous structure was seriously damaged, the protein bodies were highly broken, and a lot of nonuniform-sized and irregular fragments were seen. After 3 days of germination, small pits and holes were discovered on the surface of some starch granules which may be due to the partial starch hydrolysis by internal enzymes during germination (Wu et al. 2013).

The budding is initiated by the process of germination which leads to the activation of hydrolytic enzymes and breakdown of high molecular weight polymers; therefore, oligosaccharides, amino acids, and other biofunctional substances are formed. Germination significantly influences the physicochemical properties of brown rice which vary according to variety and germination conditions. Literature has reported that variations in germination process are desired for the particular properties (Patil and Khan 2011; Cho and Lim 2016). Wu et al. (2013) investigated the significant changes in physicochemical properties of brown rice flour during the process of germination. Germination led to a dramatic decrease in the contents of total starch, amylose, and amylopectin, viscosity profile, as well as the storage and loss modulus of brown rice flour.

Significant physicochemical property changes of brown rice flour were observed under different germination conditions. Increasing steeping time (from 24 to 72 h) or decreasing the pH of steeping water (in the range of pH 3–7) caused an increase in reducing sugar, free gamma-aminobutyric acid, and α-amylase activity but a decrease in total starch content and viscosities. The principal component analysis results also confirmed that pH of steeping water and the steeping time had significant effects on the physicochemical properties of germinated brown rice flour. The modification of these quality characteristics by germination may offer alternative rice flour that is more nutritious and less viscous for the exploitation in food industry (Charoenthaikij et al. 2009).

Various studies have reported the germination induced important changes in starch structures and chemical compositions, including starch yields and amylose contents (Xu et al. 2012; Pinkaew et al. 2016). Both the starch degradations and the reductions of the amylose contents affected the pasting properties of germinated brown rice starches. Germination also affected amylopectin branch chain length distributions, which led to the changes in the pasting properties of germinated brown rice starches (Pinkaew et al. 2016).

After a regular long-grain brown rice was soaked in distilled water at room temperature for 12 h and then germinated at 30 °C for 24 h with 65% relative humidity under biological oxygen demand, the germinated brown rice starch had a much lower number average molecular weight than that of ungerminated brown rice starch. Thus, the germinated brown rice starch showed a higher polydispersity than that of ungerminated brown rice starch, which indicated that starch molecules became more nonuniform (Xu et al. 2012). In addition, after the three Chinese brown rice cultivars were soaked in deionized water at 25 °C and germinated in the dark for 5 days, the average molecular weight values differed significantly among three rice cultivars. The average molecular weight values of isolated ungerminated brown rice starches from medium-grained waxy, medium-grained japonica, and long-grained indica were 2.83, 2.39, and 1.80×10^8 g/mol, respectively, while the germinated brown rice starches at the first to fifth days of germination from those three rice cultivars were in the range of $2.84–2.86 \times 10^8$ g/mol, $2.41–2.48\times 10^8$ g/mol, and $1.78–1.87 \times 10^8$ g/mol, respectively (Wu et al. 2013). The crystallinity of brown rice starch was also affected by germination. The differences in major reflections and relative crystallinity between ungerminated brown rice and germinated Malaysian brown indica rice starches have been reported (Musa et al. 2011). Germination has been reported to affect the thermal properties of brown rice starch. The decreases in the gelatinization temperatures and enthalpy of germinated long-grain brown rice starch have been observed (Xu et al. 2012; Wu et al. 2013).

Pasting properties and in vitro starch digestibility of germinated brown rice were investigated by Chung et al. (2012). Steeping in water (30 °C, 24 h) raised the moisture content and germination percentage of brown rice. Pasting viscosity was substantially decreased but gelatinization temperatures and enthalpy were decreased only marginally by germination (30 °C, 48 h). The digestibility of starch in brown rice was increased by germination. The contents of rapidly digestible starch, slowly digestible starch, and resistant starch in the cooked brown rice were 47.3%, 40.8%, and 11.9%, respectively, but changed to 57.7%, 39.1%, and 3.2%, respectively, upon germination.

Significant changes on hydration and pasting properties of brown rice flour were found during germination. The starch degradation by enzyme activity could be evidenced with the decrease in viscosity and water-binding capacity (Cornejo and Rosell 2015). Foaming capacity and stability of rice flours increased after germination. Germination resulted to changes in pasting and thermal characteristics of rice flours. Germinated rice flours had better physicochemical properties with reduced phytic acid and starch contents, which can be utilized as functional ingredients in the preparation of rice-based products (Chinma et al. 2015).

Influence of Germination on Cooking and Textural Properties of Brown Rice

Although normal brown rice has high nutritional value, its popularity is low due to poor cooking properties. However, germinated brown rice is easily cooked, and the texture is softer than that of brown rice. Therefore, germinated brown rice could become a popular healthy food. Despite its nutritional value and beneficial physiological effects, brown rice is not widely consumed because it has undesirable cooking properties, low organoleptic quality, and harsh texture (Wu et al. 2013). Numerous studies have demonstrated that germination improves texture and acceptability of brown rice and also enhances nutrient and phytochemical bioavailability (Tian et al. 2004). During germination, changes occur in biochemical, nutritional, and sensory characteristics resulting in the degradation of carbohydrates and proteins which significantly affects the cooking and texture properties of rice.

Germinated brown rice offers an excellent appearance, improved shelf life, and handling ease. Unlike white rice, germinated brown rice provides more sweetness. Jiamyangyuen and Ooraikul (2008) reported that effects of germination on cooking and textural properties of cooked rice were more pronounced when rice was soaked and germinated for a longer period. They reported that germinated rice requires less cooking time and water absorbed by kernels during germination resulted in size expansion. From the sensory evaluation, they showed that the cooked germinated rice is sweeter, softer, swelled, and cohesive than cooked regular brown rice. Germinated brown rice has a softer texture than normal brown rice due to the reaction between phytic acid and minerals during the birth of the sprout which indicates that it can be easily cooked and is easier to digest.

The hardness value of cooked germinated rice significantly decreased with increasing germination time. The lower hardness values were correlated with the decrease of starch content. The starch content of germinated rice decreased with increasing germination time. These results indicate the strong effect of the germination time on the cooked rice texture. Germination process modifies the microstructure of brown rice. The starch granules were more loosely packed as the germination time was longer. This modification of microstructure led to lower hardness of cooked dried germinated rice and higher number of fissured kernels after drying. From the sensory assessment, the texture of germinated rice was better than that of rice without germination (Chungcharoen et al. 2015).

Germination is an efficient method that provides a soft brown rice texture after cooking. During germination, hydrolytic enzymes are produced and activated (Banchuen et al. 2009). These enzymes destroy hydrogen bonds at random points in the high-molecular-weight polymers particularly starch, to lower molecular weight polymers, i.e., glucose and maltose. The decay of starch enables the change of textural properties. A good quality of finished germinated rice depends on germination time and drying temperature. Germination time involves directly the decomposition of starch. Prolonging germination time provides more decomposition of starch and higher level of bioactive components. In addition, the decay of starch during germination may affect the product quality during drying because of the weaker strength of kernels which will produce a large number of fissured kernels and influence the texture (Chungcharoen et al. 2015).

The metabolic changes of dry seeds require sufficient energy, which is mostly from the degradation of residual starch that composes 90% of brown rice. α-Amylase, which randomly breaks the α-1,4 bonds in starch, is considered the principal enzyme for starch degradation in brown rice. However, in the early stage of germination, glucose produced from sucrose and maltose during starch degradation by β-amylase is the main source of energy because the α-amylase becomes active in the late stage (Palmiano and Juliano 1972). Moongngarm and Saetung (2010) reported that α-amylase was produced during the protein synthesis in the late stage (seedling growth) rather than preexisting in seeds, although the dry seeds contain a small amount of the enzyme. On the other hand, β-amylase, which sequentially breaks down the starch to maltose units, is largely present in brown rice endosperm and becomes activated with the start of imbibitions. In addition to these two main enzymes, a debranching enzyme that digests the α-1,6 bond of starch also becomes active during germination. As a result of amylolytic actions, germination causes a decrease in the amount of amylose in brown rice, which is one of the key factors for determining the quality and eating texture of cooked rice. Jiamyangyuen and Ooraikul (2008) observed the influence of germination on the eating and sensory properties of brown rice and reported that cohesiveness and softness in cooked germinated brown rice increased by germination, whereas the stickiness and blandness decreased. They also reported that germinated brown rice was easier to cook and required less cooking time. These phenomena are likely due to the decreased amylose content by the action of amylolytic enzymes during germination and the cracks on rice kernels formed during germination and drying.

Influence of Germination on Nutritional Properties of Brown Rice

The metabolic activity of brown rice increases as soon as it is hydrated during soaking. Complex biochemical changes occur during germination in various parts of the seed. Because no external nutrients are added during the germination process, only water and oxygen are consumed by the germinating seed, desirable nutritional changes mainly stem from the decomposition of complex compounds into more simple forms and their transformation into essential constituents as well as the breakdown of nutritionally undesirable constituents (Chavan et al. 1989; Cho and Lim 2016). Germination caused significant changes in chemical compositions, bioactive compounds, and amino acids as compared to nongeminated brown rice. The germination process activated the enzymes which could improve some chemical components and remarkably increase the reducing sugar and total sugars (Moongngarm and Saetung 2010).

Germination process is an inexpensive and effective technology for improving cereal quality. Germination process leads to structural modification and synthesis of new compounds with high bioactivity and can increase the nutrition value and stability of grains. Protein, magnesium, phosphorus, potassium, and antioxidant properties of rice increased after germination while phytic acid and total starch contents decreased (Chinma et al. 2015). Roohinejad et al. (2009) reported that gamma-aminobutyric acid content after germination in Malaysian brown rice seeds ranged between 0.01 and 0.1 mg/g. The quantity of glutamic acid and protein content varied between 10.1–15.2 mg/g and 6.99–10.17%, respectively. A significant positive correlation exists between the concentration levels of protein and glutamic acid. The developed germinated brown rice showed the rich source of nutritional and functional components as compared to ungerminated brown rice.

Kayahara et al. (2001) observed that not only existing nutrients are increased but new components are also released from the inner change due to germination. The nutrients which have increased significantly include gamma-aminobutyric acid, lysine, vitamin E, dietary fiber, niacin, magnesium, vitamin B_1, and vitamin B_6. The other nutrients that increased in germinated brown rice were inositols, ferulic acid, tocotrienols, potassium, zinc, and γ-oryzanol. In particular, the amount of gamma-aminobutyric acid in germinated brown rice was noticed to be ten times more as compared to milled white rice and two times more than that of brown rice. Trachoo et al. (2006) found that germination of rice grains increased many nutrients such as vitamin B, reducing sugar, and total protein contents of germinated brown rice were higher than those of brown rice and white rice. Choi et al. (2006) reported that upon 24-h germination, the increased amounts of these nutrients relative to those in the nongerminated brown rice were 3.4 times for fructose, 2.75 times for reducing sugars, and 7.97 times for gamma-aminobutyric acid. The increase of nutrients was observed higher in case of germinated brown rice made by 24-h germination as compared to that made by 48-h germination. Among nutrients, a significant increase was observed in gamma-aminobutyric acid contents.

Changes in active components during brown rice germination have also been reported by several researchers. Pang and Zhang (1994) reported that the increase in the contents of ascorbic acid, glutathione, and riboflavin is linear with germination time. Ohtsubo et al. (2005) found that germinated brown rice contains more total dietary fiber (145%), soluble dietary fiber (120%), and insoluble dietary fiber (150%) than brown rice after 72 h of germination. Cáceres et al. (2017) reported that germination increases the γ- oryzanol, total phenolic content, and antioxidant activity of brown rice.

Germinated brown rice is a good source of the phenolics associated with antioxidant effects. Germination significantly increased by 63.2% and 23.6% the total phenolic and flavonoid contents, respectively. The percentage contribution of bound phenolics to total was 42.3% before and decreased slightly to 37.6% after germination. The percentage contribution of bound flavonoids to total, 51.1%, was the same before and after germination. The change in the amounts of free and bound forms indicated that transformations could occur during the germination process. The levels of ferulic, coumaric, syringic, and caffeic acid significantly increased. The ratio of bound ferric reducing antioxidant power to total was basically constant, while germination increased the ratio of bound oxygen radical absorbance capacity to total (Ti et al. 2014).

Product Development from Germinated Brown Rice

There is a huge potential for the development of products high in nutritional and functional components. In order to increase the nutrient density of rice-based products, it is important to use germinated brown rice as raw material. Germinated brown rice is a rich source of bioactive components and minerals. Value-added products have been developed from germinated brown rice either alone or in combination with other ingredients resulted in better-quality products. Germinated brown rice is used to make various products in food industry such as rice balls, bread, doughnuts, cookies, rice burger, etc. The utilization of germinated brown rice as an ingredient in bakery products such as cookies, breads, and cakes is one of the approaches to improve their nutritive and bioactive qualities (Chung et al. 2014; Chalermchaiwat et al. 2015).

Germinated brown rice flour and germinated glutinous brown rice flour have been used as functional food ingredients in breadmaking, due to its high content of bioactive compounds (Charoenthaikij et al. 2012). Germinated rice flour showed appropriate functionality for being used as raw ingredient in gluten-free breadmaking. The germination time of the rice has a significant effect on flour properties and the resulting bread quality. Specifically, flours obtained after 24 h of germination led to an improvement in bread texture, which might be ascribed to the increase of amylase activities as well as starch degradation, which agrees with hydration and pasting results. The bread color also improved as a result of nonenzymatic browning reaction. However, excessive germination deteriorated the product as a result of extensive amylolysis (Patil and Khan 2011; Cornejo and Rosell 2015).

The effect of germination conditions on the nutritional benefits of germinated brown rice flour bread has been determined at different germination conditions. When comparing different germination times (0, 12, 24 and 48 h), germination for 48 h provides germinated brown rice flour bread with nutritionally superior quality on the basis of its higher content of protein, lipids, and bioactive compounds, increased antioxidant activity, and reduced phytic acid content and glycaemic index, although a slight decrease in in vitro protein digestibility was detected. Overall, germination seems to be a natural and sustainable way to improving the nutritional quality of gluten-free rice breads (Cornejo et al. 2015).

Morita et al. (2007) used various additives for making germinated brown rice breads and evaluated suitable combinations of germinated brown rice for breadmaking to provide germinated brown rice bread with high nutritional properties. The 30% of the wheat flour substitution with germinated brown rice showed the improved bread quality with sensory acceptability.

Ohtsubo et al. (2005) prepared the puffed germinated brown rice which contained the notable amount of oryzanol, inositol, total ferulic acid, and total dietary fibers. Further, they observed that the products prepared by the co-extrusion of germinated brown rice contained more free amino acids, such as gamma-aminobutyric acid, alanine, aspartic acid, and glutamic acid, than white and brown rice. The puffed germinated brown rice was exploited for the development of bread and the wheat bread prepared with 30% puffed germinated brown rice contained showed gamma-aminobutyric acid and free sugars, such as maltose, than ordinary wheat bread. The developed bread was shown to be sweeter and equivalently palatable as a result of the organoleptic test.

Substituting wheat flour with germinated brown rice flour for sugar-snap cookies resulted in increasing the residual moisture content and decreasing the hardness. Moreover, using heat-moisture-treated germinated brown rice flour substantially made the cookies darker and improved the storage stability of the cookies by retarding moisture loss and hardening during storage. The heat-moisture-treated germinated brown rice is an effective alternative of wheat flour for cookie preparation providing improved quality and nutritional value (Chung et al. 2014).

The incorporation of germinated brown rice flour to wheat flour dramatically changes the cooking and textural quality of cooked noodles, resulting in increased cooking loss and softness. However, heat-moisture treatment of germinated brown rice apparently modifies the pasting properties of the mixtures of wheat and germinated brown rice flours, and the noodles made from the flour containing the treated germinated brown rice showed substantially improved cooking and textural qualities comparable to the control wheat noodle. Consequently, heat-moisture treatment allowed the germinated brown rice to be effectively used as a replacement of wheat flour for noodle making with additional health-promoting and nutritive values. However, heat-moisture treatment often induces darkening and flavor change due to the Maillard reaction, which should be carefully considered for the use of treated germinated brown rice (Chung et al. 2012).

Extrusion has become a very important process in the manufacture of snack foods due to its ease of operation and ability to produce a variety of desirable sizes, shapes, and textures. The extruded snacks were developed from germinated brown rice flour. Changes in feed moisture and screw speed significantly affected physical properties of brown rice-based extrudates. Decreasing feed moisture increased destruction of free gamma-aminobutyric acid in extrudates. DPPH radical-scavenging activities, ferric reducing antioxidant power, and total phenolic content of extrudates were significantly lower than those of the non-extruded germinated brown rice flour (Chalermchaiwat et al. 2015). Extruded rice product was developed from germinated brown rice by Gujral et al. (2012). The results have shown that the product contains notable source of phenolic compounds. Furthermore, the extrudates from germinated brown rice were used to make an instant pudding, which upon evaluation scored higher as compared to the pudding from control brown rice.

Health Benefits of Germinated Brown Rice

Particular attention has been given to functional foods that play an important role in disease prevention or in slowing the progress of chronic diseases through the provision of essential nutrients by regular consumption (Viuda-Martos et al. 2010). Recently, the health benefits of germinated brown rice have attracted an increasing number of researchers, and the relationship between a germinated brown rice diet and the prevention of certain diseases has been identified. Germinated brown rice has gained significant attention during the last decade as a tool of enhancing eating quality and potential health-promoting functions. Consumption of germinated brown rice is receiving increasing attention supported by scientific evidence on its beneficial health effects reducing the risk of diseases such as obesity, cardiovascular diseases, type-2 diabetes, neurodegenerative diseases, and osteoporosis, and germinated brown rice has been identified as a natural and inexpensive substitute of conventional white rice to improve nutritive and health status of a large population that currently eat rice as staple food. Germinated brown rice has become popular among health-conscious consumers, such as diabetics, hypercholesterolemics, and hypertensive patients, because of its bioactive compounds (Ito et al. 2005; Wu et al. 2013; Wunjuntuk et al. 2016).

Hyperlipidemia comprised of a heterogeneous group of disorders whose characteristic expression is an elevation in the plasma concentration of serum cholesterol. Oh et al. (2003) observed that germinated brown rice water extract administration improved the serum lipid profiles of mice suffering from obesity. Miura et al. (2006) found that both germinated brown rice suppressed hypercholesterolemia induced by hepatoma growth by means of upregulating cholesterol catabolism. Roohinejad et al. (2010) observed that changes in blood cholesterol could be modulated using germinated brown rice rather than nongerminated brown rice, suggesting that germinated brown rice is a more efficient functional diet food that affects high blood cholesterol. The excellent hypolipidemic action of germinated brown rice may be due to the high concentration of different biologically active components, such as gamma-aminobutyric acid, dietary fibers, γ-oryzanol, or other antioxidants. Roohinejad et al. (2009) observed that the antihypercholesterolemic effect of γ-oryzanol is partially due to its sterol moiety, which is partly split off from the ferulic acid part in the small intestine by cholesterol esterase.

Germinated brown rice supplementation has the greatest impact on increasing antioxidant enzyme activity and vitamin E level and on reducing lipid peroxidation in hypercholesterolemia rabbit, thereby preventing the formation of atherosclerotic plaques. Moreover, germinated brown rice diet can also reduce the level of hepatic enzymes (Esa et al. 2013). Germinated brown rice is a more efficient functional diet food in reducing the risk of atherogenesis and coronary artery disease through its protection against low-density lipoprotein oxidation. The results of numerous epidemiological and clinical studies have been extremely convincing in showing that a moderate or higher intake of dietary fiber can effectively lower cardiovascular disease risk through its action on low-density lipoprotein cholesterol (Estruch et al. 2009; Kendall et al. 2010). Gamma-aminobutyric acid, which increases significantly after germination, was reported to suppress an increase in serum low-density lipoprotein cholesterol levels caused by chronic ethanol administration in mice (Oh et al. 2003).

Imam et al. (2014) evaluated the effects of germinated brown rice in the dietary management of cardiovascular disease. Diet-induced hypercholesterolemic rats were fed with germinated brown rice, in comparison with normal, high-fat-diet rats. Germinated brown rice reduced weight gain and improved lipid parameters and low-density lipoprotein, partly through transcriptional regulation of hepatic lipoprotein lipase, peroxisome proliferator-activated receptor gamma, adiponectin, and ATP-binding cassette. The results suggest germinated brown rice could ameliorate cardiovascular disease risk by modulating lipid metabolism and oxidative stress.

High blood pressure or hypertension, one manifestation of cardiovascular disease, continues to be a major cause of morbidity and death. Germinated brown rice has been demonstrated to have a role in lowering blood pressure. Ebizuka et al. (2009) examined the effects of pregerminated brown rice on hypertension and the biochemistry of blood in spontaneously hypertensive rats fed a diet consisting of 40% pregerminated brown rice. The results of their observation showed that pregerminated brown rice had a significant and strong antihypertensive effect in spontaneously hypertensive rats. The antihypertensive effect of germinated brown rice may be due to the complex actions of various components abundantly found in germinated rice, including gamma-aminobutyric acid, dietary fiber, γ-oryzanol, and ferulic acid. Ferulic acid, whose content increases greatly after germination, had a beneficial effect in lowering the blood pressure in streptozotocin-induced diabetic rats and spontaneously hypertensive rats (Ardiansyah et al. 2008).

Parboiled germinated brown rice is produced by steaming germinated paddy of Thai rice variety. Wunjuntuk et al. (2016) determined the anti-fibrotic and anti-inflammatory effects of parboiled germinated brown rice in rats induced by carbon tetrachloride. Rats in parboiled germinated brown rice + carbon tetrachloride group showed the hepatoprotective effects by significantly decreasing the markers of liver inflammation, fibrosis, and hydroxyproline content, as well as the score of necro-inflammation and fibrosis. In addition, parboiled germinated brown rice suppressed the expression level of α-smooth muscle actin significantly as compared with carbon tetrachloride group. These results suggested that regular consumption of parboiled germinated brown rice could reduce liver inflammation and fibrosis.

Cancer remains one of the major causes of mortality worldwide. There is convincing evidence suggesting that germinated brown rice predominantly contains biologically active substances that produce the chemopreventive and antitumor properties of rice. Germinated brown rice, which maintains the majority of bioactive constituents in rice bran and germ and even contains more biofunctional components than nongerminated brown rice, may be a good choice for a staple diet with respect to cancer prevention. Effects of germinated brown rice on the proliferation and apoptosis of cancer cells were reported in literature (Tian et al. 2004; Cho and Lim 2016). Oh and Oh (2004) found that germinated brown rice extracts with enhanced levels of gamma-aminobutyric acid inhibited leukemia cell proliferation and stimulated cancer cell apoptosis. Gamma-aminobutyric acid, which is abundant in germinated brown rice, was also reported to suppress tumor growth in small airway epithelia. Schuller et al. (2008) suggest that a germinated brown rice diet with high levels of gamma-aminobutyric acid might prevent pulmonary adenocarcinoma in smokers. Gamma-aminobutyric-acid-enriched fractions extracted from germinated brown rice also displayed an inhibitory action on the proliferation of some cancer cells and a stimulatory action on immune responses (Oh and Oh 2004).

Furthermore, gamma-aminobutyric acid has other physiological functions, such as blood vessel strengthening, modulation of insulin secretion, prevention of blood cholesterol amplification, alleviation of emotional unrest, improvement from stroke, and enhancement of liver and kidney functions (Oh et al. 2003).

The incidence of diabetes is increasing worldwide, both in developed and developing countries. The beneficial effects of germinated brown rice on fasting blood glucose levels, postprandial blood glucose levels, and insulin response in humans have been reported in literature. It was reported that both nondiabetic patients and hyperglycemic patients with uncontrolled diabetes who ate diets supplemented with germinated brown rice had lower postprandial blood glucose levels than those who ate white rice supplemented diets (Ito et al. 2005). Regarding the effects of the long-term consumption of germinated brown rice, study evaluated the blood glucose and lipid concentrations in impaired fasting glucose or type-2 diabetes patients who subsisted for 6 weeks on white rice or germinated brown rice diet separated by a 2-week washout interval in a crossover design. The results suggested that the consumption of germinated brown rice as a staple food in patients with type-2 diabetes was useful in improving blood glucose and lipid levels (Hsu et al. 2008). Several researchers have also evaluated the effects of a germinated brown rice diet on diabetic complications using streptozotocin-induced diabetic rats. Hagiwara et al. (2004) observed that germinated brown rice ameliorated the blood glucose, type-1 plasminogen-activator, and lipid peroxide concentrations in diabetic rats in comparison with diabetic rats fed with white rice, suggesting that germinated rice intake may ameliorate diabetic vascular complications such as ischemic heart disease. Usuki et al. (2007) reported that compared with the brown rice diet, except for its blood glucose-lowering effect, the germinated brown rice diet also mitigated diabetic neuropathy and was much more effective in easing peripheral nerve dysfunction. The results suggested that germinated brown rice is a good food ingredient in preventing diabetes and its associated complications.

The reduction in blood glucose levels and the incidence of diabetic vascular complications in diabetic patients fed a diet rich in germinated rice might result from the substantially higher dietary fiber content of germinated rice bran (Seki et al. 2005). Gamma-aminobutyric acid produced during germination of brown rice or the vitamins and minerals in the bran and germ layer of germinated brown rice may also be associated with the effect of lowering blood glucose or preventing diabetic complications (Usuki et al. 2007).

Conclusion

Germinated brown rice as one of healthy foods becomes more and more popular in the world. Several combinations have been carried out to optimize the germination conditions and maximize the beneficial attributes of germinated brown rice. The germination process induces significant changes in composition which have a considerable effect on properties of rice. Germination process generally results in improved levels of phytochemicals such as gamma-aminobutyric acid, γ-oryzanol, ferulic acid, antioxidant activity, minerals, and vitamins. The modification of quality characteristics by germination may offer alternative nutritious rice to the food industry. Consumption of germinated brown rice is receiving increasing interest among consumers and supported by scientific evidence on its beneficial health effects by reducing the risk of diseases such as obesity, cardiovascular diseases, type-2 diabetes, and neurodegenerative diseases.

References

Ardiansyah, Ohsaki Y, Shirakawa H, Koseki T, Komai M (2008) Novel effects of a single administration of ferulic acid on the regulation of blood pressure and the hepatic lipid metabolic profile in stroke-prone spontaneously hypertensive rats. J Agric Food Chem 56(8):2825–2830

Banchuen J, Thammarutwasik P, Ooraikul B, Wuttijumnong P, Sirivongpaisal P (2009) Effect of germinating processes on bioactive component of Sangyod Muang Phatthalung rice. Thai J Agric Sci 42(4):191–199

Cáceres PJ, Peñas E, Martinez-Villaluenga C, Amigo L, Frias J (2017) Enhancement of biologically active compounds in germinated brown rice and the effect of sun-drying. J Cereal Sci 73:1–9

Capanzana MV, Buckle KA (1997) Optimisation of germination conditions by response surface methodology of a high amylose rice (*Oryza sativa*) cultivar. LWT Food Sci Technol 30(2):155–163

Chalermchaiwat P, Jangchud K, Jangchud A, Charunuch C, Prinyawiwatkul W (2015) Antioxidant activity, free gamma-aminobutyric acid content, selected physical properties and consumer acceptance of germinated brown rice extrudates as affected by extrusion process. LWT Food Sci Technol 64(1):490–496

Charoenthaikij P, Jangchud K, Jangchud A, Piyachomkwan K, Tungtrakul P, Prinyawiwatkul W (2009) Germination conditions affect physicochemical properties of germinated brown rice flour. J Food Sci 74(9):C658–C665

Charoenthaikij P, Jangchud K, Jangchud A, Prinyawiwatkul W, No HK (2012) Composite wheat–germinated brown rice flours: selected physicochemical properties and bread application. Int J Food Sci Technol 47(1):75–82

Chavan JK, Kadam SS, Beuchat LR (1989) Nutritional improvement of cereals by sprouting. Crit Rev Food Sci Nutr 28(5):401–437

Chinma CE, Anuonye JC, Simon OC, Ohiare RO, Danbaba N (2015) Effect of germination on the physicochemical and antioxidant characteristics of rice flour from three rice varieties from Nigeria. Food Chem 185:454–458

Cho DH, Lim ST (2016) Germinated brown rice and its bio-functional compounds. Food Chem 196:259–271

Choi HD, Park YK, Kim YS, Chung CH, Park YD (2004) Effect of pretreatment conditions on γ-aminobutyric acid content of brown rice and germinated brown rice. Korean J Food Sci Technol 36(5):761–764

Choi ID, Kim DS, Son JR, Yang CI, Chun JY, Kim KJ (2006) Physico-chemical properties of giant embryo brown rice (Keunnunbyeo). J Appl Biol Chem 49(3):95–100

Chung HJ, Cho DW, Park JD, Kweon DK, Lim ST (2012) In vitro starch digestibility and pasting properties of germinated brown rice after hydrothermal treatments. J Cereal Sci 56(2):451–456

Chung HJ, Cho A, Lim ST (2014) Utilization of germinated and heat-moisture treated brown rices in sugar-snap cookies. LWT Food Sci Technol 57(1):260–266

Chungcharoen T, Prachayawarakorn S, Tungtrakul P, Soponronnarit S (2015) Effects of germination time and drying temperature on drying characteristics and quality of germinated paddy. Food Bioprod Process 94:707–716

Cornejo F, Rosell CM (2015) Influence of germination time of brown rice in relation to flour and gluten free bread quality. J Food Sci Technol 52(10):6591–6598

Cornejo F, Caceres PJ, Martínez-Villaluenga C, Rosell CM, Frias J (2015) Effects of germination on the nutritive value and bioactive compounds of brown rice breads. Food Chem 173:298–304

Ebizuka H, Ihara M, Arita M (2009) Antihypertensive effect of pre-germinated brown rice in spontaneously hypertensive rats. Food Sci Technol Res 15(6):625–630

Esa NM, Kadir KKA, Amom Z, Azlan A (2013) Antioxidant activity of white rice, brown rice and germinated brown rice (in vivo and in vitro) and the effects on lipid peroxidation and liver enzymes in hyperlipidaemic rabbits. Food Chem 141(2):1306–1312

Estruch R, Martinez-Gonzalez MA, Corella D, Basora-Gallisá J, Ruiz-Gutierrez V, Covas MI, … Pena MA (2009) Effects of dietary fibre intake on risk factors for cardiovascular disease in subjects at high risk. J Epidemiol Community Health 63(7):582–588

Gujral HS, Sharma P, Kumar A, Singh B (2012) Total phenolic content and antioxidant activity of extruded brown rice. Int J Food Prop 15(2):301–311

Hagiwara H, Taiichiro SEKI, Ariga T (2004) The effect of pre-germinated brown rice intake on blood glucose and PAI-1 levels in streptozotocin-induced diabetic rats. Biosci Biotechnol Biochem 68(2):444–447

Hsu TF, Kise M, Wang MF, Ito Y, Yang MD, Aoto H, … Yamamoto S (2008) Effects of pre-germinated brown rice on blood glucose and lipid levels in free-living patients with impaired fasting glucose or type 2 diabetes. J Nutr Sci Vitaminol 54:163–168

Imam MU, Ishaka A, Ooi DJ, Zamri NDM, Sarega N, Ismail M, Esa NM (2014) Germinated brown rice regulates hepatic cholesterol metabolism and cardiovascular disease risk in hypercholesterolaemic rats. J Funct Foods 8:193–203

Ito Y, Mizukuchi A, Kise M, Aoto H, Yamamoto S, Yoshihara R, Yokoyama J (2005) Postprandial blood glucose and insulin responses to pre-germinated brown rice in healthy subjects. J Med Investig 52(3, 4):159–164

Jiamyangyuen S, Ooraikul B (2008) The physico-chemical, eating and sensorial properties of germinated brown rice. Journal of Food Agriculture and Environment 6(2):119–124

Kayahara H, Tsukahara K, Tatai T (2001) Flavour, health and nutritional quality of pre-germinated brown rice. In: Spanier AM (ed) Food flavours and chemistry: advances of the new millennium. Royal Society of Chemistry, Cambridge, pp 546–551

Kendall CW, Esfahani A, Jenkins DJ (2010) The link between dietary fibre and human health. Food Hydrocoll 24(1):42–48

Komatsuzaki N, Tsukahara K, Toyoshima H, Suzuki T, Shimizu N, Kimura T (2007) Effect of soaking and gaseous treatment on GABA content in germinated brown rice. J Food Eng 78(2):556–560

Lin JH, Singh H, Chang YT, Chang YO (2011) Factor analysis of the functional properties of rice flours from mutant genotypes. Food Chem 126(3):1108–1114

Lu ZH, Zhang Y, Li LT, Curtis RB, Kong XL, Fulcher RG, … Cao W (2010) Inhibition of microbial growth and enrichment of γ-aminobutyric acid during germination of brown rice by electrolyzed oxidizing water. J Food Prot 73(3):483–487

Mir SA, Bosco SJD, Shah MA, Mir MM, Sunooj KV (2016) Variety difference in quality characteristics, antioxidant properties and mineral composition of brown rice. J Food Meas Charact 10(1):177–184

Miura D, Ito Y, Mizukuchi A, Kise M, Aoto H, Yagasaki K (2006) Hypocholesterolemic action of pre-germinated brown rice in hepatoma-bearing rats. Life Sci 79(3):259–264

Miyoshi K, Sato T, Takahashi N (1996) Differences in the effects of dehusking during formation of seeds on the germination of seeds of indica and japonica rice (*Oryza sativa* L.) Ann Bot 77(6):599–604

Moongngarm A (2011) Influence of germination conditions on starch, physicochemical properties, and microscopic structure of rice flour. Int Conf Biol Environ Chem 1:78–82

Moongngarm A, Saetung N (2010) Comparison of chemical compositions and bioactive compounds of germinated rough rice and brown rice. Food Chem 122(3):782–788

Morita N, Maeda T, Watanabe M, Yano S (2007) Pre-germinated brown rice substituted bread: dough characteristics and bread structure. Int J Food Prop 10(4):779–789

Musa AS, Umar M, Ismail M (2011) Physicochemical properties of germinated brown rice (*Oryza sativa* L.) starch. Afr J Biotechnol 10(33):6281–6291

Oh CH, Oh SH (2004) Effects of germinated brown rice extracts with enhanced levels of GABA on cancer cell proliferation and apoptosis. J Med Food 7(1):19–23

Oh SH, Soh JR, Cha YS (2003) Germinated brown rice extract shows a nutraceutical effect in the recovery of chronic alcohol-related symptoms. J Med Food 6(2):115–121

Ohtsubo KI, Suzuki K, Yasui Y, Kasumi T (2005) Bio-functional components in the processed pre-germinated brown rice by a twin-screw extruder. J Food Compos Anal 18(4):303–316

Palmiano EP, Juliano BO (1972) Biochemical changes in the rice grain during germination. Plant Physiol 49(5):751–756

Pang Z, Zhang S (1994) Studies on the changes of active components during the germination of rice seed. Nat Prod Res Dev 7(1):26–31

Patil SB, Khan MK (2011) Germinated brown rice as a value added rice product: a review. J Food Sci Technol 48(6):661–667

Pinkaew H, Thongngam M, Wang YJ, Naivikul O (2016) Isolated rice starch fine structures and pasting properties changes during pre-germination of three Thai paddy (*Oryza sativa* L.) cultivars. J Cereal Sci 70:116–122

Roohinejad S, Omidizadeh A, Mirhosseini H, Saari N, Mustafa S, Mohd Yusof R, … Yazid M (2009) Effect of pre-germination time of brown rice on serum cholesterol levels of hypercholesterolaemic rats. J Sci Food Agric 90(2):245–251

Roohinejad S, Omidizadeh A, Mirhosseini H, Saari N, Mustafa S, Mohd Yusof R, … Yazid M (2010) Effect of pre-germination time of brown rice on serum cholesterol levels of hypercholesterolaemic rats. J Sci Food Agric 90(2):245–251

Roohinejad S, Omidizadeh A, Mirhosseini H, Saari N, Mustafa S, Meor Hussin AS, … Abd Manap MY (2011) Effect of pre-germination time on amino acid profile and gamma amino butyric acid (GABA) contents in different varieties of Malaysian brown rice. Int J Food Prop 14(6):1386–1399

Saikusa T, Horino T, Mori Y (1994) Accumulation of γ-aminobutyric acid (GABA) in the rice germ during water soaking. Biosci Biotechnol Biochem 58(12):2291–2292

Schuller HM, Al-Wadei HA, Majidi M (2008) Gamma-aminobutyric acid, a potential tumor suppressor for small airway-derived lung adenocarcinoma. Carcinogenesis 29(10):1979–1985

Seki T, Nagase R, Torimitsu M, Yanagi M, Ito Y, Kise M, … Ariga T (2005) Insoluble fiber is a major constituent responsible for lowering the post-prandial blood glucose concentration in the pre-germinated brown rice. Biol Pharm Bull 28(8):1539–1541

Thitinunsomboon S, Keeratipibul S, Boonsiriwit A (2013) Enhancing gamma-aminobutyric acid content in germinated brown rice by repeated treatment of soaking and incubation. Food Sci Technol Int 19(1):25–33

Ti H, Zhang R, Zhang M, Li Q, Wei Z, Zhang Y, … Ma Y (2014) Dynamic changes in the free and bound phenolic compounds and antioxidant activity of brown rice at different germination stages. Food Chem 161:337–344

Tian S, Nakamura K, Kayahara H (2004) Analysis of phenolic compounds in white rice, brown rice, and germinated brown rice. J Agric Food Chem 52(15):4808–4813

Trachoo N, Boudreaux C, Moongngarm A, Samappito S, Gaensakoo R (2006) Effect of germinated rough rice media on growth of selected probiotic bacteria. Pak J Biol Sci 9:2657–2661

Usuki S, Ito Y, Morikawa K, Kise M, Ariga T, Rivner M, Robert KY (2007) Effect of pregerminated brown rice intake on diabetic neuropathy in streptozotocin-induced diabetic rats. Nutr Metab 4(1):25

Viuda-Martos M, López-Marcos MC, Fernández-López J, Sendra E, López-Vargas JH, Pérez-Álvarez JA (2010) Role of fiber in cardiovascular diseases: a review. Compr Rev Food Sci Food Saf 9(2):240–258

Watanabe M, Maeda T, Tsukahara K, Kayahara H, Morita N (2004) Application of pregerminated brown rice for breadmaking. Cereal Chem 81(4):450–455

Wu F, Chen H, Yang N, Wang J, Duan X, Jin Z, Xu X (2013) Effect of germination time on physicochemical properties of brown rice flour and starch from different rice cultivars. J Cereal Sci 58(2):263–271

Wunjuntuk K, Kettawan A, Rungruang T, Charoenkiatkul S (2016) Anti-fibrotic and anti-inflammatory effects of parboiled germinated brown rice (Oryza sativa 'KDML 105') in rats with induced liver fibrosis. J Funct Foods 26:363–372

Xu J, Zhang H, Guo X, Qian H (2012) The impact of germination on the characteristics of brown rice flour and starch. J Sci Food Agric 92(2):380–387

Zhang Q, Xiang J, Zhang L, Zhu X, Evers J, van der Werf W, Duan L (2014) Optimizing soaking and germination conditions to improve gamma-aminobutyric acid content in japonica and indica germinated brown rice. J Funct Foods 10:283–291

Zhang Q, Jia F, Zuo Y, Fu Q, Wang J (2015) Optimization of cellulase conditioning parameters of germinated brown rice on quality characteristics. J Food Sci Technol 52(1):465–471

Chapter 12
Value-Added Products from Brown Rice

Shabir Ahmad Mir, Manzoor Ahmad Shah, and Idrees Ahmed Wani

Introduction

Cereal products occupy an important place in human nutrition and are consumed daily by the majority of the population. However, celiac disease patients are unable to consume the cereal products that contain gluten proteins. Gluten-free products were initially introduced for people who have intolerance to some specific peptides comprised in the gluten proteins (Catassi and Fasano 2008). Nevertheless, there is an increasing demand for gluten-free foods motivated by health concern. Rice is one of the most appropriate cereal flours for preparing gluten-free products due to its several quality properties such as hypoallergenic, colourless and bland taste and has high amount of easily digested carbohydrates (Gujral and Rosell 2004).

Rice is one of the most important cereals of the world. Most people in Asia and tropical and subtropical countries use rice as a major staple food. Brown rice is the dehusked paddy which retains the embryo and bran layers of the grain. Brown rice is rich in minerals, vitamins and phytochemicals mostly present in the bran layer contributing to the biological activities including antioxidant, anticarcinogenic and antiallergic activities, antiatherosclerosis and amelioration of iron-deficiency anaemia to sustain the human health (Hudson et al. 2000; Min et al. 2012). Brown rice is not

S.A. Mir (✉)
Department of Food Technology, Islamic University of Science and Technology,
Awantipora, Jammu and Kashmir, India
e-mail: shabirahmir@gmail.com

M.A. Shah
Department of Food Science and Technology, Pondicherry University, Puducherry, India
e-mail: mashshft@gmail.com

I.A. Wani
Department of Food Science and Technology, University of Kashmir,
Srinagar, Jammu and Kashmir, India
e-mail: idwani07@gmail.com

© Springer International Publishing AG 2017
A. Manickavasagan et al. (eds.), *Brown Rice*,
DOI 10.1007/978-3-319-59011-0_12

preferred mostly by consumers unlike polished rice, because it requires longer cooking time and has an undesirable and a firm texture. However, in recent years the nutritional and functional significance of brown rice has been recognized, and its consumption is being encouraged against the prevailing occurrence of diseases reported in connection with consumption of polished rice (Liu 2004; Srinivasan et al. 2007).

There is an increasing demand for processed foods such as popped, extruded and snack products; hence substitution of brown rice flour in the formulation of value-added products is the need of the hour. Further, the utilization of brown rice could be exploited in the processing of gluten-free products, breakfast cereals, bakery products and noodles.

Why Consume Brown Rice Products?

The important reason for use of brown rice flour is its nonallergenic nature. There is a small proportion of the population who are allergic to wheat gluten. The use of rice as a substitute for wheat in the diets of wheat-intolerant patients has long been recognized. Rice, being the least allergenic of all cereals, is often the first cereal recommended for infants. In addition, one important point in favour of the use of rice flour lies in its versatility. Firstly, rice varieties differ widely in their inherent quality, which enables different use values to be imparted to the flour. Secondly, one can use either brown rice or parboiled or otherwise pregelatinized rice, producing flours with different properties. These flours can impart a varied flavour as well as texture to the product made from modified flour. So the brown rice flour has a wide diversity in terms of use for diverse products (Gallagher et al. 2004; Mir et al. 2017a).

Composite flours have been widely used in the preparation of value-added products, which increase their nutritional and functional properties. The use of brown rice is of great importance to the rice industry particularly for the development of products (Tiwari et al. 2011) and due to various health benefits including anticarcinogenic and hypocholesterolemic effects and have been used as ingredient to various food products (Wang et al. 1999; Jayadeep et al. 2009).

As the consumer demand is changing day by day due to health concerns and they mostly prefer health beneficial value-added products, brown rice has been nowadays mostly used for the said purpose and fulfils the need of the consumer. Brown rice flour is preferred for better nutrition value and is considered as cost-effective and promising approach to alleviate malnutrition and other health-related problems.

Popped Rice

Popped rice is a popular snack product prepared from rice due to its light weight and crispness. Popped rice is one of the famous products prepared from whole rice kernel and is very popular snack product in many countries as breakfast cereal. It is prepared from hydrothermal-treated rice and by heating in sand, oil or microwave oven (Hoke et al. 2005). Brown rice nowadays is commonly used for popping purpose due to nutritional properties. During popping, the size of rice kernels increases, and a fully heat-treated crisp, porous and ready-to-eat product is created. Researchers have reported that the variety and composition of the grain such as amylose content, protein content, moisture content and degree of gelatinization affects the popping quality of grains significantly (Maisont and Narkrugsa 2010; Mir et al. 2016a). Popped rice is also commonly used in cereal drinks, ready-to-eat breakfast cereal and infant food as main ingredient. Popped rice is not only a staple in the diet as a major source of carbohydrate but it also contributes beneficial nutrients including phytochemicals, vitamins, minerals and dietary fibre which have been linked to reduce disease risk (Maisont and Narkrugsa 2009).

Popping process was optimized for brown rice based on expansion ratio. A central composite design with interactive effect of three independent variables, including salt content (1–2.5 g/100 g raw material), moisture content (13–17 g/100 g raw material) and popping temperature (210–240 °C), was used to study their effects on the expansion ratio of brown rice. The experimental values of expansion ratio were ranged from 5.24 to 6.85. The optimal condition including salt content (1.75 g/100 g raw material), moisture content (15 g/100 g raw material) and popping temperature (225 °C) was predicted for a maximum expansion ratio of 6.79, which was then proved to be 6.85 through the experiment. Raw and popped brown rice was investigated for physical properties including hardness; L*, a* and b* value; length/breadth ratio; bulk density; and minerals, which showed the significant differences. The optimized popped rice sample was evaluated for structural, spectroscopic and thermal properties, which showed the significant difference from raw rice (Mir et al. 2015).

Puffing exhibited significant effect on the physical and antioxidant properties of brown rice. The puffing process degraded the starch integrity, which affected the physical properties of grain significantly. The degraded starch during puffing treatments led to changes in X-ray diffraction pattern and decrease in peak intensities of Raman spectrum. The retrograded starch formed during parboiling facilitated to change the grain into expanded form. The total phenolic content, total flavonoid content, reducing power and DPPH radical scavenging activity decreased upon puffing treatments. The popped brown rice showed the presence of notable amount of minerals like calcium, iron, potassium, magnesium, manganese, phosphorus, sulphur and zinc (Mir et al. 2016a).

Extruded Products

The utilization of rice is of great importance to the food industry particularly in the development of extruded value-added products (Ohtsubo et al. 2005). Brown rice is used as raw material for the development of various snack products. Rice flour has become an attractive ingredient in the extrusion industry due to its unique attributes and ease of digestion (Kadan et al. 2003).

Brown rice is used for the production of gluten-free pasta with improved nutrition. Brown rice pasta produced with an extrusion temperature of 120 °C and screw speed at 120 rpm had a similar quality to that made from other gluten-free flour. The extrusion temperature and screw speed also significantly affected the cooking quality and textural properties of brown rice pasta. The cooking loss and hardness of pasta were 6.7% and 2387.2 g, respectively (Wang et al. 2016).

From a nutritional point of view, the use of brown rice is considered an important gain, especially for consumers suffering from celiac disease. The gluten-free pastas made with brown rice and corn meal were developed. An increase of the feed moisture content (up to 43.4 g/100 g) caused an increase in cooking loss, but for corn meal flour (up to 46.8 g/100 g), this increase caused an increase in cooking time. The pasta with the highest brown rice flour content (87 g/100 g) had higher contents of lipids and fibre, as well as better mineral and amino acid profiles. The pasta produced with a ratio of 40:60 (corn meal/brown rice), with 30 g/100 g of moisture content and at 70 °C received better evaluations with respect to texture (sensory and instrumental) compared with pasta made with a ratio of 13.2:86.8 (corn meal/brown rice) and with 35 g/100 g moisture content at 80 °C (da Silva et al. 2016).

The development of an extrudate from germinated brown rice by a twin screw extruder was investigated by Ohtsubo et al. (2005). They prepared germinated brown rice by soaking in water for 72 h at 30 °C followed by drying to 13–15% moisture content at 15 °C in a low-humidity artificial weather-control room. Total dietary fibre, total ferulic acid and amino butyric acid content of germinated brown rice were found to be higher than those of ordinary brown rice or polished rice. They also found that the puffed germinated brown rice contained more oryzanol, inositol, total ferulic acid and total dietary fibres than the unpuffed white rice. The products prepared by the co-extrusion of germinated brown rice (90%) and beer yeast (10%) contained more free amino acids, such as amino butyric acid, glycerin, alanine, aspartic acid and glutamic acid, than white rice, brown rice and puffed germinated brown rice.

Brown rice extrudes were developed from brown rice by Sharma et al. (2012). The brown rice was germinated for 24 and 48 h, dried and milled into grit. The grit from controlled (ungerminated) and germinated brown rice was extruded at 100 and 120 °C. The total phenolic content of the control and germinated brown rice varied from 0.803 to 0.992 mg/g ferulic acid equivalent and germination increased total phenolic content by 8.8–12.0%. The antioxidant activity varied from 6.96% to 15.86%, and germination increased the antioxidant activity by 18.2–37.2%. Upon extrusion at 100 °C, the total phenolic content decreased by over 50%. A further decrease of 6–15% in total phenolic content was observed when the extrusion temperature was increased from 100 °C to 120 °C. Similar decrease in the antioxidant activity was observed upon extrusion and rise in extrusion temperature. Significant increases in water solubility, water absorption capacity and expansion percentage of extrudates were observed upon extrusion. The extrudates from germinated brown rice were used to make an instant pudding, which upon evaluation scored higher as compared to the pudding from control brown rice.

Mir et al. (2017a) investigated the technological and nutritional properties of snacks developed from brown rice and chestnut flour. Pasting properties significantly varied among the flour ratios and the higher levels of brown rice flour led to the decrease in pasting properties. Colour parameters and L*, a* and b* values of snacks increased with increase in brown rice flour. Expansion ratio of snacks is decreased, while hardness increased with progressive increase of brown rice flour. Mineral analysis showed that blended snacks were rich in potassium, phosphorus, sulphur, magnesium, etc. Flavour value of snacks was increased with the increase in chestnut flour level in the blend. The study demonstrated that brown rice-chestnut flour-based snacks have good potential for consumer acceptance and are regarded as health-promoting functional food, especially for celiac disease patients.

Pastor-Cavada et al. (2011) developed the extruded products based on brown rice, corn and legume. Results showed that expansion, solubility and specific mechanical energy consumption were higher for rice blends than for corn blends, while density followed an inverse trend. Addition of legumes produces a decrease of expansion and an increase in solubility in both rice containing. The performance of each mixture during extrusion and the physical properties of the extruded products were considered to be in the range of those expected for snack-type products. They have a better nutritional quality than a traditional extrudate.

Bread

Bread is mostly consumed as major dietary source of calories. Increasing numbers of diagnosed cases and growing awareness make the availability of gluten-free breads an important socioeconomic and health issue. The development of gluten-free breads has attracted great attention as a result of better diagnoses of relationship between gluten-free products and health. Brown rice is one of the important raw materials used for gluten-free breads either alone or in combination with other flours. Brown rice is also rich source of nutritional components which enhanced the value of product. The market demand for gluten-free products, especially rice based, is increasing day by day due to growing number of celiac disease cases (Mir et al. 2016b).

The impact of brown rice flour incorporation, at three different levels of 5%, 10% and 15% to the wheat flour preparations on rheological properties of wheat-based dough and quality of wheat-based flat bread, was investigated. The brown rice flour incorporation mainly affected the chemical properties of flours, the rheological characteristics of dough and quality and shelf life of bread. The protein-related properties of flours principally experienced reduction; however, the ash content had an increase, along with brown rice flour incorporation. The rheological properties of dough were affected considerably by brown rice flour substitution, wherein the sample containing 5% brown rice flour was closest to brown rice flour-free dough (control). Regarding the yielded bread, brown rice flour addition affirmatively affected sensorial properties and firmness quality evaluation, wherein the bread made from dough with composite flour fortified with 5% brown rice flour was scored the best. The findings from instrumental firmness quality assessment were confirmed as the bread containing 5% brown rice flour remained softer and demanded lowest force to be compressed over the storage period. Overall, results showed that adding brown rice flour up to 5% can be used in baking of flat bread since it meets the desired quality parameters (Khoshgozaran-Abras et al. 2014).

Breads were prepared with various combinations of maize, brown rice and wheat flours. The bread volume decreased, whereas bread weight and moisture content increased with the increasing level of maize and brown rice flour. The crumb and crust colour of breads were improved with addition of 8% maize and 8% brown rice flour in bread formulation. The protein content and other nutrients of breads were increased by addition of maize and brown rice flours. The analysis of bread containing 8% maize and 8% brown rice flours showed protein 9.76%, fat 4.10%, ash 2.10%, crude fibre 5.16%, sugar 2.26% and total carbohydrates 46.91%. Bread having 8% maize and 8% brown rice flour had most acceptable flavour, texture, colour and overall acceptability when compared with other bread parameters of maize and brown rice flour (Islam et al. 2012).

Veluppillai et al. (2010) developed the bread from wheat flour and malted rice flour. Several formulation of bread was prepared from the blend of wheat and malted rice flour, aiming to find a blend with better nutritional quality and consumer acceptability. The substitution of wheat flour with malted rice showed that 35% level gave the best results according to the physical and sensory qualities of bread. The final blended bread formulation was compared with the wheat bread. The malted rice wheat blend bread contains higher content of soluble dietary fibre (0.62%), insoluble dietary fibre (3.95%), total dietary fibre (4.57%) and free amino acid content (0.62 g/kg) as compared to wheat bread (0.5%, 2.73%, 3.23% and 0.36 g/kg, respectively). Demirkesen et al. (2013) developed the gluten-free bread from different tigernut flour/rice flour blends (0:100, 5:95, 10:90, 15:85, 20:80 and 25:75). Breads prepared with tigernut flour/rice flour blend of 10:90 and 20:80 have shown the most acceptable texture and specific volume values.

The gluten-free bread was developed from germinated brown rice. Germinated rice flour showed appropriate functionality for being used as raw ingredient in gluten-free breadmaking. The germination time of the rice has a significant effect on flour properties and the resulting bread quality. Specifically, flours obtained after 24 h of germination led to an improvement in bread texture, which might be ascribed to the increase of amylase activities as well as starch degradation, which agrees with hydration and pasting results. Also bread colour improved as a result of non-enzymatic browning reaction. However, excessive germination deteriorated the product as a result of extensive amylolysis. Germinated rice flour of more than 24 h of germination was not suitable for breadmaking (Cornejo et al. 2015).

Watanabe et al. (2004) found that the substitution of brown rice or germinated brown rice to wheat flour lowered specific volume of bread more than the control bread without brown rice or germinated brown rice with increasing amounts of substitution. The improving effect was more obvious for 10 or 20% germinated brown rice than for brown rice. Among the functional components in brown rice and germinated brown rice breads, amino butyric acid was unexpectedly decomposed from the final bread. Therefore, germinated brown rice improved the bread quality when substituted for wheat flour.

Morita et al. (2007) used various additives for making germinated brown rice breads and evaluated suitable combinations of germinated brown rice and additives for breadmaking to provide germinated brown rice bread with high functional properties. The 30% of the wheat flour was substituted with germinated brown rice (GBR 30), and they found that combined additions of phytase, hemicellulase and sucrose fatty acid ester to GBR 30 improved the bread qualities with more suitable dough properties as compared with sample without additives. During fermentation, the amounts of gas leaked from the GBR 30 dough were suppressed by the additions. Phytase and hemicellulase hydrolysed the phytate and hemicellulose in germinated brown rice, and the maturity and extensibility of the GBR 30 dough were caused by the activated yeast with formed phosphate and decomposed bran, making the large loaf volume and softness of breadcrumbs during storage. Consequently, they observed that the combined additions with phytase, hemicellulase and sucrose fatty acid ester to GBR 30 improved the dough and bread qualities.

Cakes

Rice cake is an important bakery product. Cakes of good quality are produced from brown rice as it's higher in nutrients and dietary fibre than white rice. Because of consumer interest in low calorie and healthy foods, as well as marketing efforts on the part of major food companies, brown rice cake is rapidly gaining widespread consumer acceptance nowadays.

Cakes have been developed from rice flour which has wide range of properties and known for their sensory properties. The rice cakes were using both long-grain and medium-grain brown rice. The effects of tempering moisture, heating temperature and heating time on various rice cake quality attributes were investigated. A greater specific cake volume was obtained at lower tempering moisture, higher heating temperature and longer heating time for long-grain brown rice. For medium-grain rice, higher tempering moisture yielded a larger specific cake volume. The lightness of rice cakes correlated well with their expansion; the less expanded cake always had a lighter colour. Medium-grain brown rice also produced cakes that were much more fragile than those produced from long-grain brown rice (Huff et al. 1992).

Puffed rice cakes were produced from a blend of ground black rice and medium-grain brown rice by using a rice cake machine. Effects of moisture content, heating temperature and heating time on quality of the black rice cake were investigated.

The specific volume of black rice cakes showed an increasing trend with increasing tempering moisture, heating temperature and heating time. The hardness of puffed black rice cake decreased as tempering moisture and heating time increased and was influenced more by medium-grain brown rice content than heating temperature. In general, the black rice cake lightness and yellowness decreased steadily with increasing black rice content, while tempering moisture, heating temperature and heating time did not significantly affect the colours of black rice cake. On the contrary, the redness increased with increasing black rice contents (Lee et al. 2008).

Puffed rice cakes were produced also from a blend of medium-grain brown rice and black rice using a rice cake machine. Specific volume of rice cakes was higher at 18% tempering moisture. As heating time and temperature increased, specific volume of rice cake also increased. Lightness of developed cakes was increased with decreasing of heating temperature and time (Kim et al. 2001).

Noodles

Rice quality plays an important role in noodle quality. High amylose rice varieties make much better noodles compared to low amylose varieties. They provide a better ordered structure, appropriate strength and lower density (higher swelling). Rice noodles are one of the commonly known rice products consumed in Southeast Asia (Cham and Suwannaporn 2010). Chung et al. (2012) investigated the effect of heat-moisture treatment of germinated brown rice on the cooking and texture quality of the noodles containing mixtures of wheat and germinated brown rice flours. With the increase in germinated brown rice flour percentage, hardness and tensile strength of the noodles decreased, whereas water absorption and cooking loss increased. Pasting properties of the flour mixtures significantly decreased by increasing the amount of germinated brown rice. The noodle containing the heat-moisture-treated germinated brown rice flours showed the lower cooking loss and higher hardness value and tensile strength. However, substitution of wheat flour with heat-moisture-treated germinated brown rice flour leads to the darker-coloured noodle because of thermal discolouration of germinated brown rice.

Quality characteristics of Korean wheat noodles prepared from brown glutinous rice flour with and without aroma were studied by Kee et al. (2000). When compared with the control, aromatic brown glutinous rice samples produced noodles with a greater degree of lightness and a less intensity of yellowing. Replacement of up to 20% of Korean wheat flour by aromatic brown glutinous rice flour and brown glutinous rice flour in noodle had similar cooked properties such as weight gain, volume and water absorption as compared with the control. From the result of sensory evaluation, composite flours (addition up to 30% aromatic brown glutinous rice flour and up to 20% brown glutinous rice flour) and control were rated with a relatively high-quality score for appearance, taste and overall eating quality.

Cookies

Cookies represent baked products containing three major ingredients: flour, sugar and fat. In contrast to bread dough, cookie dough is characterized with low water content. The major attributes that affect cookie quality are texture, flavour and appearance. There have been attempts to increase the health beneficial components of cookies by adding various whole grain or brown rice. Utilization of germinated brown rice in sugar-snap cookies and effect of heat-moisture treatment of the germinated brown rice were investigated. Sugar-snap cookies were prepared with white rice, brown rice, germinated brown rice and the treated germinated brown rice flours, as substitutes for wheat flour (30–100 g/100 g). All cookies containing rice flours, regardless of germination and heat-moisture treatment, required significantly less force to compress than did the wheat flour cookie, and this softening effect was increased as the level of rice flour substitution increased. The cookies made with the germinated brown rice flour displayed inferior physical characteristics compared to those with wheat flour, but the cookies containing the treated germinated brown rice flour showed improved physical properties with lower moisture content and higher spread factor than those containing untreated germinated brown rice flour. The cookies containing the treated germinated brown rice flours showed relatively a low degree of firming during the ambient storage. The overall results showed that the cookies with acceptable quality and improved nutrition could be prepared by partial or complete replacement of wheat flour with the heat-moisture-treated germinated brown rice flour (Chung et al. 2014). Heat-moisture treatment is one of the thermal treatments for cereals and starch which has been used to change their inherent structure and subsequently the physical properties for utilization.

The quality characteristics of cookies containing brown rice flour, which has a greater variety of functional components than wheat flour, were studied. The results of the pasting properties show that the inclusion of brown rice flour to the wheat flour mixture did not affect the pasting temperature for up to 30% inclusion. The total dietary fibre and total polyphenol content increased, and colour of the cookies became darker with increasing brown rice flour content. According to the results from texture analysis, the hardness decreased, and the crispness increased significantly with increasing brown rice flour content. From the acceptance test, the aroma and texture of the cookies with added brown rice flour were significantly lower than those of the wheat flour cookies. However, the appearance, taste and overall acceptance of the cookies with added brown rice flour did not differ significantly from those of the wheat flour cookies. The brown rice flavour of the cookies with brown rice flour was significantly stronger than that of the wheat flour cookie, but there were no significant differences among the cookies with brown rice flour in it. The crispness of the cookies increased significantly with the inclusion of brown rice flour in the mixture, especially for the mixture with 30% brown rice flour which had the highest value of crispness among the cookies. The graininess and brown colour of the cookies increased significantly with increasing brown rice flour content, especially for the mixture with 30% brown rice flour which had the highest values among the cookies (Lee and Oh 2006).

Mir et al. (2017b) developed and evaluated the properties of brown rice-based cracker incorporated with apple pomace. For the production of desirable product, carboxymethyl cellulose was used as the main additive. The developed product showed the good consumer acceptability. The crackers were rich source of dietary fibre; minerals especially potassium, phosphorus, sodium, sulphur and zinc; and antioxidants which are good for celiac disease patients.

Idli

Nowadays whole grain consumption is increased through reformulation of regularly consumed traditional products by partial or complete replacement with whole grains. Idli is a popular Indian breakfast dish consumed by many people all over the world and is made from milled rice and black gram. The wide consumption of idli makes it ideal as a model for studying acceptability of a food reformulated with brown rice. Manickavasagan et al. (2013) evaluated the acceptability of idli regarding texture, colour and sensory properties when white rice was replaced with brown rice at five replacement levels (0%, 25%, 50%, 75% and 100%). Instrumental hardness and gumminess were proportional to the level of brown rice replacement, while springiness and cohesiveness did not vary by replacement level. Liking ratings for overall acceptability were similar at the three lowest levels of brown rice replacement. Although brown rice replacement reduced the liking score for various sensory attributes, especially for 75% and 100% replacements, more than 90% of the panellists preferred brown rice-blended idli as their first choice.

Conclusion

Brown rice is a good source of bioactive components which serve as a functional food for healthy life style. In recent years, the nutritional and functional significance of brown rice has been recognized and its consumption being encouraged. The demand for healthy foods among consumers has increased because of their nutritional and therapeutic significance. The formulated brown rice-based products were observed to be rich in functional components which are beneficial for the human health. The products developed from the brown rice have good potential for consumer acceptance and are regarded as health-promoting functional foods. The introduction of brown rice-based products in the market would increase the diversity of functional products and, even more importantly, of functional foods suitable for celiac disease patients.

References

Catassi C, Fasano A (2008) Celiac disease. In: Arent EA, Dal Bello F (eds) Gluten-free cereal products and beverages. Elsevier, London, pp 1–27

Cham S, Suwannaporn P (2010) Effect of hydrothermal treatment of rice flour on various rice noodles quality. J Cereal Sci 51(3):284–291

Chung HJ, Cho A, Lim ST (2012) Effect of heat-moisture treatment for utilization of germinated brown rice in wheat noodle. LWT-Food Sci Technol 47(2):342–347

Chung HJ, Cho A, Lim ST (2014) Utilization of germinated and heat-moisture treated brown rices in sugar-snap cookies. LWT-Food Sci Technol 57(1):260–266

Cornejo F, Caceres PJ, Martínez-Villaluenga C, Rosell CM, Frias J (2015) Effects of germination on the nutritive value and bioactive compounds of brown rice breads. Food Chem 173:298–304

da Silva EMM, Ascheri JLR, Ascheri DPR (2016) Quality assessment of gluten-free pasta prepared with a brown rice and corn meal blend via thermoplastic extrusion. LWT-Food Sci Technol 68:698–706

Demirkesen I, Sumnu G, Sahin S (2013) Quality of gluten-free bread formulations baked in different ovens. Food Bioprocess Technol 6(3):746–753

Gallagher E, Gormley TR, Arendt EK (2004) Recent advances in the formulation of gluten-free cereal-based products. Trends Food Sci Technol 15(3):143–152

Gujral HS, Rosell CM (2004) Improvement of the breadmaking quality of rice flour by glucose oxidase. Food Res Int 37(1):75–81

Hoke KAREL, Housova J, Houska M (2005) Optimum conditions of rice puffing. Czech J Food Sci 23(1):1–11

Hudson EA, Dinh PA, Kokubun T, Simmonds MS, Gescher A (2000) Characterization of potentially chemopreventive phenols in extracts of brown rice that inhibit the growth of human breast and colon cancer cells. Cancer Epidemiol Biomarkers Prev 9(11):1163–1170

Huff HE, Hsieh F, Peng IC (1992) Rice cake production using long-grain and medium-grain brown rice. J Food Sci 57(5):1164–1167

Islam MZ, Ud-Din MS, Haque MA (2012) Studies on the effect of brown rice and maize flour on the quality of bread. J Bangladesh Agri Univ 9(2):297–304

Jayadeep A, Singh V, Rao BS, Srinivas A, Ali SZ (2009) Effect of physical processing of commercial de-oiled rice bran on particle size distribution, and content of chemical and bio-functional components. Food Bioprocess Technol 2(1):57–67

Kadan RS, Bryant RJ, Pepperman AB (2003) Functional properties of extruded rice flours. J Food Sci 68(5):1669–1672

Kee HJ, Lee ST, Park YK (2000) Preparation and quality characteristics of Korean wheat noodles made of brown glutinous rice flour with and without aroma. Korean J Food Sci Technol 32(4):799–805

Khoshgozaran-Abras S, Azizi MH, Bagheripoor-Fallah N, Khodamoradi A (2014) Effect of brown rice flour fortification on the quality of wheat-based dough and flat bread. J Food Sci Technol 51(10):2821–2826

Kim JD, Lee JC, Hsieh FH, Eun JB (2001) Rice cake production using black rice and medium-grain brown rice. Food Sci Biotechnol 10(3):315–322

Lee MH, Oh MS (2006) Quality characteristics of cookies with brown rice flour. J Korean Soc Food Cult 21(6):685–694

Lee JC, Kim JD, Hsieh FH, Eun JB (2008) Production of black rice cake using ground black rice and medium-grain brown rice. Int J Food Sci Technol 43(6):1078–1082

Liu RH (2004) Whole grain phytochemicals and health. J Cereal Sci 46(3):207–219

Maisont S, Narkrugsa W (2009) Effects of some physicochemical properties of paddy rice varieties on puffing qualities by microwave. Kasetsart J Nat Sci 43(3):566–575

Maisont S, Narkrugsa W (2010) Effects of salt, moisture content and microwave power on puffing qualities of puffed rice. Kasetsart J Nat Sci 44(2):251–261

Manickavasagan A, Reicks M, Singh V, Sawsana A, Intisar AM, Lakshmy R (2013) Acceptability of a reformulated grain-based food: implications for increasing whole grain consumption. Food Sci Human Wellness 2(3):105–112

Min B, Gu L, McClung AM, Bergman CJ, Chen MH (2012) Free and bound total phenolic concentrations, antioxidant capacities, and profiles of proanthocyanidins and anthocyanins in whole grain rice (*Oryza sativa* L.) of different bran colours. Food Chem 133(3):715–722

Mir SA, Bosco SJD, Shah MA, Mir MM, Sunooj KV (2015) Process optimization and characterisation of popped brown rice. Int J Food Prop 19(9):2102–2112

Mir SA, Bosco SJD, Shah MA, Mir MM (2016a) Effect of puffing on physical and antioxidant properties of brown rice. Food Chem 191:139–146

Mir SA, Shah MA, Naik HR, Ahmad I (2016b) Influence of hydrocolloids on dough handling and technological properties of gluten-free breads. Trends Food Sci Technol 51:49–57

Mir SA, Bosco SJD, Shah MA (2017a) Technological and nutritional properties of gluten-free snacks based on brown rice and chestnut flour. J Saudi Soc Agric Sci. http://dx.doi.org/10.1016/j.jssas.2017.02.002

Mir SA, Bosco SJD, Shah MA, Santhalakshmy S, Mir MM (2017b) Effect of apple pomace on quality characteristics of brown rice based cracker. J Saudi Soc Agric Sci 16(1):25–32

Morita N, Maeda T, Watanabe M, Yano S (2007) Pre-germinated brown rice substituted bread: dough characteristics and bread structure. Int J Food Prop 10(4):779–789

Ohtsubo K, Suzuki K, Yasui Y, Kasumi T (2005) Bio-functional components in the processed pre-germinated brown rice by a twin-screw extruder. J Food Compos Anal 18(4):303–316

Pastor-Cavada E, Drago SR, González RJ, Juan R, Pastor JE, Alaiz M, Vioque J (2011) Effects of the addition of wild legumes (*Lathyrus annuus* and *Lathyrus clymenum*) on the physical and nutritional properties of extruded products based on whole corn and brown rice. Food Chem 128(4):961–967

Sharma P, Gujral HS, Singh B (2012) Antioxidant activity of barley as affected by extrusion cooking. Food Chem 131(4):1406–1413

Srinivasan M, Sudheer AR, Menon VP (2007) Ferulic acid: therapeutic potential through its antioxidant property. J Clin Biochem Nutr 40(2):92–100

Tiwari U, Gunasekaran M, Jaganmohan R, Alagusundaram K, Tiwari BK (2011) Quality characteristic and shelf life studies of deep-fried snack prepared from rice brokens and legumes by-product. Food Bioprocess Technol 4(7):1172–1178

Veluppillai S, Nithyanantharajah K, Vasantharuba S, Balakumar S, Arasaratnam V (2010) Optimization of bread preparation from wheat flour and malted rice flour. Rice Sci 17(1):51–59

Wang M, Hettiarachchy NS, Qi M, Burks W, Siebenmorgen T (1999) Preparation and functional properties of rice bran protein isolate. J Agric Food Chem 47(2):411–416

Wang L, Duan W, Zhou S, Qian H, Zhang H, Qi X (2016) Effects of extrusion conditions on the extrusion responses and the quality of brown rice pasta. Food Chem 204:320–325

Watanabe M, Maeda T, Tsukahara K, Kayahara H, Morita N (2004) Application of pregerminated brown rice for breadmaking. Cereal Chem 81(4):450–455

Part V
Storage

Chapter 13
Extending Shelf Life of Brown Rice Using Infrared Heating

Chandrasekar Venkitasamy and Zhongli Pan

Introduction

Rice is the most widely consumed staple food for a large percentage of the human population in the world, particularly in Asia. Most of the rough rice produced is consumed as white (milled) rice. After harvest, rough rice is normally dried with hot air or ambient air and then stored in silo or warehouse for an extended duration to meet the supply needed for daily consumption of our society with ensured food safety and quality (Patindol et al. 2005). Recently, there has been an increasing trend in the consumption of brown (unmilled) rice due to its higher nutrition value than that of white rice. Meanwhile, avoiding polishing and whitening by consuming brown rice could reduce the energy use in milling process by as much as 65% (Javier 2004). There is also a desire to store brown rice before milling rather than rough rice, which saves significant amount of space and allows quick and easy access to rice by people during emergencies. Both brown rice stored as bulk before milling and in retail packages have a short shelf life problem because the lipid in rice bran on the whole grain can cause quality deterioration due to lipid oxidation. In addition, the lipid-rich bran layer on the whole grain is susceptible to microbial and insect damage.

C. Venkitasamy
Department of Biological and Agricultural Engineering, University of California, Davis, One Shields Avenue, Davis, CA 95616, USA
e-mail: cvenkitasamy@ucdavis.edu

Z. Pan (✉)
Department of Biological and Agricultural Engineering, University of California, Davis, One Shields Avenue, Davis, CA 95616, USA

Healthy Processed Foods Research Unit, USDA-ARS-WRRC, 800 Buchanan St, Albany, CA 94710, USA
e-mail: zhongli.pan@ars.usda.gov

© Springer International Publishing AG 2017
A. Manickavasagan et al. (eds.), *Brown Rice*,
DOI 10.1007/978-3-319-59011-0_13

Brown rice produced by dehulling of rice (without milling) is superior to white rice in terms of nutrients and biologically active phytochemicals, such as minerals, oils, vitamins, iron, dietary fiber, thiamine, tocopherols, tocotrienols, γ-oryzanol, and γ-aminobutyric acid, and it has been acknowledged as a healthy food (Juliano and Bechtel 1985; Chen et al. 2015). Thiamine and oil contents of brown rice are about five times that of white rice, while fiber, niacin, phosphorous, potassium, iron, sodium, and riboflavin are about two to three times greater than white rice (Champagne 1994). These nutrients are typically in the rice bran layer of the brown rice (Chen et al. 2015). Brown rice has a low carbohydrate than white rice. Moreover, brown rice has the bran layer and germ intact, giving it a nutty flavor and rich texture.

Storing rice in brown rice form before milling is more popular in Japan, South Korea, and Taiwan than the other regions in the world. However, to extend the shelf life of brown rice, the store facilities are equipped with climate control to keep the brown rice at a low temperature for reducing the oxidation, which is costly, particularly during summer time. It is ideal to store rice at ambient temperature if the rice bran layer can be stabilized by inactivating the lipase enzyme. Storage of brown rice at ambient temperature may save more than 30% of storage space and reduce 20% of weight during storage and transportation compared to rough rice storage (Malekian et al. 2000; Thakur and Gupta 2006). In addition, the hulling process used for brown rice production may break up the cells in the outer layer, releasing lipase enzyme, which accelerates the breakdown of oil in the bran layer, liberating free fatty acids (FFAs) causing rancidity and off flavor of brown rice during storage which reduces the sensory quality (Das et al. 2012).

It has been reported that brown rice can be stored for 3–6 months at ambient temperature (23–34 °C) if the grain is dried to 14% moisture content (Champagne 1994; Javier 2004; Das et al. 2012). Brown rice packed in modified atmospheres such as carbon dioxide flushing and under vacuum also showed a slow deterioration in quality and extended its shelf life (Sharp and Timme 1986; Genkawa et al. 2008; Ory et al. 1980). At the retail market, brown rice is also packaged in heat-sealed cans and bags (Sharp and Timme 1986) to effectively minimize the rancidity and maintain stability of brown rice.

The susceptibility of brown rice to become rancid due to lipid hydrolysis and oxidation has limited the storage and marketing of the brown rice. The lipids in brown rice are readily hydrolyzed by lipases, which releases free fatty acids. Several approaches have been utilized to stabilize brown rice or rice bran against lipolytic hydrolysis: (a) inactivating lipase by heating rough rice with moist gas (Van Atta et al. 1952) or by parboiling or precooking (McCabe 1976); (b) removing the kernel oil, which serves as the substrate for lipase, through organic solvent extraction (Kester 1951); and (c) denaturation and inactivation of lipase and lipase-producing bacteria and mold by liquid ethanol and ethanol vapors (Champange et al. 1991; Champagne and Hron 1992). The technologies employed in industry include (a) extrusion, (b) microwave treatment, (c) ohmic heating, (d) dry heat treatment, (e) γ-irradiation, (f) parboiling, (g) toasting, and (h) infrared (IR) drying (Pan et al. 2008; Yilmaz et al. 2014; Chen et al. 2015; Bergonio et al. 2016; Ding et al. 2016, 2017). The most popular technique for stabilizing rice bran is high temperature and pressure extrusion.

An ideal and economical way for extending the shelf life of brown rice and rice bran is achieving stabilization during drying since it is an essential step in rice postharvest. However, the hot air temperature used in current drying practice is too low to achieve inactivation of lipase. IR drying was studied to effectively inactivate the lipase enzymes in rough rice and brown rice and to reduce free fatty acid formation, thus improving long-term storage stability (Ding et al. 2015). In addition to inactivation of lipase, IR drying of rice has significant other advantages over conventional hot and ambient air drying of rice. Since IR radiation can penetrate into a rice kernel, the heat transfer to the grain is faster and more energy-efficient than with convective air-drying methods (Ginzberg 1969). IR drying provides faster and uniform heating of kernels during drying and thus reduces fissuring, resulting in increased total rice yield (TRY), head rice yield (HRY), and milling quality compared to hot air drying. Infrared drying also provides disinfestation action for common insects, including lesser grain borers, Angoumois grain moths, and dark-brown grain beetles (Ginzberg 1969; Pan et al. 2008). This chapter summarizes the mechanism of IR heating, advantages of IR drying of food materials, IR emitter types and investigations performed on IR drying of rough rice, and its influence on the milling yield, rice quality, and storage characteristics of rough rice, brown rice, and rice bran quality.

Infrared Heating of Food Materials

IR heating has been gaining popularity in several food processing unit operations, including drying, peeling, baking, roasting, blanching, pasteurization, sterilization, disinfection, and disinfestation, cooking, and popping (Pan et al. 2016). It offers many advantages over conventional hot air and steam heating methods, including:

1. A consistent and uniform distribution of IR energy provides better product quality.
2. High energy efficiency is achieved as IR energy is transferred from IR emitters by radiation to product surface without requiring any heating medium.
3. High intensity of IR heat can reduce the needed processing time.
4. High degree of automation allows precise control of process parameters.
5. Compact design and different types of IR emitters permit a space-saving footprint.
6. It provides water saving and reduces waste generation in blanching and peeling of foods and is environmentally friendly (Pan and Atungulu 2010a; Pan et al. 2016).

Infrared Radiation

IR radiation is a part of the electromagnetic spectrum as shown in Fig. 13.1 (Modest 1993; Pan and Atungulu 2010b) and has both a spectral and directional dependence. IR has wavelengths ranging from 0.7 to 1000 µm (Siegel and Howell 2002) and is classified into three different categories, near-IR (NIR), mid-IR (MIR), and far-IR (FIR), with corresponding spectral range of 0.75–1.4 µm, 1.4–3.0 µm, and

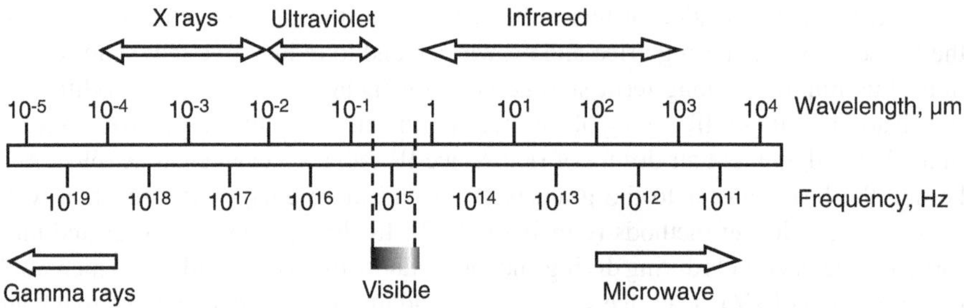

Fig. 13.1 Electromagnetic wave spectrum (Source: Pan and Atungulu 2010b)

3.0–1000 μm, respectively (Krishnamurthy et al. 2008; Sakai and Hanzawa 1994; Pan et al. 2016). The IR radiation band relevant to heating of foods, including drying, was provided by Sandu (1986). Water has very strong absorption of IR radiation at around 2.7–3.3, 6.0, and greater than 12.5 μm. The O–H bonds in water absorb IR energy and start to rotate with the same frequency as the incident radiation. This transformation of IR radiation to rotational energy causes the evaporation of water (Nindo and Mwithiga 2010).

In IR heating, the temperature of IR emitter is directly related to the wavelength of the radiation. The relationship between the wavelength and temperature can be described by the basic laws for blackbody radiation, including Planck's law, Wien's displacement law, and Stefan–Boltzmann's law (Sakai and Hanzawa 1994; Dangerskog and Osterstrom 1979). The high temperature of NIR can result in rapid increase of product temperature and potentially reduce processing time, but risk overheating the product, causing product quality deterioration. If the operating temperature of the emitter is too low, it may not produce adequate radiant energy to meet the energy requirement for a thermal process of foods (Pan et al. 2016). Therefore, it is critical to select IR emitters with appropriate wavelengths and emissive power for different food processes.

Infrared Absorption Characteristics of Food Materials

Food materials are considered as complex matrices consisting of different biochemical macromolecules, biological polymers, inorganic salts, and water. During IR heating, the food absorbs radiation at certain wavelength and also reflects and transmits radiation at other wavelengths. The absorbed radiation energy produces heat within the object. The interaction between IR radiation and the optical and physical properties of food materials is important in the design of an IR heating system or optimizing IR-based thermal processes. During IR heating, the IR absorption properties of foods depend on three factors: chemical composition, physical dimension, and properties of food products (Sandu 1986; Krust et al. 1962; Ginzberg 1969). For surface treatments, it is ideal that radiation energy is absorbed at the product surface with minimal penetration, while for other applications such as blanching and dehydration, selecting the electromagnetic wavelengths with strong penetration capability to the products is desirable.

Selection of wavelength range in IR heating is critical to ensure IR heating efficiency and final product quality. Otherwise, excessive heating of material surface and inadequate heating of product interior may be observed when the IR-emitting wavelength was not consistent with the optimal absorption wavelengths of food materials (Lentz et al. 1995). It was found that the maximum absorption of IR radiation by medium-grain rough rice occurred at a wavelength of 2.9 μm (Bekki 1991). The penetration depths of NIR on various food products were summarized by Ginzburg (1969). The penetration capacity and reflectance increases as the wavelength of the radiation decreases during IR heating. This indicates that NIR has a higher penetrating capability than FIR. FIR radiation has been more advantageous for food thermal processing by considering the total energy input and easy control, as well as most of the components in food absorb radiant energy in the region of FIR.

Infrared Emitters for Food Processing Applications

IR emitters are divided into three categories based on wavelength: short-wave, medium-wave, and long-wave emitters. The short-wave emitters operate at very high emissive power and temperatures. They are not suitable for most of food processing applications due to the possibility of overheating without appropriate temperature control. Medium-wave IR emitters emit a radiation spectrum of wavelengths ranging from 1.4 to 3.0 μm and are used for drying and curing of food products. Long-wave IR emitters are useful for low-temperature process applications in combination with convection heating (Pan et al. 2005). The thermal efficiency, heat transfer rate, and effect of radiation on heated materials are closely related to the range of wavelength emitted by the IR emitters (Siegel and Howell 2002).

IR radiation can be produced by systems utilizing either electricity or hydrocarbon gases. The corresponding emitters are classified into electrical and gaseous types based on energy sources. Electrical IR emitters have a filament that can be heated to a high temperature. They are manufactured with standardized energy values or powers, such as 60, 100, 150, 200, and 500 watts. A number of electrical emitters can be mounted on a panel at required spacing between each emitter to give the desired energy intensity. These emitters consist of a reflector, incandescent lamp, quartz tube, and resistance elements such as metallic tube, ceramic tube, and nonmetallic rod. Electrical IR emitters normally have conversion efficiency from 78% to 85%, whereas gas-fired IR emitters have a conversion efficiency of 40–46% (Ramaswamy and Marcohe 2005). Electric emitters can reach a power intensity of up to 400 kW/m^2, while gas-fired emitters normally do not exceed a power intensity of 22 kW/m^2 (Nindo and Mwithiga 2010). In general, electric heaters are expensive and can conveniently manipulate the power intensity of a given system according to the design requirement (Pan et al. 2016).

Gas-fired IR emitters use propane or natural gas as fuel and are available in different sizes for industrial application. The advantages of gas-fired emitters include uniform heating of the target with high thermal efficiency and low operating cost; they are also more reliable and durable than electrical emitters and are independent of electricity. There are different types of gas-fired IR emitters, such as direct flame IR radiator, ceramic burner, metal fiber burner, high-intensity porous burner, and catalytic gas-fired IR emitter. Among these, the catalytic gas-fired IR emitter is a relatively new diffusion-type heater that operates on a catalytic exothermic chemical oxidation–reduction principle. It converts methane or propane gas to moisture and carbon dioxide in the presence of platinum and oxygen and thus releases IR energy. The catalytic IR emitter emits radiation heat over a wide range of wavelengths without any visible flame. These emitters require preheating of the catalyst before the combustion process is initiated. Typically, these emitters have a power density varying from 6 to 28 kW/m^2 and radiation efficiency ranging from 30% to 75% (Das and Das 2010). The heating efficiency of gas-fired emitters depends on the size of the heater and the separation distance between the heater and the object being heated (Lia et al. 2004, Nindo and Mwithiga 2010). Detailed descriptions of emitter types, designs, selections, and characteristics were reviewed by Das and Das (2010).

Infrared Drying and Dehydration of Food Materials

In IR drying, also called thermal radiation drying, heat is transferred to the drying materials by radiant energy. In natural radiation drying (solar drying), radiation from the sun is tapped either directly or indirectly for drying purposes. Artificial IR drying involves the use of IR radiation generators or emitters. IR drying has been investigated as a potential method for obtaining high-quality dried foodstuffs, including fruits, vegetables, and grains (Abe and Afzal 1997; Pan et al. 2008, 2011, 2016). IR radiation drying is fundamentally different from convective drying because the material is dried directly by absorption of IR energy rather than transfer of heat from the air (Bal et al. 1970). IR energy is transferred from the heating element to the product surface without heating the surrounding air. The radiation impinges on the exposed material, penetrates it, and is converted to sensible heat. The penetration capability depends on the properties of the treated material and the temperature of the radiation source. The penetration provides more uniform heating in individual rice kernels and reduces the moisture gradient during heating and drying. In addition, due to radiation heating, the temperature of a rice kernel is not limited by the wet bulb temperature of the surrounding air and would become high in a short timeframe.

The merits and demerits of IR drying are described in detail by Nindo and Mwithiga (2010). IR has several advantages over the conventional drying, including the high heat transfer rates, use of alternative energy source, increased energy efficiency, short drying time, easy control of material temperature, a reduced necessity for air flow across the product, high degree of process control parameters, space saving along with clean working environment, better product quality, and ability to achieve complete disinfestation and disinfection of foods (Afzal et al. 1999; Tan et al. 2001; Pan et al. 2008; Shih et al. 2008; Khir et al. 2014; Dostie et al. 1989; Navari et al. 1992; Sakai and Hanzawa 1994; Mongpreneet et al. 2002; Nowak and Lewicki 2004; Krishnamurthy et al. 2008; Gabel et al. 2006; Wang et al. 2014). In IR drying, there is a very little absorption of heat by the space separating the product and emitter unless the intervening medium is saturated with water vapor. No direct contact with material is required as in conduction drying. IR radiation, similar to visible radiation (light), can be focused to increase heating intensity, provide fast treatments, or target a particular area (Nindo and Mwithiga 2010). Energy efficiency of IR drying can be improved by placing IR source enclosed in a chamber with a highly reflective surface to take advantage of the multiple reflections within the enclosure. To achieve maximum uniformity of radiant flux density on the surface of foodstuffs being irradiated, there should be proper spacing between individual IR generators (if more than one unit is used) and the distance between the radiation sources and the foodstuff (Il'yasov and Krasnikov 1991).

IR drying is not always easily applicable. Since foods usually come in complex shapes and sizes, the application of IR in such situations may be limited because energy impinging on the material will be different from place to place (Nindo and Mwithiga 2010). Normally, IR modules are used to dry a thin layer of foods (Nindo and Mwithiga 2010). The products to be dried can be in motion facilitated by a conveyor belt or a vibrating plate. Because IR radiation does not depend on the surrounding medium, the product surface can be quickly heated up to desired temperatures. However, allowing the temperature to rise too high and/or heating for extended periods of time is detrimental to the product quality (Sharma et al. 2005; Gabel et al. 2006; Kumar et al. 2005). To prevent high product temperatures especially for heat-sensitive materials, intermittent heating and IR heating in combination with hot air drying or freeze drying have been practiced (Wang et al. 2016; Ding et al. 2015, 2016, 2017: Atungulu and Pan 2010). Since, IR can be used to quickly heat rough rice in a single or thin layer to a relatively high temperature, it should be possible to use the sensible heat from the heated rice to remove more moisture during cooling, which could make the overall IR rough rice drying process more energy efficient (Pan et al. 2008).

Combination of IR and hot air drying has been used as an efficient and rapid drying method compared to hot air drying alone. The energy and operating costs of the combined drying mode are lower than convective drying systems for several food and agricultural products, particularly for the drying of heat-sensitive products (Kocabiyik 2010). Use of tempering or holding treatment by maintaining the temperature of heated product after IR heating followed by natural cooling was found to reduce the moisture content of rice by up to 2.2% by using the sensible heat from heated rice without using additional energy and achieved high rice milling quality (Pan et al. 2011). In addition, IR heating followed by tempering and natural cooling inactivates lipase enzyme and significantly reduces the generation of free fatty acids in rough and brown rice (Ding et al. 2015). Controlling the IR radiation intensity or shortening the drying frequency by using intermittent IR drying has been used to minimize the changes in color and nutrients of dried product (Chua et al. 2004). Several studies proved that using IR for drying of food results in a better-quality product and saving of drying time and energy (Riadh et al. 2015).

Infrared Drying of Rice

Infrared Drying of Rough Rice

The earliest use of IR for drying rough rice was reported in the early 1960s (Schroeder and Roseberg 1960; Pan et al. 2011). A high drying rate of rice was achieved by spreading the rice in a single layer. It has been reported that only 7 min were required to reduce the moisture content (MC) from 20% to 14.8% (d.b.) using NIR radiation heating, compared to 30 min for hot air drying (Rao 1983). Using IR to preheat the rough rice to 60 °C followed by hot air drying at 49 °C for 2–3 min removed approximately 2% MC in single pass. Bekki (1991) found that the maximum absorption of IR radiation by medium-grain rough rice occurred at a wavelength of 2.9 μm. IR drying results in uniform heating of kernels in the drying process and thus reduces fissuring, thereby maintaining the yield and milling quality of rice (Khir et al. 2014; Pan et al. 2011). IR drying of rice has been resulted in a high drying rate, good milling quality, effective disinfestation and disinfection of rough rice, inactivation of lipase enzyme, improvement in shelf life of both rough and brown rice, and effective stabilization of rice bran (Pan et al. 2008, 2011, Khir et al. 2011, 2014, Wang et al. 2016).

Catalytic IR heating of rough rice having high and low harvest MCs in single layer was performed by Pan et al. (2008, 2011) to study the effect of IR heating time, IR intensity, bed thickness on heating and drying rates, and milling characteristics of rice followed by tempering and cooling treatments. The technical feasibility of simultaneous drying and disinfestation of rice using IR heating has been evaluated. A catalytic emitter with dimensions of 30 × 60 cm, surface temperature of 650 °C, and corresponding peak wavelength of 3.1 μm was used for the study. The average IR intensity at the rough rice bed surface was 5348 W/m^2. A single layer of rice of 250 g at a loading rate of 2 kg/m^2 was placed on the drying bed. The tempering was conducted by keeping the rice samples in closed containers placed in an incubator at the same temperature as the heated rice for 4 h immediately following the heating. The tempered and non-tempered samples were each cooled using natural cooling (slow cooling) or forced-air cooling at ambient temperature. The results of this study related to drying rates and milling characteristics of rice followed by tempering and cooling treatments are summarized in the following sections.

Rough Rice Temperature and Heating Rate

The average rice temperature during IR heating increased with the increase in heating time and IR intensity and decreased with increase in bed thickness (Pan et al. 2008, 2011). The rice sample temperatures for different heating durations at different initial MCs, radiation intensities, and drying bed thicknesses are shown in Fig. 13.2a, b. The rice temperature increased with increase in heating duration and radiation intensity for samples with the same initial MC and drying bed thickness. No significant differences in rice temperature were observed between the low and

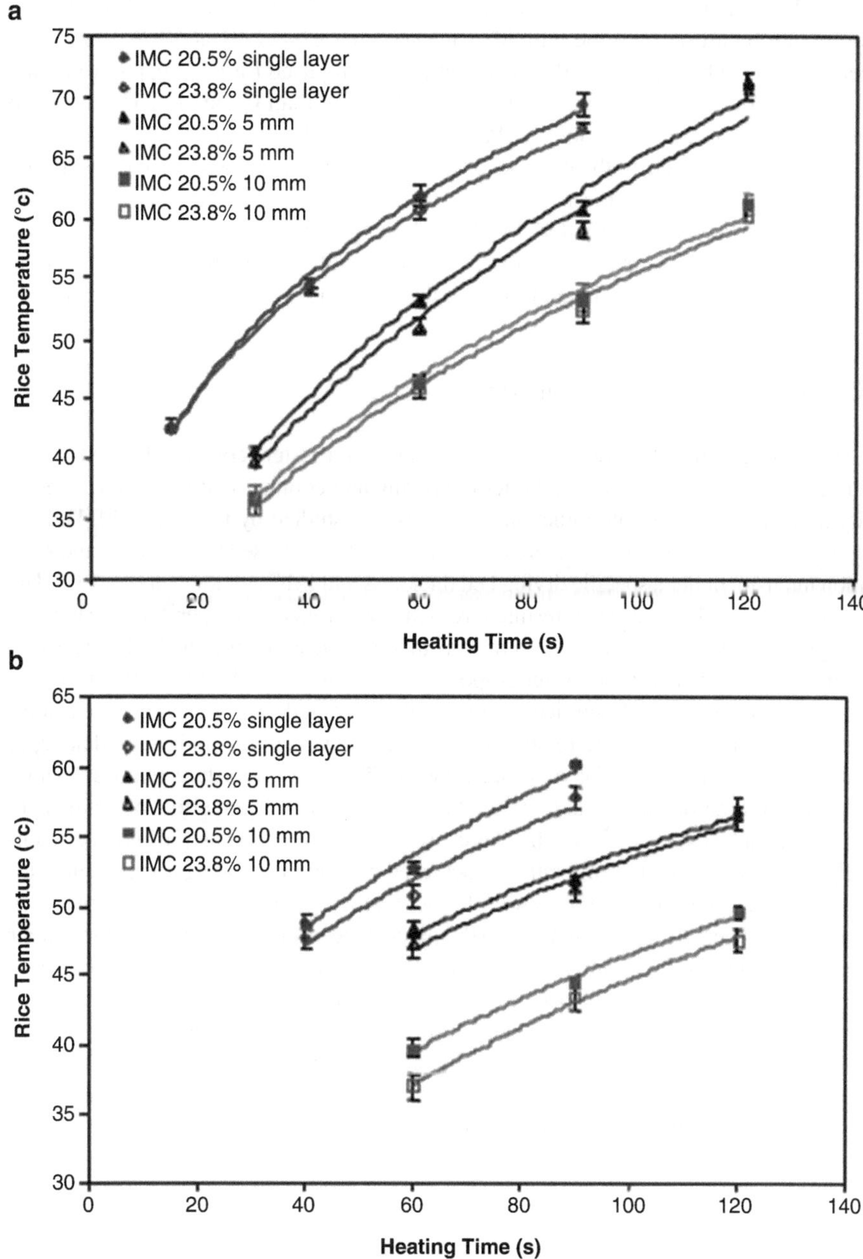

Fig. 13.2 Relationship between rice temperatures and heating time at drying bed thickness and initial moisture contents under radiation intensity of (**a**) 5348 W/m^2 and (**b**) 4658 W/m^2 (Source: Pan et al. 2011)

high MC rice after IR heating; however, the low MC rice samples had slightly higher temperatures than the high MC rice samples, especially at 60, 90, and 120 s heating time, which could be due to less energy being used for heating the water and a lower evaporative cooling effect in the low MC rice than in the high MC rice with the constant radiation heat supply (Pan et al. 2011). The heating rate of rice during IR heating decreased with increase of the drying bed thickness as increased bed thickness increases the mass of the rice per batch. In another study, Das et al. (2004a) used a vibration-aided IR dryer having a radiation intensity of 3100–4290 W/m^2 to dry rough rice grain with bed depths of 12–16 mm and found that the grain depth had an insignificant effect on the drying rate at a given IR intensity.

Moisture Removal and Drying Rate

The moisture removal for rice samples with an initial moisture content (IMC) of 20.5% and 23.8% during IR heating and after tempering and cooling treatment with different drying bed thicknesses and radiation intensities was studied by Pan et al. (2011). Rice moisture removal during IR heating increased with the increased heating time and radiation intensity under a specific drying bed thickness and IMC as shown in Fig. 13.3. The moisture removal increase was resulted from the increased rice temperature due to more energy being absorbed by the rice kernels with longer heating time and higher radiation intensity which caused more water evaporation compared to a shorter heating time and low radiation intensity. When rice samples with different depths were heated to a similar temperature, the moisture removal during heating was similar, indicating that depth was not a limiting factor for moisture removal. Therefore, moisture removal mainly depended on the rice temperature (Pan et al. 2011). In significant effect of grain depth on drying rate was also observed by Das et al. (2004a & b). Pan et al. (2011) experimentally proved that a high drying rate was achieved by using IR heating alone, even without counting the moisture removal during tempering and cooling, compared to conventional heated air drying of 0.1 to 0.2 percentage points per min due to the low air temperature used (Kunze and Calderwood 2004). The influence of initial moisture content, rice temperature, drying bed thickness, tempering, and cooling methods on moisture diffusivity of rough rice under IR heating was studied by Khir et al. (2011). Rough rice moisture diffusivities under IR heating and cooling were significantly affected by rice temperature and tempering treatment, respectively. High heating rate and moisture diffusivity were achieved with IR heating of rough rice than hot air drying.

Milled Rice Quality

IR drying of rough rice resulted in high moisture removal and milling quality by heating the rice samples to about 60 °C followed by tempering and natural cooling (Amaratunga et al. 2005; Pan et al. 2008, 2011; Khir et al. 2011). Meeso et al. (2004) achieved a significant improvement in the head rice yield and whiteness of rough rice using infrared radiation heating as a result of partial gelatinization of rice kernels.

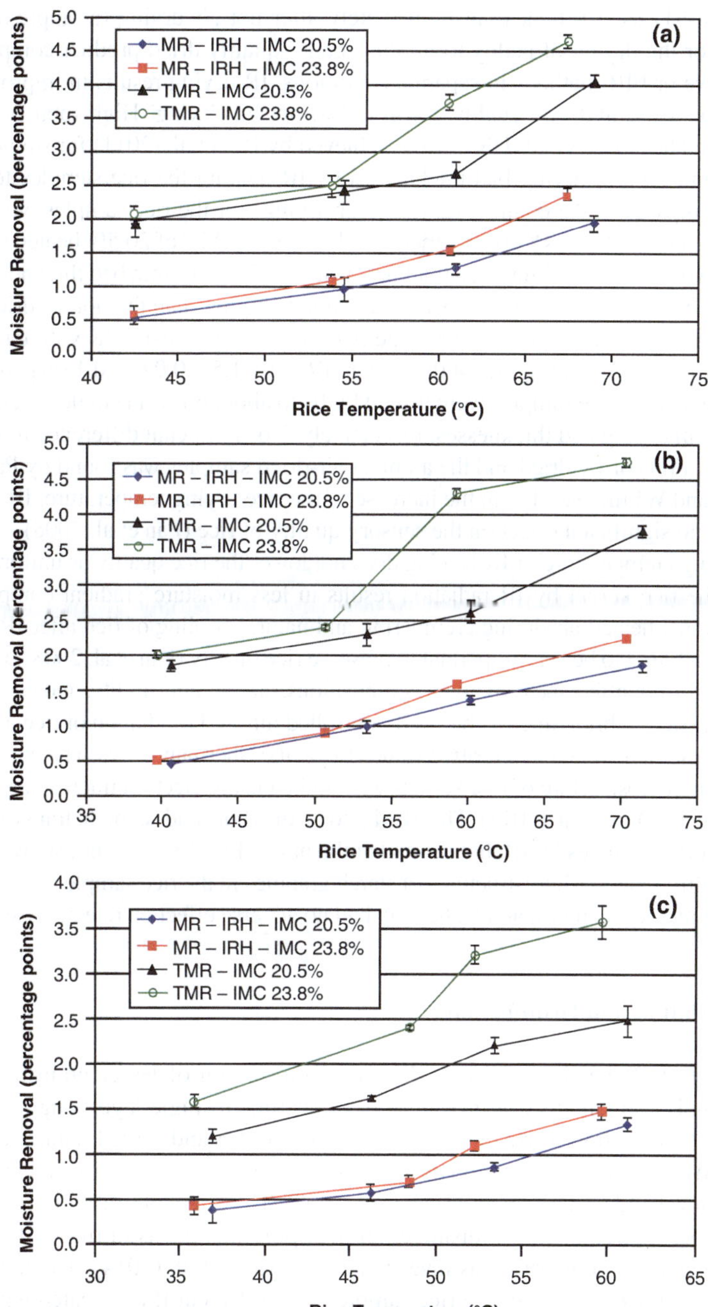

Fig. 13.3 Moisture removal of rice with initial moisture contents (MCs) of 20.5% and 23.8% caused by infrared heating and cooling after tempering treatment with (**a**) single-layer, (**b**) 5 mm, and (**c**) 10 mm drying bed thicknesses under radiation intensity of 5348 W/m². *MR* moisture removal, *IRH* infrared heating, *TMR* total moisture removal (Source: Pan et al. 2011)

Bekki (1991) used FIR heating immediately after hot air drying (at approximately 40 °C) for the drying of paddy to produce better-quality dried product compared to convective or FIR methods in separate conditions. IR drying results in improvement of milling yield and rice quality compared to hot air drying. High total rice yield (TRY) and head rice yield (HRY) were achieved by Pan et al. (2011) for single-layer, 5 mm, and 10 mm drying bed thicknesses by IR heating the rice samples to about 60 °C compared to the controls (ambient air drying) for the rice samples with IMCs of 20.5% and 23.8%. TRYs of IR-dried rough rice with IMC of 20.5% heated to 61 °C (single layer), 60.2 °C (5 mm bed thickness), and 61.2 °C (10 mm bed thickness) were found to be 0.65, 0.88, and 0.59 percentage points higher than the hot air-dried samples. Moreover, compared to control, the rice samples dried using IR with tempering and natural cooling had significantly higher HRY (by 1.52, 0.94, and 1.06 percentage points) when the rice samples were heated by IR to about 60 °C in single-layer, 5 mm, and 10 mm drying bed thicknesses, respectively. No significant difference in whitening index (WI) of IR-dried and the ambient air-dried samples was found by Pan et al. (2011), and WI increased with the increase of the rice drying temperature. IR heating also has no significant effect on the sensory quality of rice (Pan et al. 2008).

The high temperature of IR heating does not lower the rice quality as uniform heating of the rice kernel by IR radiation results in less moisture gradient compared to conventional heated-air drying. Tempering and natural cooling of rice after IR drying have been found to be very important to preserve rice quality (Pan et al. 2008, 2011). IR drying does not affect the proximate compositions, starch digestibility, and energy values of the rice. A brief, intense, but well-controlled infrared irradiation of rice increases the wholesomeness and its nutritive value (Keya and Sherman 1997). The effect of IR drying on microstructure of rice kernels was studied using electron microscopy (SEM) technique by Ding et al. (2015). The SEM photos of whole and segment transverse sections of dried samples by IR, hot air, and ambient air (Fig. 13.4) did not show any significant differences. Microstructures of starch granules in the rice samples were similar for the three rice samples, and IR heating had no adverse effect on rice microstructure.

Disinfestation and Disinfection

Influence of IR heating of rough rice on disinfestation of lesser grain borer and Angoumois grain moths was investigated. IR heating of single-layer rough rice samples to 60 °C, followed by tempering at 60 °C for 4 h and forced natural cooling, achieved complete disinfestation of moths and grain borers in addition to high moisture removal and milling quality (Pan et al. 2008). The effect of infrared (IR) heating and tempering treatments on disinfection of *Aspergillus flavus* in freshly harvested rough rice and storage rice was investigated by Wang et al. (2014). The log reductions of *A. flavus* spores for the rice samples with different IMCs heated with IR to 60 °C followed by tempering for different time periods are presented in Fig. 13.5.

The *A. flavus* populations in IR-dried rough rice significantly decreased ($\alpha < 0.05$) as tempering time increased. With the same tempering time, the *A. flavus* population significantly decreased ($\alpha < 0.05$) as the IMC of rough rice increased. Therefore,

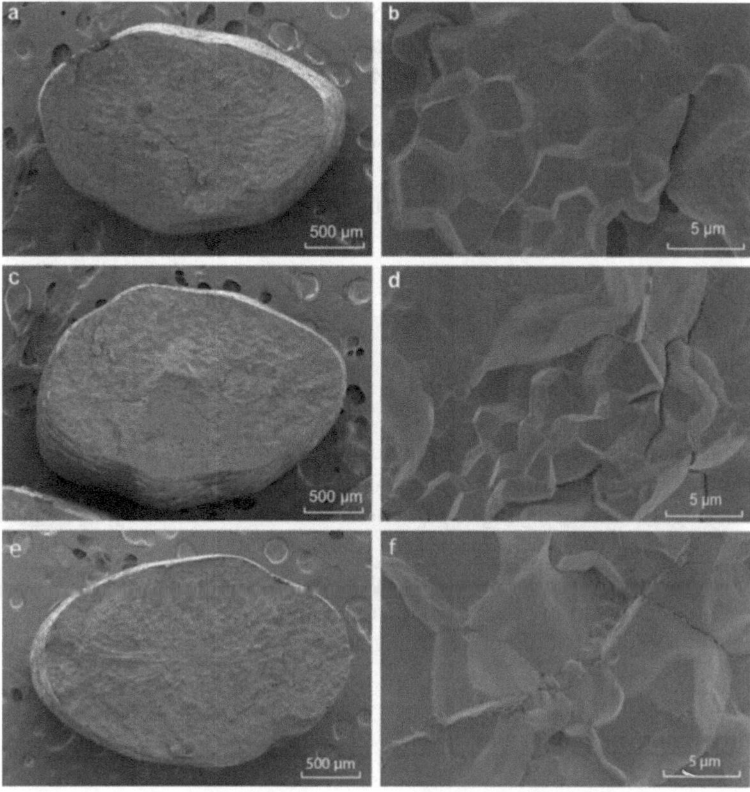

Fig. 13.4 Scanning electron micrograms of the rice dried with ambient air (**a**, **b**), infrared radiation (**c**, **d**), and hot air (**e**, **f**) (Source: Ding et al. 2015)

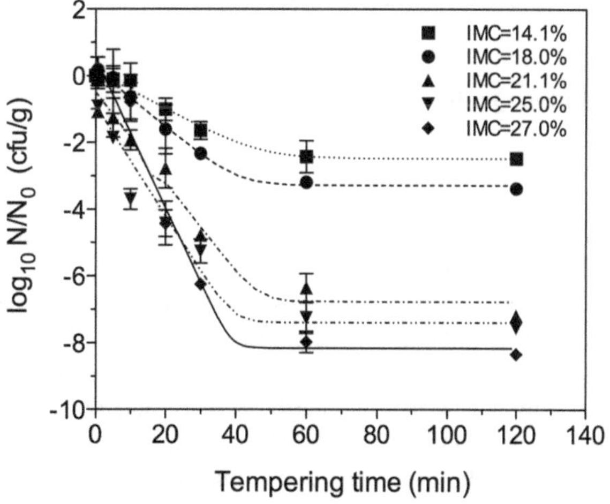

Fig. 13.5 Disinfection and fitting curves for *Aspergillus flavus* spores for rough rice with different initial moisture contents (Source: Wang et al. 2014)

heating rice samples to 60 °C followed by tempering for 120 min was more than adequate to inactivate *A. flavus* (based on a 5-log reduction requirement) in rice with an IMC above 21.1% (Wang et al. 2014). Measured quantity of moisture can be added to bring the rice with low IMC to 21.1% before IR drying to inactivate the *A. flavus* populations. From these studies, it can be concluded that IR heating can be used for simultaneous drying and disinfestation of freshly harvested rough rice. For dry rice, the disinfection can be achieved by rewetting the dry rice to sufficient moisture.

Infrared Drying of Parboiled Rice

Rice Temperature and Heating Rate

IR heating of parboiled rice has been found to provide faster heating of rice to increase the temperature of rice to the drying temperature quickly compared to the hot air drying. Results of drying of parboiled rice of two varieties by Tirawanichakul et al. (2012) using IR at two intensities (1000 and 1500 W) and hot air under the temperature range of 60–100 °C at an inlet air flow rate of 1.0 ± 0.2 m/s are shown in Fig. 13.6. The IR drying increased the rice temperature to the target temperature quickly compared to that of hot air drying and reduced the drying time.

Moisture Removal and Drying Rate

The IR heating resulted in high drying rate or moisture removal compared to hot air drying and reduced the drying time for parboiled rice because energy in the form of electromagnetic wave (IR) can be transferred and absorbed directly to the rice grain kernel by heat radiation with a low heat loss to the surrounding. The drying rate increased with the increase in the intensity of IR radiation (Tirawanichakul et al. 2012). The increased drying rate decreased the drying time by almost half compared to hot air drying.

Milling Quality

IR drying of parboiled rice had higher HRY compared to hot air-dried and ambient-dried samples. This is because high infrared power can penetrate more deeply leading to higher rate of heat transfer to kernel resulting in greater starch gelatinization (Tirawanichakul et al. 2012). However, contradicting results on HRY were reported by Das et al. (2004a) for IR drying of parboiled rice using higher radiation intensities of 1509–5514 W/m². About 6–8% reduction in HRY was observed when the intensity increased from 1509 to 5514 W/m². The HRY reduction increased with increase in IR intensity. The reduction in HRY was attributed to a marked increase in drying rate as the intensity level increased. The rapid moisture removal contributed to a higher state of stress in the rice kernel at the time of drying, causing a high percent of breakage.

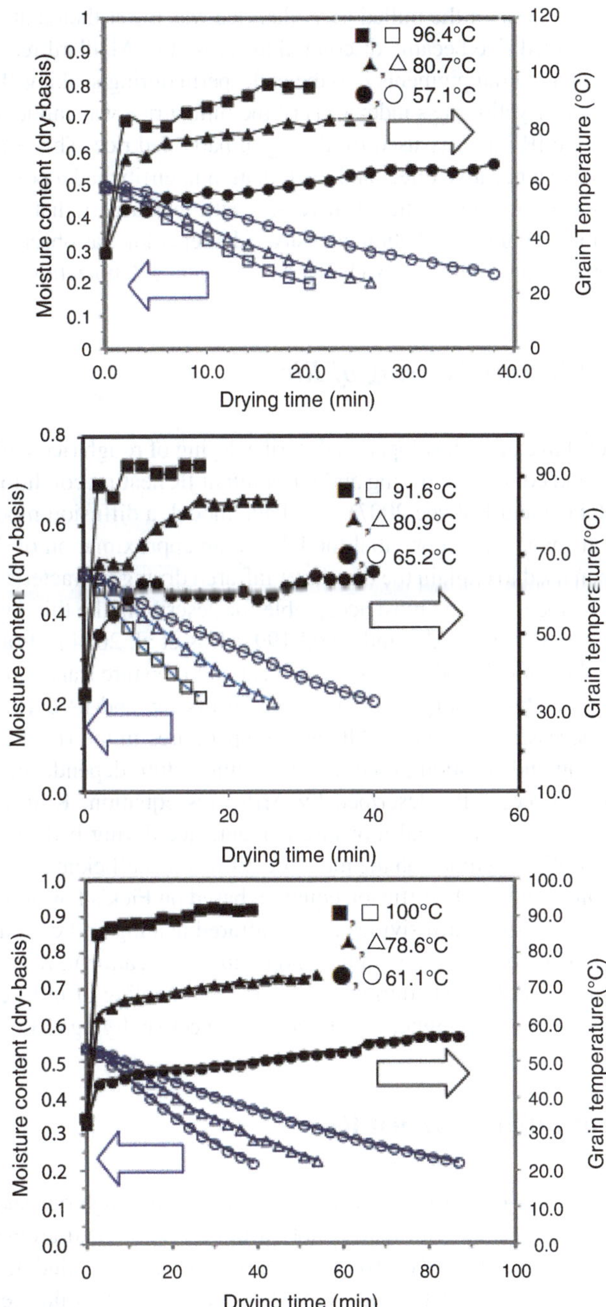

Fig. 13.6 Moisture content and grain temperature against drying time of parboiled rice drying with IR 1000 W (*top*), IR 1500 W (*middle*), and hot air (*bottom*) at inlet drying temperatures of 57.1–100 °C (Source: Tirawanichakul et al. 2012)

The whiteness value of the milled parboiled rice was lower than that of the milled rough rice. Parboiled rice became discolored as a result of Maillard reaction and the diffusion of hull and bran pigments into the endosperm during soaking (Bhattacharya 2004). The color or yellowness index (YI) of the milled rice was influenced by grain bed depth and the IR intensity used for drying of parboiled rice. The grain bed depth was found to have a negative effect and the radiation intensity had a positive effect on YI. At a particular bed depth, the YI increased with increased radiation intensity or increased heating. At higher radiation intensity, a darker color was observed. Radiation intensity and bed depth significantly changed the percent gelatinization of the kernels.

Modeling of Infrared Drying of Rice

Several models have been developed to describe drying of rough rice with IR drying, combined IR, and convective drying and intermittent IR heating conditions (Abe and Afzal 1997; Cihan and Kahvec 2007). The Page model, a diffusion model based on spherical grain shape, an exponential model, and an approximation of the diffusion model had been used to explain the thin-layer infrared drying characteristics of rough rice. The Page model was the most acceptable for describing the thin layer of rough rice under infrared drying (Abe and Afzal 1997; Das et al. 2004a). Prakash (2011) developed mathematical models to describe heat and moisture transport phenomena in the rice kernels during sorption, convective air drying, and IR drying processes. The models successfully predicted MC and temperatures in the rice kernels during air and IR drying and sorption processes. The temperature dependence of moisture diffusivity was successfully described by Arrhenius equation. Khir et al. (2011) investigated the effects of initial moisture content, rice drying bed thickness, temperature, and cooling methods on the moisture diffusion coefficient and the moisture diffusivity using an unsteady diffusion equation based on Fick's law and slope methods. Rough rice moisture diffusivities under infrared heating and cooling were dramatically influenced by rice temperature and tempering treatment, respectively. The moisture diffusion coefficients during the heating and cooling of infrared dried rice with tempering were much higher than those of convective drying (Khir et al. 2011).

Shelf Life of Infrared Dried Rice

Storage stability of IR-dried rough rice stored in the form of rough rice and brown rice was studied and compared with that of hot air-dried (43 °C) and ambient air-dried rough rice (Ding et al. 2015, 2016, 2017). Storage characteristics including the changes in color, microstructure, free fatty acid, peroxide value, and iodine value gelatinization, pasting, cooking, and textural properties of rough rice were determined to detect any notable degradation of lipids in rough and brown rice during storage. IR drying was performed by heating rough rice to 60 °C under radiation intensity of 4685 W/m^2 in the IR dryer (Fig. 13.7), followed by 4 h tempering and natural cooling to 16.1% (dry

Fig. 13.7 Rice drying in a custom-designed catalytic IR heating unit (Source: Pan et al. 2016)

basis) moisture content. The rough rice samples dried by the three methods were stored in the form of rough rice and brown rice at 35 ± 1 °C with relative humidity of $65 \pm 3\%$ for 10 months using an experimental plan shown in Fig. 13.8 (Ding et al. 2015).

Storage of Infrared Dried Rough Rice

Milled Rice Quality

IR drying produced slightly higher TRY than hot air and ambient air drying (Ding et al. 2015). The TRY of IR-dried rice was slightly higher than other drying methods for the entire 10-month storage. The HRY values of rice dried with IR in initial 4 months of storage were slightly higher than rice dried with hot and ambient air-drying methods. At the 10 months of storage, the HRY of IR-dried rice was similar to the HRY from other two methods. The whiteness index of milled rice from IR drying and hot air drying was slightly higher than ambient air-dried samples. The whiteness index did not change during the first 4 months of storage of rough rice, and beyond 4 months of storage, the whiteness values were found to decrease. The IR drying significantly increased the whiteness of the white rice compared to hot air and ambient air drying. This is because high temperature of IR drying results in partial gelatinization of starch and collapse of the starch granules producing a transparent and glossy kernels. Moreover, a high drying temperature decreased the activity of enzymes and reduces the enzymatic browning. Shorter time of IR drying compared to other methods weakened the nonenzymatic browning reaction. No significant differences ($p > 0.5$) were found in the YI of rice dried using three drying methods. However, YI values increased during storage. The increase in yellowness of rice could be caused by lipid oxidation, Maillard reaction, and leaching of the

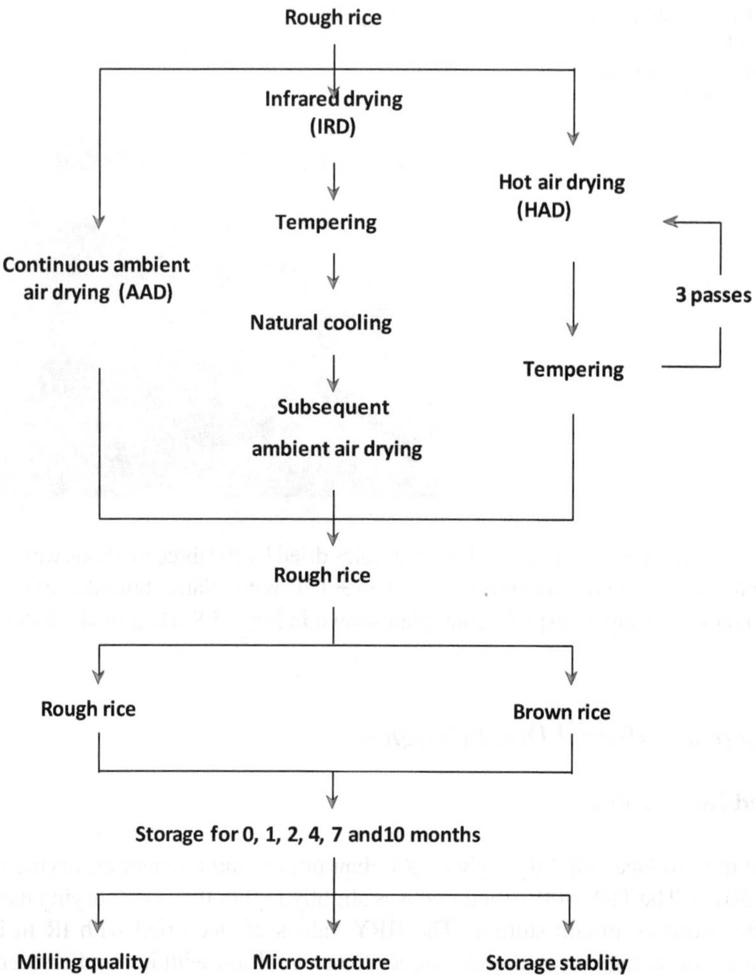

Fig. 13.8 Experimental procedure for rough rice drying and storage (Source: Ding et al. 2015)

yellow color from pigments present in rice hull and bran into the endosperm after a long period of storage. IR drying may stabilize the microstructure of the rough rice and prevents any changes in microstructure during storage. The high temperature during IR heating followed by tempering process could anneal the starch and denature protein in rice and make their structure more stable than the samples dried with hot air and ambient air drying. IR-dried rice has been reported to have better gelatinization, pasting, cooking, and textural properties compared to hot air- and ambient air-dried rice (Ding et al. 2016).

Fig. 13.9 Changes in free fatty acid (FFA) concentrations of stored rough rice dried by three methods. *IRD* IR drying, *HAD* hot air drying, *AAD* Ambient air drying (Source: Ding et al. 2015)

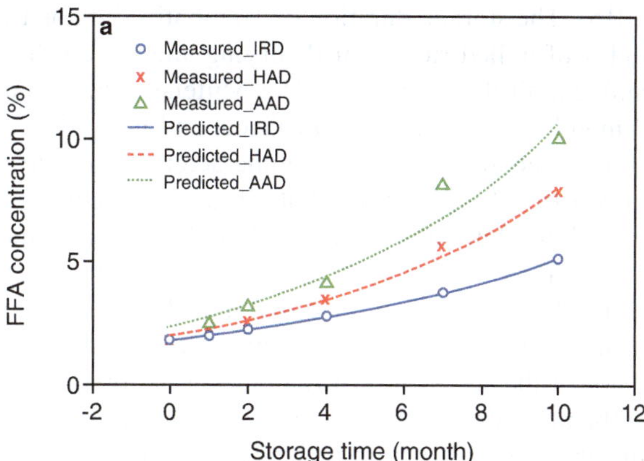

Storage Stability

Storage stability of rice can be assessed from the free fatty acid (FFA) content, peroxide value (PV), and iodine value (IV) of rice samples. The FFA and POV levels of lower than 10% and 20 Meq/kg, respectively, are considered as indicators of stable rice with accepted sensory qualities. Ding et al. (2015) studied the FFA of rough rice samples dried with IR, hot air, and ambient air over the storage period of 10 months (Fig. 13.9). FFA value of IR-dried rice was lower than the hot air- and ambient air-dried rice samples during the 0–4 months of storage. After 4 months of storage, FFA concentrations of rough rice samples dried with IR drying were significantly ($p < 0.05$) lower than those dried with hot air and ambient air.

The studies on peroxide value (PV) and iodine value (IV) showed no significant difference among rice dried using three drying methods. In general, all the stored rice samples had similar changing trend for PV and IV over the storage period. PV of all samples increased after the first month and then decreased to the least at month 4. After that, PV increased again at month 7 and became stable till month 10. All the rice dried by IR, hot air, and ambient air did not have significant difference in iodine value at month 0, and it had a similar steady descend over the storage period (Ding et al. 2015). IR drying of rice stabilizes the microstructure and causes effective inactivation of lipase and reduces the lipid hydrolysis to improve the long-term storage stability and significantly extends the period of safe storage.

Storage of Brown Rice

Milled Rice Quality

Brown rice stored after IR drying produced slightly higher TRY and HRY than those dried with hot air. No significant differences were found in TRY and HRY of IR-dried and ambient air-dried rice samples stored as brown rice (Ding et al.

2015). The storage duration of 10 months did not affect the HRY. The whiteness index of milled rice from IR drying and hot air drying was slightly higher than ambient air-dried samples. The whiteness index did not change during the first 4 months of storage of brown rice, and beyond 4 months, the whiteness values were found to decrease. Color characteristics of the white rice milled from stored brown rice dried with IR, hot air, and ambient air were studied for the storage period of 10 months using L*, a*, and b* and also the yellowness index (YI) values to describe the color change in stored brown rice (Fig. 13.10). The results showed that IR drying increased the whiteness of milled rice without significant changes in yellowness. Moreover, the IR drying also reduced the changes in color of the stored brown rice during the first 4 months of storage compared to ambient air drying. Ding et al. (2017) used light transverse micrographs to show the changes in structure of white rice milled from stored brown rice dried with IR, hot air, and ambient air during 10 months of storage period. The micrographs showed that IR-dried rice had a more stable structure compared to hot air- and ambient air-dried rice kernels.

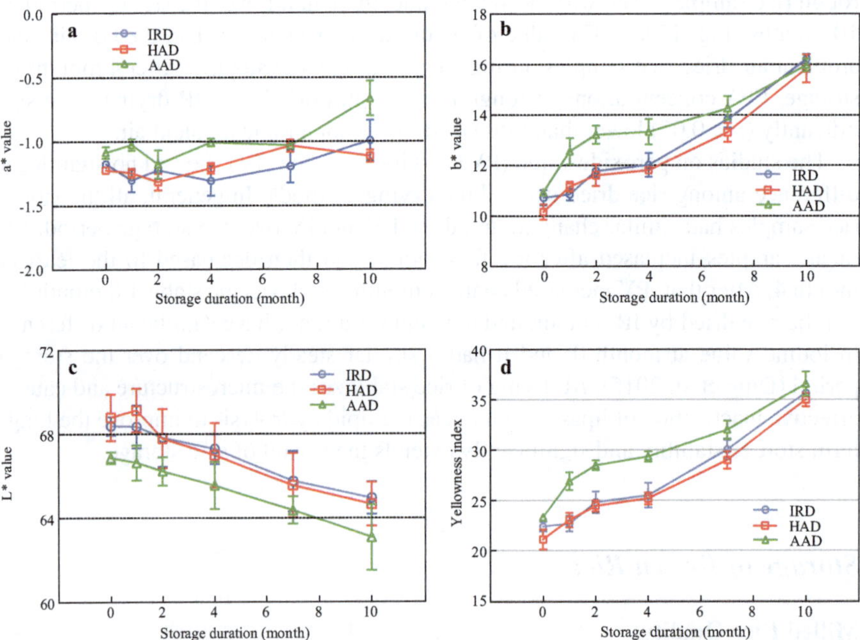

Fig. 13.10 Color characteristics of the white rice milled from stored brown rice dried with infrared (IRD), hot air (HAD), and ambient air (AAD) during storage. The figures (**a**), (**b**), (**c**), and (**d**) show the a*, b*, L*, and yellowness index values of the rice samples, respectively (Source: Ding et al. 2017)

Storage Stability

Ding et al. (2015) studied the concentrations of free fatty acids (FFA) of brown rice over the storage period of 10 months prepared from rough rice dried using IR, hot air, and ambient air (Fig. 13.11). FFA value of IR-dried rice was lower than the hot air- and ambient air-dried rice samples during the 0–4 months of storage. After 4 months of storage, FFA concentrations of rough rice samples dried with IR drying were significantly ($p < 0.05$) lower than those dried with hot air and ambient air. The PV and IV values of stored brown rice samples showed no significant difference among rice dried using three drying methods. IR heating had no negative effects on oxidation of unsaturated fatty acids to saturated fatty acids (Ding et al. 2015). PV of all samples increased after the first month and then decreased to the least at 4 months of storage. PV increased again at month 7 and became stable till month 10. The IV values steadily decreased from 0 to 10 months of storage.

Storage of brown rice prepared from IR-dried rough rice did not have adverse effects on the cooking quality of milled rice. The cooking characteristics water uptake ratio (WUR), volume expansion ratio (VER), and solid loss percentage (SLP) of the milled rice samples dried with IR showed no significant difference ($p > 0.05$) with those dried with hot air and ambient air drying. WUR and VER of rice samples from three drying methods slightly increased in the first 4 months, following a considerable increase until the seventh month; in the end there was a slight decrease between seventh and tenth months of storage period.

No significant difference in the hardness values of the cooked white rice samples dried with IR, hot air, and ambient air was noticed. The hardness of all cooked rice samples was in the same level during the first 4 months of storage, and then it significantly increased during the prolonged storage beyond 4 months. Adhesiveness of cooked rice samples dried using IR and hot air was slightly higher than those dried with ambient air, and the adhesiveness of all rice samples slightly decreased during the storage.

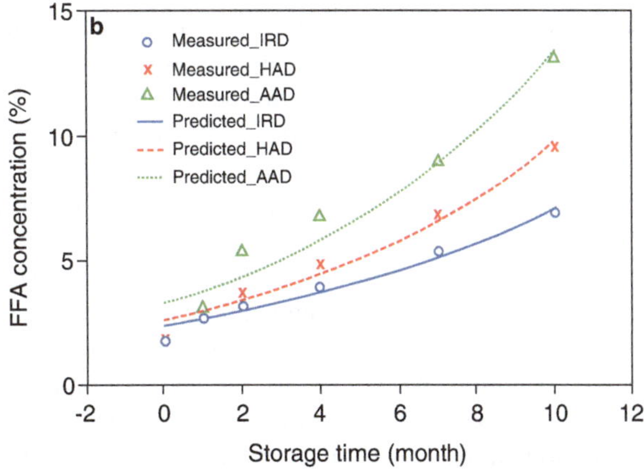

Fig. 13.11 Changes in free fatty acid (FFA) concentrations of stored brown rice prepared from rough rice dried by three methods. *IRD* IR drying, *HAD* hot air drying, *AAD* ambient air drying (Source: Ding et al. 2015)

The thermal and pasting properties of stored brown rice dried using IR, hot air, and ambient air were investigated by studying the onset (T_o), peak (T_{op}), conclusion (T_c), range of gelatinization temperature (W_p), and the gelatinization enthalpy (ΔH) of rice flour prepared from the milled white rice from stored brown rice (Ding et al. 2017). IR-dried rice had a slightly lower T_o, T_c, W_p, and ΔH values than hot air- and ambient air-dried rice, whereas rice from all three drying methods had similar gelatinization characteristics. IR drying reduced the changes in the cooking and textural characteristics by stabilizing the microstructure and properties of proteins and starch granules located in the surface layer of rice kernels. All the parameters (T_o, T_p, T_c, W_p, and ΔH) of the stored brown rice congruously increased in the first 4 or 7 months and then slightly decreased. The results showed that IR drying could be used as a feasible efficient drying technique that can maintain the physicochemical properties of brown rice during storage and improve its storage characteristics (Ding et al. 2017).

The accelerated storage study of rough rice and brown rice by storing at 35 ± 1 °C with relative humidity of $65 \pm 3\%$ showed that FFA concentration of hot air-dried rough rice was increased to 7.84% (Fig. 13.9) after 10 months of storage, whereas the FFA concentration of IR-dried rice stored in the form of brown rice was increased to only 6.91% (Fig. 13.11) after 10 months of storage. These results show that IR-dried rice can be stored in brown rice form at ambient temperature with even longer shelf life than that of rough rice dried with ambient and hot air.

Stabilization of Rice Bran by IR Drying of Rice

Rice bran is an important by-product of rice milling that constitutes 8–10% of the weight of brown rice. Rice bran is a nutritionally valuable by-product of paddy milling containing significant amounts of bioactive phytochemicals such as tocopherols, tocotrienols, γ-oryzanol, and other plant sterols (Yilmaz et al. 2014). Utilization of rice bran is severely restricted by the activity of endogenous enzymes, such as lipase, which can hydrolyze the triglyceride into glycerol and free fatty acids (FFAs). Immediately after milling, the FFA level begins to increase and the bran soon becomes unsuitable for further processing to produce oil for human consumption or as animal feed owing to its reduced pH, rancid flavor, and soapy taste. Lipases are the major cause of this deterioration in addition to oxidases (indirectly). In order to better utilize the rice bran from rice milling process, the oil must be extracted quickly after milling process to control FFAs at a low level. However, this practice may not be feasible due to restriction of facility and production schedule. Therefore, it is a normal practice that the bran undergoes an energy-intensive extrusion stabilization process to inactivate the potent lipase enzyme immediately after milling to extend storage time at ambient conditions for controlling the FFA concentration at a low level (less than 10%) before oil extraction (Wang et al. 2016).

IR irradiation can be used for the inactivation of lipases in order to extend the shelf life of rice bran. Rice bran was stabilized effectively by passing it in the rice bran stabilization system (Fig. 13.12) by varying the heating time from 1 to 10 min and IR power from 200 to 900 W (Yilmaz et al. 2014).

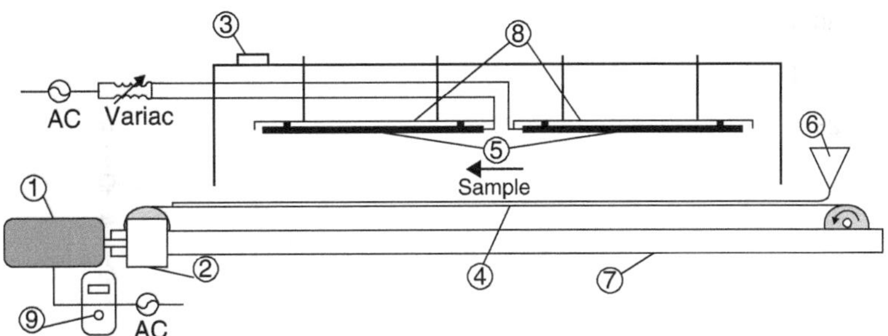

Fig. 13.12 Schematic diagram of prototype IR rice bran stabilization system: 1, electric motor; 2, gear reducer; 3, fan; 4, conveyor; 5, IR emitter; 6, hopper; 7, frame; 8, reflector; 9, electronic variator (Source: Yilmaz et al. 2014)

The FFA content of raw and IR-stabilized rice bran samples was monitored every 15 days during 6 months of storage. FFA content of rice bran stabilized at 600 W IR power for 5 min remained below 5% for 165 days without a significant change in γ-oryzanol content or fatty acid composition. However, a significant decrease in tocopherol content was observed in stabilized rice bran compared with raw bran.

IR drying can be used for simultaneous rough rice drying and stabilization of bran. The effectiveness of IR drying for simultaneous rough rice drying and rice bran stabilization was evaluated by packing rice bran samples in plastic ziplock bags obtained from milling of IR-dried and hot air (control)-dried rice samples and storing the bran for 38 days at temperature of 20 ± 1 °C and RH of 46 ± 3% (Wang et al. 2016). FFA concentrations were determined for rice bran samples produced from milling rough rice samples with different initial MCs and stored for different durations. The FFA concentration of control samples increased sharply to more than 10% in less than 7 days of storage, whereas the FFA concentration of IR-dried samples gradually increased to 10% after 17, 19, and 30 days of storage for samples with initial MC of 20.06, 25.53, and 32.5 g moisture/100 g dry solid, respectively (Fig. 13.13). The lipase activity declined with increase in tempering time and initial moisture content of the rough rice. IR heating of freshly harvested rough rice to 60 °C followed by tempering treatment for 3 h or longer has effectively inactivated the lipase and extended the storage stability of rice bran for up to 38 days after milling. At the same time, high drying rate and good milling quality have been achieved by IR drying of rough rice.

Conclusion

The consumption of brown rice has been increasing due to its higher nutrition value, nutty flavor, and rich texture. It is also desirable to store rice in brown rice form before milling which has been practiced in controlled climate with low temperature in some countries, which could save space and provide convenience. However, the short shelf life of brown rice in bulk storage before milling and in retail packing has been a significant

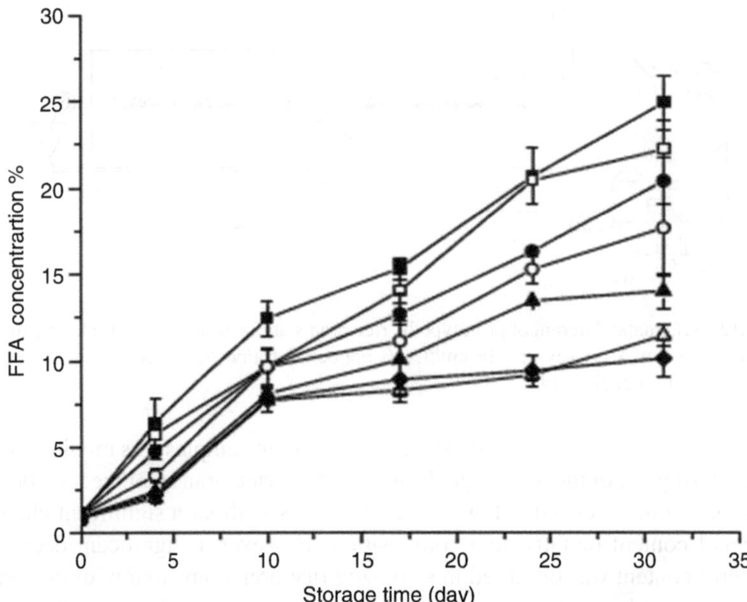

Fig. 13.13 Total FFA concentration in treated rice bran milled from IR-dried rice with initial moisture content 32.5 g moisture/100 g dry solid (Source: Wang et al. 2016)
■ control; □ IR; ● tempering, 1 h; ○ tempering, 2 h; ▲ tempering, 3 h; △ tempering, 4 h; ◆ tempering, 5 h

concern. The effective and economic way to extend the shelf life of brown rice is to inactivate the lipase enzyme to minimize the lipid oxidation. The currently used low-temperature hot air drying cannot stabilize the rice against deterioration. Recently, IR drying has been reported that it effectively inactivated the lipase enzymes during rough rice drying. The IR drying resulted in high drying rate, good milling quality, effective disinfestation and disinfection of rough rice, and improvement in shelf life of both rough and brown rice due to partial lipase inactivation in the bran layer. More importantly, the IR-dried rice can be stored in brown rice form at ambient temperature with even longer shelf life than that of rough rice dried with ambient and hot air. The rice bran from milling of IR-dried rice has sufficient shelf life for further value-added utilization without need for additional stabilizing treatment, which make the product more competitive.

References

Abe T, Afzal TM (1997) Thin-layer infrared radiation drying of rough rice. J Agric Eng Res 67(4):289–297
Afzal TM, Abe T, Hikida Y (1999) Energy and quality aspects during combined FIR-convection drying of barley. J Food Eng 42:177–182

Amaratunga K, Pan Z, Zheng X, Thompson JF (2005) Comparison of drying characteristics and quality of rough rice dried with infrared and heated air. American Society of Agricultural Engineers Meetings Papers. ASAE paper no. 056005, St. Joseph, pp 1–10

Atungulu GG, Pan Z (2010) Infrared radiative properties of food materials. In: Pan Z, Atungulu GG (eds) Infrared heating for food and agricultural processing. CPC Press, Boca Raton, pp 19–40

Bal S, Wratten FT, Chesnen JL, Faulkner MD (1970) An analytical and experimental study of radiant heating of rice grain. Trans. ASAE 13(5):644–652

Bekki E (1991) Rough rice drying with a far-infrared panel heater. Journal of Japan Society Agriculture Machinery 53(1):55–63

Bergonio KB, Lucatin LGG, Corpuz GA, Ramos NC, Duldulao JBA (2016) Improved shelf life of brown rice by heat and microwave treatment. J Microbiol Biotechnol Food Sci 5(4):378–385

Bhattacharya KR (2004) Parboiling of rice. In: Champagne ET (ed) Rice chemistry and technology. AACC Int., Saint Paul, pp 329–404. ISBN-13: 978–1–891127-34-2

Champagne ET (1994) Brown Rice stabilization. In: Marshall WE, Wadsworth JI (eds) Rice science and technology. Marcel Dekker, Inc., New York, pp 17–35

Champagne ET, Hron RJ Sr (1992) Stabilizing brown rice to lipolytic hydrolysis by ethanol vapors. Cereal Chem 69(2):152–156

Champange ET, Hron RJ Sr, Abraham G (1991) Stabilizing brown rice products by aqueous ethanol ethanol extraction. Cereal Chem 67:267–271

Chen Y, Jiang W, Jiang Z, Chen X, Cao J, Dong W, Dai B (2015) Changes in physicochemical, structural, and sensory properties of irradiated brown japonica rice during storage. J Agric Food Chem 63:4361–4369

Chua KJ, Chou SK, Mujumdar AS, Ho JC, Hon CK (2004) Radiant-convective drying of osmotic treated agro-products: effects on drying kinetics and product quality. Food Control 15:145–158

Cihan A, Kahvec K (2007) Modeling of intermittent drying of thin layer rough rice. J Food Eng 79(1):293–298

Dangerskog M, Osterstrom L (1979) Infra-red radiation for food processing I. A study of fundamental properties of infrared radiation. Lebensm Wiss Technol 12(4):237–242

Das I, Das SK (2010) Emitters and infrared heating system design. In: Pan Z, Atungulu GG (eds) Infrared heating for food and agricultural processing. CRC Press, Boca Raton, pp 57–88

Das I, Das SK, Bal S (2004a) Drying performance of a batch type vibration aided infrared dryer. J Food Eng 64(1):129–133

Das I, Das SK, Bal S (2004b) Specific energy and quality aspects of infrared (IR) dried parboiled rice. J Food Eng 62:9–14

Das A, Das S, Subudhi H, Mishra P, Shrama S (2012) Extension of shelf life of brown rice with some traditionally available materials. Indian J Tradit Knowl 11(3):553–555

Ding C, Khir R, Pan Z, Zhao L, Tu K, El-Mashad H, McHugh TH (2015) Improvement in shelf life of rough and brown rice using infrared radiation heating. Food Bioprocess Technol 8(5):1149–1159

Ding C, Khir R, Pan Z, Wood DF, Tu K, El-Mashad H, Berrios J (2016) Improvement in storage stability of infrared dried rough rice. Food Bioproc Tech: Int J 9:1010–1020

Ding C, Khir R, Pan Z, Venkitasamy C, Tu K, El-Mashad H, Berrios J (2017) Influence of infrared radiation drying on storage characteristics of brown rice. Submitted to J Cereal Sci

Dostie M, Seguin JN, Maure D, Ton-That QA, Chatingy R (1989) Preliminary measurement on the drying of thick porous materials by combinations of intermittent infrared and continuous convection heating. In: Mujumdar SA, Roques MA (eds) Drying. Hemisphere Press, New York

Gabel M, Pan Z, Amaratunga KSP, Harris L, Thompson JF (2006) Catalytic infrared dehydration of onions. J Food Sci 71(9):E351-E357

Genkawa T, Uchino T, Inoue A, Tanaka F, Hamanaka D (2008) Development of a low-moisture-content storage system for brown rice: storability at decreased moisture contents. Biosyst Eng 99(4):515–522

Ginzberg AS (1969) Application of infrared radiation in food processing. Leonard Hill Books, London

Il'yasov SG, Krasnikov VV (1991) Physical principles of infrared radiation of foodstuffs. Hemisphere Publishing Corp, New York

Javier EQ (2004) Let's promote brown rice to combat hidden hunger. Rice Today, January, 38

Juliano BO, Bechtel DB (1985) The rice grain and its gross composition. In: Juliano BO (ed) Rice Chemistry and technology, 2nd edn. American Association of Cereal Chemists, Eagan, pp 17–57

Kester EB (1951) Stabilization of brown rice. US Patent 2,538,007

Keya EL, Sherman U (1997) Effects of a brief, intense infrared radiation treatment on the nutritional quality of maize, rice, sorghum, and beans. Food Nutr Bull 18(4):382–387

Khir R, Pan Z, Salim A, Hartsough BR, Mohamed S (2011) Moisture diffusivity of rough rice under infrared radiation drying. LWT-Food Sci Tech 44(4):1126–1132

Khir R, Pan Z, Thompson JF, Salim A, Hartsough BR, Salah M (2014) Moisture removal characteristics of thin layer rough rice under sequenced infrared radiation heating and cooling. J Food Process Preserv 38(1):430–440

Kocabiyik B (2010) Combined infrared and hot air drying. In: Pan Z, Atungulu GG (eds) Infrared heating for food and agricultural processing. CRC Press, Boca Raton, pp 101–116

Krishnamurthy K, Khurana HK, Jun S, Irudayaraj J, Demirci A (2008) Infrared heating in food processing: an overview. Compr Rev Food Sci Food Saf 7:2–13

Krust PW, McGlauchlin LD, Mcquistan RB (1962) Elements of infrared technology. Wiley, New York

Kumar DGP, Hebbar HU, Sukumar D, Ramesh MN (2005) Infrared and hot-air drying of onions. J Food Process Preserv 29:132–150

Kunze OR, Calderwood DL (2004) Rough rice drying: Moisture adsorption and desorption. In Rice: Chemistry and Technology, 223–268. St. Paul, Minn. American Association of Cereal Chemists

Lentz RR, Pesheck PS, Anderson GR, DeMars J, Peck TR (1995) Method of processing food utilizing infrared radiation. US Patent 5,382,441

Lia BX, Lua YP, Liua LH, Kudo K, Tana HP (2004) Analysis of directional radiative behavior and heating efficiency for a gas-fired radiant burner. J Quant Spectrosc Radiat Transf 92:51–59

Malekian F, Rao RM, Prinyawiwatkul W, Marshall WE, Windhauser M, Ahmedna M (2000) Lipase and lipoxygenase activity, functionality and nutrient losses in rice bran during storage (Bull., no. 870, pp. 1–68). Baton Rouge: LSU Agricultural Centre.

McCabe D (1976) Process for preparing a quick-cooking brown rice and the resulting product. US Patent 3,959,515

Meeso NA, Madhiyanon T, Soponronnarit S (2004) Influence of FIR irradiation on paddy moisture reduction and milling quality after fluidized bed drying. J Food Eng 65(2):293–301

Modest MF (1993) Radiative heat transfer. McGraw-Hill International Editions, New York

Mongpreneet S, Abe T, Tsurusaki T (2002) Accelerated drying of welsh onion by far infrared radiation under vacuum conditions. J Food Eng 55:147–156

Navari P, Andrieu J, Gevaudan A (1992) Studies on infrared and convective drying of non-hygroscopic solids. In: Mujumdar SA, Roques MA (eds) Drying. Elsevier Science, Amsterdam, pp 685–694

Nindo CI, Mwithiga G (2010) Infrared drying. In: Pan Z, Atungulu GG (eds) Infrared heating for food and agricultural processing. CRC Press, Boca Raton, pp 89–100

Nowak D, Lewicki PP (2004) Infrared drying of apple slices. Innovative food Sci. Emerging Technol 5(3):353–360

Ory RL, Delucca AJ, St. Angelo AJ, Dupuy HP (1980) Storage quality of brown rice as affected by packaging with and without carbon dioxide. J Food Protection 43:929–932

Pan Z, Atungulu GG (2010a) Infrared dry blanching. In: Pan Z, Atungulu GG (eds) Infrared heating for food and agricultural processing. CRC Press, Boca Raton, pp 169–201

Pan Z, Atungulu GG (2010b) Infrared heating for food and agricultural processing. CRC Press, Boca Raton

Pan Z, Olson DA, Amaratunga KSP, Olsen CW, Zhu Y, McHugh TH (2005) Feasibility of using infrared heating for blanching and dehydration of fruits and vegetables. ASAE paper no. 056086. American Society of Agricultural Engineers, USA

Pan Z, Yang J, Brandl M, McHugh HT, Bingol G (2008) Infrared heating for improved safety and processing efficiency of dry roasted almonds. Report for Almond Board of California, pp 1–20

Pan Z, Khir R, Bett-Garber KL, Champagne ET, Thompson JF, Salim A, Hartsough BR, Mohamed S (2011) Drying characteristics and quality of rough rice under infrared radiation heating. Trans ASABE 54(1):203–210

Pan Z, Venkitasamy C, Li X (2016). Infrared processing of foods. Elsevier reference module in food sciences. doi:10.1016/B978-0-08-100596-5.03105-X

Patindol J, Wang YJ, Jane JI (2005) Structure-functionality changes in starch following rough rice storage. Starch/Stärke 57:197–207

Prakash B (2011) Mathematical modeling of moisture movement within a rice kernel during convective and infrared drying. PhD dissertations. University of California Davis, Department of Biological and Agricultural Engineering. Available online from ProQuest Dissertations & Theses Global

Ramaswamy HS, Marcohe M (2005) Food processing principles and applications. CRC Press, Boca Raton

Rao ON (1983) Effectiveness of infrared radiation as a source of energy for paddy drying. J. Agric. Eng. 20(2):71–76

Riadh MM, Ahmad SAB, Marhaban MH, Soh AC (2015) Infrared heating in food drying: an overview. Dry Technol 33(3):322–335

Sakai N, Hanzawa T (1994) Applications and advances in far-infrared heating in Japan. Trends Food Sci Technol 5:357–362

Sandu C (1986) Infrared radiative drying in food engineering: a process analysis. Biotechnol Prog 2:109–119

Schroeder HW, Roseberg DW (1960) Effect of infrared intensity on vixero visero rough rice. Rice J 61(1):28–43

Sharma GP, Verma RC, Pathare PB (2005) Thin layer infrared radiation drying of onion slices. J Food Eng 67(1):361–366

Sharp R, Timme L (1986) Effects of storage time, storage temperature, and packaging method on shelf life of brown rice. Cereal Chem 63(3):247–251

Shih C, Pan Z, McHugh TH, Wood D, Hirschberg E (2008) Sequential infrared radiation and freeze-drying method for producing crispy strawberries. Trans ASABE 51(1):205–216

Siegel R, Howell JR (2002) Thermal radiation heat transfer, 3rd edn. Hemisphere, Washington, 868 pp

Tan M, Chus KJ, Mujumdar AS, Chou SK (2001) Effect of osmotic pre-treatment and infrared radiation on drying rate and color changes during drying of potato and pineapple. Drying Tech 19(19):2193–2207

Thakur AK, Gupta AK (2006) Water absorption characteristics of paddy, brown rice and husk during soaking. J Food Eng 75(2):252–257

Tirawanichakul S, Bualuang O, Tirawanichakul Y (2012). Study of drying kinetics and qualities of two parboiled rice varieties: hot air convection and infrared irradiation. Songklanakarin Journal of Science and Technology, 34(5):557–568

Van Atta GR, Kester EB, Olcott HS (1952) Method of stabilizing rice products. US Patent 2,585,978

Wang B, Khir R, Pan Z, EI-Mashad H, Atungulu GG, Ma H, McHugh TH, Qu W, Wu B (2014) Effective disinfection of rough rice using infrared radiation heating. J Food Prot 77(9):1538–1545

Wang T, Khir R, Pan Z, Yuan Q (2016) Simultaneous rough rice drying and rice bran stabilization using infrared radiation heating. Accepted by LWT – Food Science and Technology

Yilmaz N, Tuncel BN, Kocabiyik H (2014) Infrared stabilization of rice bran and its effects on gamma-oryzanol content, tocopherols and fatty acid composition. J Sci Food Agric 94:1568–1576

Chapter 14
Storage Entomology of Brown Rice

M. Loganathan, J. Alice R.P. Sujeetha, and R. Meenatchi

Introduction

Rice (*Oryza sativa L.*) plays a fundamental role in world food security and socio-economic development. It is the staple food of more than half of the world's population and provides employment for millions of rice producers, processors, and traders worldwide. The food components and environmental load of rice depends on the rice form that results from different processing conditions. Brown rice (BR), germinated brown rice (GBR), and partially milled rice (PMR) contain more health beneficial food components compared to well-milled rice (WMR) (Roy et al. 2011). Brown rice is also known as "hulled" or "unmilled" rice. Brown rice is an excellent source of manganese and a good source of the minerals selenium and magnesium. It is whole grain rice. It has a mild, nutty flavor and is more nutritious than white rice, but goes rancid more quickly because of the bran and germ, which are removed to make white rice. Any rice, including long-grain, short-grain, or glutinous rice, may be eaten as brown rice.

The complete milling and polishing that converts brown rice into white rice destroys 67% of the vitamin B3, 80% of the vitamin B1, 90% of the vitamin B6, half of the manganese, half of the phosphorus, 60% of the iron, and all of the dietary fiber and essential fatty acids (Pankaj 2008). The thiamine and oil contents of brown rice are approximately five times that of white rice, while the fiber, niacin, phosphorous, potassium, iron, sodium, and riboflavin contents are approximately 2–3 times greater (Champagne and Hron 1992). Brown rice packs a double punch by being a concentrated source of the fiber needed to minimize the amount of time cancer-causing

M. Loganathan (✉) • R. Meenatchi
Indian Institute of Food Processing Technology, Thanjavur, Tamil Nadu, India
e-mail: logu@iifpt.edu.in; meena@iifpt.edu.in

J.A.R.P. Sujeetha
National Institute of Plant Health Management (NIPHM), Hyderabad, Telungana, India
e-mail: dirpqpniphm-ap@nin.in

© Springer International Publishing AG 2017
A. Manickavasagan et al. (eds.), *Brown Rice,*
DOI 10.1007/978-3-319-59011-0_14

substances spend in contact with colon cells and being a good source of selenium, a trace mineral that has been shown to substantially reduce the risk of cancer. It is essential to store brown rice by appropriate ways to reduce the qualitative and quantitative losses caused by abiotic and biotic factors affecting them.

Storage of Brown Rice

Storing brown rice offers considerable advantages, i.e., handling a smaller quantity and lower space requirements, as the husk contributes about one-fourth of the weight and over one-third the volume of paddy (Houston 1972). Storage has an important facet causing rice to age. Aging is a natural and spontaneous phenomenon involving changes in the physical and chemical characteristics of the rice that modify the cooking, processing, eating, and nutritional qualities, and affect the commercial value of the grain. Storage is an important step for the management of biotic and abiotic factors influencing the quality of brown rice. Stored brown rice can have losses in both quantity and quality. The storage of brown rice brings about several problems, the most important of which is spoilage. Of prime importance in this respect are microorganisms, insects, rodents, and other pests, causing huge losses of quantity and quality (Neethirajan et al. 2007). The hulling process also breaks up cells in the outer layer, liberating free fatty acids that cause rancidity and off-flavor. In fact, inadequate storage brings about undesirable changes in the properties and characteristics of rice. It was also reported noticeable flavor deterioration in polished rice (cooked samples) within 2 to 4 weeks of storage at ambient conditions (Mitsuda et al. 1972).

The bran layers that are milled from brown rice to obtain white rice account for the higher nutritive content of brown rice. These bran layers also have cholesterol-reducing properties (Kahlon et al. 1989; Hegsted et al. 1990). However, the use of brown rice has been limited because the oil in the bran is susceptible to becoming rancid, leading to shortened shelf life (about three to 6 months) due to off-flavors.

Insects Infesting Brown Rice

The lesser grain borer, *Rhyzopertha dominica* (F.), rice weevil, *Sitophilus oryzae* (L.), sawtoothed grain beetle, *Oryzaephilus surinamensis*, flat grain beetle, *Laemophloeus pusillus* (Chon.), rice moth, *Corcyra cephalonica* (Staint.), Indian meal moth, and *Plodia interpunctella* (Hbn.) are the most destructive insect pests of brown rice. In addition, the red flour beetle, *Tribolium castaneum* (Herbst), a cosmopolitan grain pest, can readily feed and develop on brown rice (Kavallieratos et al. 2015). These insects feed on the kernels, at times almost completely destroying them. Insects not only cause damage to grain but also become the source of contamination. Weevil progeny were more numerous in brown rice than in white rice, associated with the deterioration of unsaturated fatty acids.

Active insects generate body heat and give off moisture, both of which are absorbed by the surrounding grain, thus producing hotspots. Insects produce heat and moisture due to their metabolic activities, which can lead to the growth of microflora and the development of hotspots in rice (Neethirajan et al. 2007). The grain returns to normal temperature after being fumigated, unless other factors are involved in the heating. Conditions other than insect infestation may, of course, cause grain to heat.

Sources of Infestation

- Infestation through migration
- Infestation through conveyance
- Infestation through storage building
- Survival from last season
- Infestation arising from mills
- Cross infestation
- Residual infestation

Residual infestation results from attack by insects which have remained in the structure of the store, vessel, or vehicle after the removal of a previously infested commodity. Brown rice collected from spillages in the processing area, warehouse, or in the holds of a bulk storage container support insect populations that are capable of reproducing in large numbers and infesting clean grain.

Parameters Influencing the Infestation of Stored Pests

Degree of milling, high temperature, high moisture content, storage time, and packaging methods influence the insect infestation and deterioration of brown rice.

Degree of Milling

Low milling degree enhances odor deterioration in brown rice. Well-milled rice keeps better than under-milled rice. However, relatively small differences in milling degree might result in noticeable differences in lipid content, hydrolytic and oxidative enzymatic activities of the surface layer, and, consequently, different stabilities.

Brown rice (not polished) and white rice (infested then polished or polished then infested) was used with different polishing process intensities and different initial weevil densities. After the hulls have been removed from the rice, some insects of minor consideration in rough rice become important. The main offenders in brown and milled rice are the sawtoothed grain beetle, red flour beetle, *T. castaneum* (Herbst.), confused flour beetle, *T. confusum* (Duv.), flat grain beetle, Indian meal moth, almond moth, rice moth, corn sap beetle, *Carpophilus dimidiatus* (F.), psocids, lesser grain borer, and rice weevil. The damage to the surface of the kernel make them not only very objectionable but costly to the industry (Lucas and Riudavets 2000). In brown rice, *Sitophilus* is the major pest and frequently does major damage. Occasionally, the lesser grain borer infests brown rice, especially when it is packaged in paper bags or cardboard cartons.

Storage Moisture and Temperature

Storage moisture and temperature are considered as important factors for insect and mold growth. Molds were observed on rice with more than 14.4% moisture content, but no mold was observed on rice with moisture content less than 12.8%. The storage of brown rice at low moisture content could be as effective as low-temperature storage of brown rice. Temperature has a positive correlation with the moisture content of brown rice. When brown rice was stored at a low temperature of 4°C for 6 months, no changes in the moisture content was observed. However, after storage at 25°C, there was a reduction in the moisture content and this moisture loss altered the microstructure of the starch granule, affecting the cooking and textural properties of brown rice. It is desirable to store brown rice at low temperatures of 4°C to avoid the creation of unfavorable environments leading to off-odors and insect attacks reducing the quality of the rice.

Storage Time

Aging is an important factor deteriorating the quality of brown rice. Many traditional materials have been tested for the storability of brown rice (Das et al. 2012). Six months' storage revealed that parad tablet and boric acid remained free-flowing and maintained a healthy look, while all others became infested with ants (Das et al. 2012). Parad tablets are used mainly in Ayurvedic medicine and people used to wrap a single tablet in muslin cloth for the safe storage of pulses and are generally used for deworming. The prolonging of shelf life might be due to the suppression of fatty acids and malondialdehyde.

Packaging

Brown rice is expected to store for only 6 months under average conditions. This is because of the presence of essential fatty acids, which quickly go rancid as they oxidize. Brown rice is more nutritious than milled rice, but there is a traditional consumer preference for white (milled) rice, which has better appearance, is translucent, and more palatable. The short shelf life has been implicated as a deterrent to larger amounts of brown rice being packaged for direct consumption (Schutz and Fridgen 1974). Ibni Hajar et al. (1997) reported that free fatty acid content is between 8.3% and 15.3%, and after 6 months, the content is between 53.0% and 65.3%. Fatty acids can be released by lipase present in the rice aleurone (bran) layer of damaged grains and by high lipase-containing bacteria and fungi adhering to rice (DeLucca et al. 1978). Infection and infestation deteriorate the quality during storage mainly because of lipolytic hydrolysis of about 3% oil present in the rice (Hunter et al. 1951). The simplest packaging technique is the use of polyethylene bags. Using this technique, brown rice and partially milled rice can be stored at ambient conditions for only 3 months due to subsequent insect infestation.

Wahid et al. (2003) studied brown rice storage with different packaging techniques, viz., vacuum, use of oxygen absorbents, and carbon dioxide, and compared them with packaging using polyethylene plastic (PE) as the control, with storage at ambient temperatures for 6 months. They reported that brown rice and partially milled rice can be stored for 6 months and 4 months, respectively, without insect infestation. An insect-infested brown rice sample was found in PE packaging (control) after 3 months of storage. All the packaging techniques used were lower in moisture content, higher in bulk density, and not significantly different in texture and flavor of cooked rice compared to the PE packaging. The packaging techniques used did not have much of a beneficial effect on the quality maintenance for partially milled rice during storage.

Hermetic Storage

The storage life of brown rice can be extended for years when packaged in air-tight containers with an inert nitrogen atmosphere. Currently, in some regions, brown rice is parboiled to increase its shelf life. In other regions, refrigeration is preferred. Parboiling reduces the quality of brown rice and refrigeration is expensive. Shelf life extension is important for transportation over long distances or overseas markets.

De Dios (2004) attempted to explore the benefits derived from hermetic storage of brown rice, milled rice at various milling degrees, and rice bran. It was observed that the oxygen concentration in rice bran, brown rice, and regular milled rice reduced to 7.6%, 10.6%, and 15.9%, respectively, after 3 to 6 months of storage. Although these atmospheres were not sufficient to obtain complete mortality of insects, the modified atmosphere did retard insect growth and development, as evidenced by weak and abnormal progenies of *R. dominica*. Rice quality was preserved throughout the storage period. The storage atmosphere is not only modified by insect metabolism but also through chemical oxygen-depleting activity of these commodities (De Dios 2004). Pioneering modern hermetic storage (Calderon and Navarro 1980; Navarro et al. 1989; Navarro and Calderon 1980) has resulted in the broad use of safe, pesticide-free hermetic storage suitable for many commodities and seeds, particularly in hot, humid climates.

Conclusion

Brown rice has a shelf life of approximately 6 months. Hence, storage plays a vital role in providing insect-free rice to consumers and to reach international markets. Different storage practices such as hermetic storage, refrigeration, or freezing can significantly extend the lifetime of rice. But under tropical conditions, freezing is costly and hermetic storage using superbugs is cheaper and viable. The use of low-density polyethylene (LDPE) as a packaging material can prevent reinfestation. The use of aluminum phosphide tablets is a common practice to protect rice from stored pests in bulk storage and warehouses. However, other parameters such as moisture content and temperature have positive effects on the growth and development of stored pests. Adopting integrated management practices and the use of alternative methods will certainly help us to manage the resistant population against stored pests of brown rice.

The research on the storage of and insect infestation in brown rice is limited. Hence, further studies are required to ensure proper storage of brown rice without insect infestation, rancidity, and with good cooking qualities.

References

Calderon M, Navarro S (1980) Synergistic effect of CO_2 and O_2 mixtures on two stored grain insect pests. In: Shejbal J (ed) Controlled atmosphere storage of grains. Elsevier, Amsterdam, pp 79–84

Champagne ET, Hron RJ (1992) Stabilizing brown rice to lipolytic hydrolysis by ethanol vapors. Cereal Chem 69:152–156

Das A, Das S, Subudhi H, Mishra P, Sharma S (2012) Extension of shelf life of brown rice with some traditionally available materials. Indian J Trad Knowl 11(3):553–555

De Dios CV, Natividad DG, Tampoc EA, Javier LG (2004) Storage behavior of rice and rice bran in hermetically sealed container. In: Donahaye EJ, Navarro S, Bell C, Jayas D, Noyes R, Phillips TW (eds) Proceedings of the international conference on controlled atmosphere and fumigation in stored products, Gold Coast, 8–13 Aug 2004. FTIC Ltd. Publishing, Israel, pp 405–425

DeLucca AJ, Plating SJ, Ory RL (1978) Isolation and identification of lipolytic microorganisms found on rough rice from two growing areas. J Food Prot 41:28–30

Hegsted M, Windhauser MM, Lester SB, Morris SK (1990) Stabilized rice bran and oat bran lower cholesterol in humans. FASAB J 4:368

Houston DF (1972) Rice hulls. In: Houston DF (ed) Rice: chemistry and technology. American Association of Cereal Chemists, St. Paul, Minnesota, pp 301–352

Hunter IR, Houston DF, Kester EB (1951) Development of free fatty acids during storage of brown (husked) rice. Cereal Chem 28:232–239

Ibni Hajar R, Ajimilah NH, Rosniyana A, Wahid S (1997) Quality changes in brown and partially milled rice during storage under modified atmosphere. In: Proceedings of the asian food technology seminar '97, pp 201–205

Kahlon TS, Saunders RM, Chow FI, Chiu MC, Betschart AA (1989) Influence of rice, oat and wheat bran on plasma cholesterol in hamsters. FASAB J 3:958

Kavallieratos NG, Athanassious SG, Arthur FH (2015) Efficacy of deltamethrin against stored-product beetles at short exposure intervals or on a partially treated rice mass. J Econ Entomol 108:1416–1421

Lucas E, Riudavets J (2000) Lethal and sublethal effects of rice polishing process on Sitophilus oryzae (Coleoptera: Curculionidae). Econ Entomol 93:1837–1841

Mitsuda HF, Kawai F, Yamamoto A (1972) Underwater & underground storage of cereal grains. Food Technol 26(3):50–56

Navarro S, Calderon M (1980) Integrated approach to the use of controlled atmospheres for insect control in grain storage. In: Shejbal J (ed) Controlled atmosphere storage of grains. Elsevier, Amsterdam, pp 73–78

Navarro S, Donahaye E, Dias R, Jay E (1989) Integration of modified atmospheres for disinfestation of dried fruits. Final scientific report of project no: I-1095-86, submitted to US-Israel Binational Agricultural Research & Development Fund (BARD), 86 pp.

Neethirajan S, Karunakaran C, Jayas DS, White NDG (2007) Detection techniques for stored-product insects in grain. Food Control 18:157–162

Pankaj (2008) Brown rice is healthier white rice. Times of India, December 25, 2014

Roy P, Orikasa T, Okadome H, Nakamura N, Shiina T (2011) Processing conditions, rice properties, health and environment. Int J Environ Res Public Health 8:1957–1976

Schutz HG, Fridgen JD (1974) Consumer preference research on rice: ranking descriptions for improved products. Food Prod Dev 8(1):75–78

Wahid S, Rosniyana A, Hashifah MA, Norin SAS (2003) Packaging techniques to prolong the shelf life of brown rice and partially milled rice at ambient storage. J Trop Agric Food Sci 31(2).225–233

Part VI
Challenges in Consumption and Marketing

Chapter 15
Hurdles in Brown Rice Consumption

V. Mohan, V. Ruchi, R. Gayathri, M. Ramya Bai, S. Shobana,
R.M. Anjana, R. Unnikrishnan, and V. Sudha

Introduction

Rice is consumed as a staple by almost half the people on Earth and is associated with community life, society, and culture (Ricepedia 2016). The consumption of rice and its production are connected to seasons, rituals, celebrations, community activities, and religious functions in rice-eating countries like India, Japan, China, Korea, Indonesia, and many others (Bhattacharya 2011). According to the archeological evidence, the use of rice in China, India, and the Indus valley is more than 8000 years old (Chang 2003). In India, it is first mentioned in the *Yajurveda* around 4530 BC and in China in 2800 BC (Achaya 2009).

Rice was consumed as hand-pounded grain and not as whole grain brown rice in ancient times. Brown rice could be processed only after the invention of appropriate shelling technology in the year 1963. Brown rice retains 100% of its bran and germ (Juliano 1993) and is rich in dietary fiber, B vitamins, tocotrienols, tocopherols, minerals, phytic acid, γ-oryzanol, γ-amino butyric acids (GABA), and phenolic acids (Dinesh Babu et al. 2009). Several studies have shown that high intake of brown rice lowers the risk of developing obesity, cardiovascular disease (CVD), type 2 diabetes, and some cancers (Sun et al. 2010; Mohan et al. 2014; Seal and Brownlee 2015; Harris and Kris-Etherton 2011).

V. Mohan (✉) • R. Anjana • R. Unnikrishnan
Department of Diabetology, Madras Diabetes Research Foundation,
Chennai, Tamil Nadu, India
e-mail: drmohans@diabetes.ind.in; dranjana@drmohans.com; unnikrishnan_ranjit@yahoo.com

V. Ruchi • R. Gayathri • M. Ramya Bai • S. Shobana (✉) • V. Sudha
Department of Foods, Nutrition & Dietetics Research, Madras Diabetes Research Foundation,
Chennai, Tamil Nadu, India
e-mail: ruchivaidya@mdrf.in; gayathri@mdrf.in; ramya@mdrf.in;
shobana@mdrf.in; s2r_7@mdrf.in

© Springer International Publishing AG 2017
A. Manickavasagan et al. (eds.), *Brown Rice*,
DOI 10.1007/978-3-319-59011-0_15

However, with the advancements in milling technology to retain better yield of head grain and shelf life, the hand-pounded rice was replaced by milled (polished) white rice. The fiber, vitamins, minerals, and phytochemicals are lost in the process of milling of brown rice (Shobana et al. 2011), resulting in an endosperm-rich grain with easily digestible carbohydrates and resultant high glycemic index (GI, the parameter that measures the glycemic property of food). Both epidemiological and clinical trials have shown that high intake of refined grains (mainly white rice) contributes to the risk of insulin resistance, obesity, type 2 diabetes, and metabolic syndrome (Mohan et al. 2009; Song et al. 2014; Anjana et al. 2015). Noncommunicable diseases (NCDs) such as CVD, type 2 diabetes, and cancer account for 68% of deaths globally (WHO 2014). It is, therefore, probably not surprising that the two major consumers of rice, China and India, rank first and second in prevalence of diabetes globally. To combat the growing epidemic of chronic diseases, the national dietary guidelines of many countries encourage the intake of whole grains like brown rice. For instance, the USDA (2016) recommends that whole grains should constitute at least half of the grains in the diet so as to reduce the risk of NCDs. The national dietary guidelines of the UK, Australia, Canada, Singapore, India, and many other countries also encourage the intake of whole grains like brown rice (FAO 2016). Despite such potential health benefits, whole grain intakes continue to be low across the world (Thane et al. 2005, 2007; Richardson 2003) such as 1–3 servings per week among UK adults (Lang et al. 2003), 0–2.7 servings per day among US adult females (Liu et al. 2003), and 40 g/d among Indian adults (Mohan et al. 2009) For the past few decades, the food industry has invested significant efforts to increase the accessibility and availability of whole grain foods like brown rice and to assist in consumer choices, for example, by highlighting nutrient content of brown rice on front-of-pack labeling along with whole grain health claims. However, there are major barriers to brown rice consumption (Jonnalagadda et al. 2011) as shown in Fig. 15.1. The present chapter discusses the barriers to consumption of brown rice and suggestions to improve and encourage its consumption.

Hurdles to Brown Rice Consumption

Market Availability of Authentic Brown Rice

The authenticity of brown rice available in the market has been questionable. Ancient Indians consumed "hand-pounded rice" that was misinterpreted as brown rice (whole grain) in the ancient Indian diets. The paddy was manually hand pounded with a stone mortar and pestle and was further winnowed and used for cooking in the past (Champagne et al. 2004). During this process minimal amounts of bran and germ (rich in micro- and macronutrients) were lost. This hand-pounded rice is many a time misunderstood as brown rice till date.

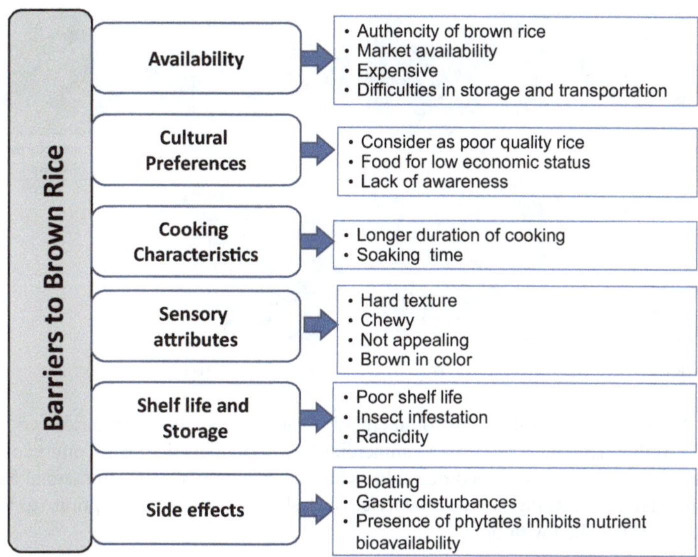

Fig. 15.1 Barriers to brown rice consumption

Lack of awareness among the population on authenticity of brown rice is a major factor that affects its promotion and availability. In India, consumption of parboiled and polished red pigmented rice (termed as *matta* rice) is common in the state of Kerala; in other parts of the country, this is misinterpreted as brown rice due to the characteristic reddish-brown color retained by the polished kernel even after polishing. This color is mainly due to the penetration of bran pigments into the endosperm, but the kernel is devoid of bran and germ constituents. In addition, polished red rice with label claims of *"ideal for diabetes and weight management," "low GI rice," "herbally treated rice suitable for diabetes,"* and *"quick cooking brown rice,"* etc. is now available in the market. The morphological attributes of some of these rice varieties have been evaluated using stereo zoom microscope and are shown in Fig. 15.2 (Shobana et al. 2015). Figure 15.2a shows commercially available so-called brown rice with bran disruption. Figure 15.2b and c shows commercially available "hand-pounded rice" and "semi brown rice" which are misleading to the consumers; both of these rice are under milled rice, and the microscopic evaluation shows clear disruption of bran and germ and is not equivalent to brown rice. Kerala polished red rice is shown in Fig. 15.2e. Efficient shelling practices are required to prepare brown rice without disruption of bran layer. Disruption of bran layer induces oxidative rancidity in brown rice and reduces shelf life. Figure 15.2f and g shows pigmented red and black-brown rice with disruption of bran. Some leading brands sell polished rice as brown rice (Fig. 15.2d). Authentic brown rice is generally not available at places easily accessible by the general public like local grocery stores or the public market, but is instead sold in high-end supermarkets (Romero 2011). Owing to its short shelf life, brown rice is purchased in small quantities as against polished white rice which

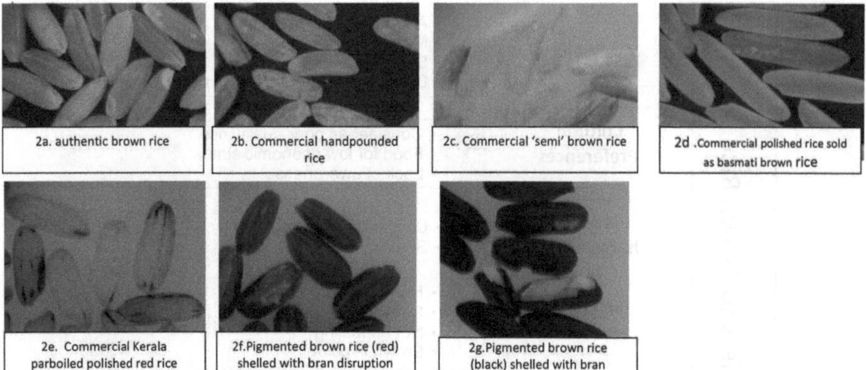

Fig. 15.2 Morphological attributes of commercially available so-called brown rice varieties in the market. (**a**) Authentic brown rice. (**b**) Commercial hand-pounded rice. (**c**) Commercial "semi" brown rice. (**d**) Commercial polished rice sold as basmati brown rice. (**e**) Commercial Kerala parboiled polished red rice. (**f**) Pigmented brown rice (*red*) shelled with bran disruption. (**g**) Pigmented brown rice (*black*) shelled with bran

can be purchased in bulk. Due to its presumed "poor appearance" and hard and chewy texture, brown rice is hardly found on restaurant menus with the possible exception of some health food restaurants (Seal and Jones 2007).

Cost is another major hurdle in the consumption of whole grain foods like brown rice. Brown rice is comparatively more expensive than white rice mainly due to its limited supply arising due to lack of demand among consumers. Also, due to oil in the bran layer, it has a shorter shelf life compared to polished rice and requires special packaging laminates (with good gas barrier properties, modified atmospheric packing such as nitrogen-flushed packing), special storage systems (modified atmospheric storage such as cocoon storage with CO2) for extended shelf life and quality control for brown rice. This in turn, makes it more expensive than brown rice (Liong 2011). Thus, more resources are needed to preserve the quality of brown rice for storage leading to increase in cost (Romero 2011).

Cultural Preferences

The food penchants that we have are as much an indication of our culture as they are about ourselves. Rice is considered as "food of God" and is used traditionally in many rituals and festivals in India, Japan, China, and other rice-eating countries. Therefore, rice comes with a spiritual and cultural baggage. White rice is generally considered as pure in appearance, as higher in quality, and as a food of affluence which in turn acts as a barrier to brown rice consumption. The comparison of various aspects of brown rice and white rice is shown in Table 15.1. Studies globally have shown that various factors (including distinct cultural background) might assist or hinder intake of brown rice. GNET (Global Nutrition Epidemiologic

Table 15.1 Comparison of brown rice and white rice

Factors	Brown rice	White rice
Availability	Selected grocery stores based on demand	Local grocery stores and restaurants
Cost	70–160 INR/kg	30–120 INR/kg
Appearance	Brown in color, bolder than polished grain	White in color, thin grain
Texture	Chewy	Soft
Soaking time	30–60 min	Soaking not required
Cooking time	Longer	Less
Side effects	Can cause bloating	No side effects reported so far
Acceptance	Poor acceptability	Favorable
Nutrient quality	Rich in dietary fiber, micro- and phytonutrients	Devoid of all bran and germ nutrients (dietary fiber, micronutrients, and phytochemicals)
Health benefits	Lowers the risk for NCDs like diabetes and CVD	Increases the risk for NCDs like diabetes and CVD

Transition), a group from Harvard School of Public Health, USA, has showed the intricacies involved in the substitution of brown rice for white rice. This study included focused group discussions (FGDs) from China, India, Nigeria, Costa Rica, Malaysia, Mexico, Puerto Rico, Tanzania, and Brazil. The outcomes showed limited awareness among the study participants about the health benefits of brown rice (Table 15.2). Though some were aware of the benefits and had consumed brown rice in the past, they had issues with its availability, cost, etc., especially in markets like India. Many studies have reported that the preference for brown rice is income dependent, with the higher income population preferring this healthier option (Hadi et al. 2012). In countries like India, white rice is being consumed in three meals, during breakfast as *dosa, idly, pongal,* or *upma,* during lunch as a whole meal that includes mixed rice or plain rice with lentils and cooked vegetables, and during dinner either as *dosa* or *idly* or plain rice with lentils and cooked vegetables as an accompaniment. As this is a tradition that has been followed in the Indian diets for many decades, it is challenging to attempt substitution of brown rice for white rice (Kumar et al. 2011; Sudha et al. 2013). During the FGDs conducted in Chennai, the statements that were recorded from the study participants were:

- *"Price is not very important, the rice should look good."*
- *"Only if the rice is white in colour will we be satisfied and then only will our family like to eat it…. There have been several occasions when we have bought rice from our village which was not the white variety. Our family members did eat it but they did not relish it and so I cannot cook it on a daily basis."*
- *"People think that since this rice does not look good, it is of cheap quality…. people test the appearance of the rice first, if they find it bulky and less white they may not like it…. it is like cow chewing food. This rice sits like a stone in the stomach, one can only eat less quantity otherwise it sits heavy and takes a long time to digest."*

Table 15.2 Overview of some consumer preference studies and its outcomes

S. No	Authors	Objective	Type of subjects	Outcome
1	Kumar et al. (2011), India	To conduct a qualitative study using focus groups to identify factors that can act as barriers to or promote acceptance of brown rice as a staple food among South Indian adults	Overweight adults; cut point for overweight $\geq$23) and adults with normal body mass index ($\geq$18.5–22.9) living in slum and non-slum residences in Chennai (n = 65)	Poor awareness about the nutritive properties of brown rice – a major barrier for acceptance Participants' perception about brown rice –inferior food for low socioeconomic group
2	Sudha et al. (2013), India	To study consumer acceptance of unmilled brown and under milled rice among urban South Indians	Overweight and normal weight adults living in slum and non-slum residences in Chennai (n = 82)	Participants preferred polished white rice Education regarding health benefits may help this population switch to brown rice Cooking quality and appearance of the rice were the major factors considered for preference and purchase
3	Zhang et al. (2010), China	Assessed the awareness and acceptability of brown rice in Chinese adults and examined the feasibility of introducing brown rice into the diet through a large, long-term randomized clinical trial to lower risk of type 2 diabetes	Thirty-two Chinese adults residing in Shanghai	Before tasting considered brown rice to be inferior in terms of taste and quality After tasting brown rice, gaining knowledge about its nutritional value showed willingness to consume brown rice Main barriers to acceptance – rough texture, inedible taste, and price Large-scale promotion is needed to change community approach toward brown rice

#	Reference, country	Objective	Study population	Findings
4	Muhihi et al. (2013), Tanzania	Evaluated factors which may promote acceptance of unrefined carbohydrates	A total of n = 45 (both men and women) employed at Muhimbili University of Health and Allied Sciences (MUHAS) in Dar es Salaam (n = 23) and Sokoine University of Agriculture (SUA) in Morogoro (n = 22)	High cost and unavailability of brown rice were strong barriers to its consumption Available only in supermarkets Expensive Educational strategies were recommended in order to increase consumption of unrefined carbohydrates
5	Muhihi et al. (2013), Tanzania	Assessed the preferences and acceptability of unrefined whole grain carbohydrate staples (i.e., brown rice, unrefined maize, and unrefined sorghum ugali) as substitutes for commonly consumed refined carbohydrates in Tanzania	Overweight and obese [body mass index (BMI) $\geq$ 25 kg/m^2] adults (n = 44) in the Dar es Salaam region, and Morogoro, a semiurban region, Tanzania	Brown rice was preferred based on smell, taste, color, appearance, texture, hardness, shine, and configuration Brown rice – rated positively by the participants, indicating that it was acceptable, palatable, and well tolerated
6	Monge-Rojas et al. (2014), Costa Rica	To identify barriers and motivators that could influence changes by substituting staple foods to a healthier one To assess the sensory perceptions of brown rice and beans in the traditional combinations among Costa Rican adults, using focus group discussions and food sensory tastings	Participants aged 40–65 years and free of major noncommunicable chronic diseases (i.e., hypertension, type 2 diabetes, cancer, or cardiovascular disease)	Brown rice reported the lowest mean hedonic liking scores Traditional habits and family support – major factors in food choices Strategies for consuming more brown rice from childhood should be enforced Transmit information of health benefits of brown rice with importance of tradition, cost-effectiveness, easy accessibility, and engaging women Masking the perceived unpleasant sensory characteristics of brown rice by incorporating them into mixed dishes may improve acceptance

- The other reason on the preference of rice also depends upon the head of the family being the main decision-maker and also the children's preferences, as they are very picky in selecting foods for consumption.

Similarly, FGDs have also been conducted in other rice-eating countries like China, Tanzania, and Costa Rica to record the perception of replacing brown rice for their usually consumed staple foods. In China, the brown rice was considered as inferior based on taste. One of the comments from the China FGD also stated that "it (brown rice) tasted like the chow for pigs." But after tasting and gaining awareness about its health benefits, the majority of participants were willing to participate in the intervention feeding trial to consume brown rice and also substitute it in their diets provided it is easily available and affordable and shelf life was improved (Kumar et al. 2011; Zhang et al. 2010; Hesse et al. 2009; Burgess-Champoux et al. 2006).

Sensory Perceptions

There was a mixed response to the sensory attributes of brown rice both in the uncooked and cooked form in various studies. Many studies reported on the chewy texture, poor appearance of both cooked and uncooked grains, and longer duration of cooking time with brown rice (Sudha et al. 2013; Zhang et al. 2010). On account of these factors, the consumption of polished white rice was preferred compared to brown rice. Indian and Chinese consumers did not rate brown rice positively (Zhang et al. 2010; Shobana et al. 2011). The main barriers to consumption of brown rice among Indians were lack of awareness about its nutritive properties and perceived inferiority (Kumar et al. 2011), while for Chinese adults, the barriers were inferior taste, quality, and price (Zhang et al. 2010). Earlier studies are methodologically comparable, thereby indicating that culture pervasively underlies all food choices. Most of the population is unaware of the health benefits of whole grains or of the recommendations regarding increased intake (Croy and Marquart 2005). Also, there is much confusion about the authenticity of brown rice available in the market. The bran portion of a brown rice may be highly colored and contain astringent, intensely flavored compounds that are not appealing to consumers. Other barriers to brown rice consumption include price, softness, texture, and cultural acceptance. However, even with limited knowledge about the health benefits of brown rice, obese adults in Tanzania and Minnesota rated brown rice positively with respect to its sensory attributes (Muhihi et al. 2013; Chase et al. 2003), indicating that some, if not all of these barriers, can be overcome.

Cooking Characteristics

Brown rice requires longer cooking time compared to milled rice, and after cooking the grains tend to be discrete and becomes hard upon standing. This is another major limiting factor that hinders brown rice consumption. Due to the presence of bran, brown rice is more resistant to penetration of water during cooking (whereas in white rice, absence of bran layer helps in easy hydration and faster cooking (Horigane et al. 2006). These properties of brown rice result in slow hydration rate and hard grain texture due to incomplete gelatinization of the starchy endosperm. The cooking time for brown rice normally varies between 30 and 50 min compared to 15 and 20 min for milled rice (Corpuz 2011). Some food technologists have identified methods to reduce the cooking time of brown rice which involves hydrothermal treatment resulting in minute fissures in the grain that allows rapid hydration and gelatinization of the grain (Corpuz 2011). Such technologies reduce cooking time but may possibly increase the glycemic response of rice (Kaur et al. 2016). It would be beneficial to develop methods that reduce cooking time without altering the glycemic properties of brown rice.

Shelf Life and Storage

Owing to its higher fat content, brown rice is susceptible to insect infestation and rancidity, which is a limiting factor for the popularization of brown rice. The bran present in brown rice is rich in oil that is vulnerable to hydrolytic and oxidative deterioration. Lipases, both within the bran and of microbial origin, hydrolyze the kernel oil. Lipases and oil are present in the seed coat and in the aleurone and germ, respectively. Dehulling of rice disrupts these layers; oil diffuses and gets into contact with lipases leading to the hydrolysis of triglycerides to free fatty acids. Lipases produced by mold and bacteria found on the kernel surfaces also have access to the bran oil during dehulling and promote its hydrolysis (Marshall and Wadsworth 1994). The process of hydrolytic and oxidative deterioration of brown rice is shown in Fig. 15.3. Lipoxygenase catalyzes the oxidation of free unsaturated fatty acids to hydroperoxides, which react to yield products such as aldehydes and hydroxy acids that impart the off flavors and odors often seen in brown rice (Corpuz 2011). Lipase-mediated rancidity may not occur if brown rice is parboiled as the enzyme gets

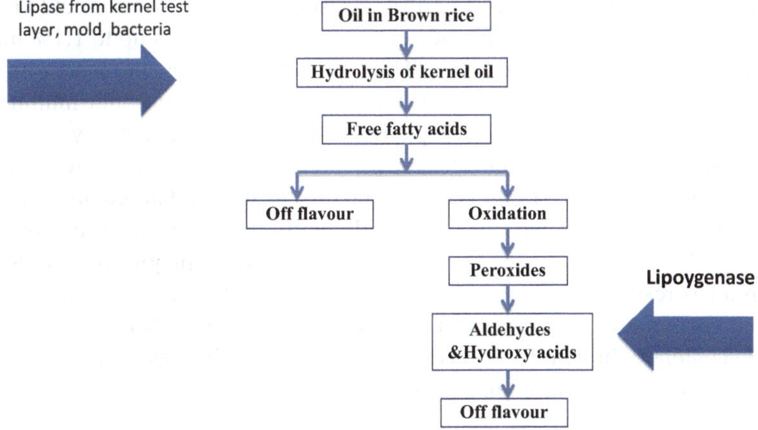

Fig. 15.3 Hydrolytic and oxidative deterioration of brown rice

inactivated during hydrothermal treatment of paddy during parboiling. However, parboiled brown rice may still become rancid due to microbial contamination promoted by its higher moisture content (Luh and Robert 1991).

The shelf life of parboiled brown rice is generally considered to be about 6 months. Purchasing small quantities of brown rice and storing it in air tight containers in a cool place may help in preservation. However, to promote brown rice consumption, it is important to develop appropriate strategies that help to improve the shelf life without altering the nutritional properties.

Side Effects

Though brown rice is nutrient dense, till date not many scientific evidences are available for its negative health effects. Due to higher fiber content, brown rice can cause bloating or feeling of fullness. Hence, among individuals with gastric disturbances like gut irritation or leaky gut, brown rice must be consumed with caution.

Callegaro and Tirapegui (1995) compared the nutritional profile of brown and white rice and concluded that anti-nutritional factors present in brown rice may adversely affect the bioavailability of nutrients. Earlier, Miyoshi et al. (1987) studied the effect of brown rice on apparent digestibility and balance of nutrients among young men on low-protein diet. The findings of the study showed that brown rice diet had more dietary fiber and the weight of fecal waste increased upon its consumption. However, on a brown rice diet, the digestibility of carbohydrates, protein, and fat decreased and so did the absorption of minerals like sodium and potassium. Nowadays high-fiber foods are preferred owing to their lower glycemic properties. The negative nitrogen balance of brown rice was higher compared to white rice. Further, the study concluded that brown rice decreased protein digestibility and led to negative nitrogen balance. However, this could be overcome by including legumes, milk, and egg in the diet.

The decreased digestibility and nutrient absorption of brown rice could be due to their phytate contents. The phytic acid content of the rice is up to 1.1% mainly (~90%) found in the aleurone layers. These phytates are known to bind with minerals like zinc, copper, iron, magnesium, niacin, and calcium and inhibit their absorption (Corpuz 2011). Further, phytic acids decrease the activity of enzyme pepsin and amylase vital for breakdown of protein and starch, respectively. Rehman and Shah (2005) observed a reduction in the phytate content of foods upon heating, and as brown rice is normally consumed after pressure cooking or steaming, the side effects due to phytates may not be of major concern. Recent research has resulted in the development of germinated brown rice (GBR); soaking and germination process during the formation of GBR helps to reduce the phytate content of brown rice to an extent and thereby enhances the positive health effects of brown rice (Wu et al. 2013).

Overcoming the Hurdles

Highly polished white rice is a preferred staple among rice-eating population, and it is very difficult to substitute brown rice over white rice due to the series of hurdles discussed earlier in the chapter. Considering these hurdles, research efforts made were successful to develop a novel Indian rice variety that would contain fiber in the form of nondigestible carbohydrates even after polishing (removal of bran and germ constituents) to mimic the functional (lower glycemic property) benefit of brown rice in addition to retaining the consumer friendly characteristics of white rice (Mohan et al. 2016).

However, considering the beneficial blend of nutritional composition and functional elements of brown rice, it is a crucial need to overcome the hurdles for its consumption. Some suggestions to overcome the hurdles to widespread use of brown rice are described in Table 15.3.

Technical Innovation

Research and development in improvising the sensory attributes and shelf life of brown rice are a crucial need. Germinated brown rice (GBR) is one such example of modified brown rice. Brown rice is soaked in water at 30 °C for 3 h and germinated for 12–20 h to get GBR. GBR has improved texture and taste compared to brown rice, and it also reduces the soaking and cooking time (Wu et al. 2013). The germination process affects pasting properties of GBR flour and would improve the bread quality when substituted for wheat flour. GBR is a whole grain as only hull is removed which causes least damage to its nutritional value. It not only contains basic nutritional components such as vitamins, minerals, dietary fibers, and essential amino acids but also contains bioactive components, such as polyphenols, γ-oryzanol, and

Table 15.3 Steps to overcome hurdles in brown rice consumption

Technical innovation	Policy advocacy	Community awareness
- Development of appropriate processing equipment - Improvement of postharvest facilities. - Promotion of germinated brown rice (GBR) - Irradiation technique may help to improve the shelf life - Food industries should come out with tasty ready to eat brown rice products	- FSSAI should set a limit for rice polishing - Enhance availability through public distribution system (PDS) - Taxes and subsidy for brown rice production as well as distribution - Eliciting cooperation of rice millers with incentives and promotions	- Promote brown rice in national health programs, nutrition societies, central and local governments, and private religious organizations - Encouragements at school, home, community, and food service managements - Changing the mindset through communication like SMS/mass media communication
Cohesive policies/integrated and affordable actions		

gamma amino butyric acid (GABA), and has a lower glycemic index. Studies have shown that substituting GBR over white rice ameliorates the hyperglycemia, boosts the immune system, lowers blood pressure, reduces risk of cardiovascular disorders, inhibits development of cancer cells, and assists in the treatment of anxiety disorders (Hsu et al. 2008; Dinesh Babu et al. 2009; Imam et al. 2014). Methods to reduce the cooking time of brown rice and also to improve the textural features of brown rice without affecting the nutritional properties of brown rice have to be devised.

It is essential for the food processing industries to initiate development of ready-to-cook and ready-to-eat health foods from brown rice to increase the range of healthy whole grain food choices. Processing techniques like irradiation in addition to already existing modified atmospheric packaging will improve the storage stability of brown rice (Chen et al. 2015). Governments should encourage these techniques by providing subsidy. Instead of refined rice flour, brown rice flour should be made available in the market for utilization in flour-based rice preparations. Moreover, the utilization of brown rice should be extended to the preparation of flakes, noodles, and such other products prepared using polished rice. Some of the common concerns regarding brown rice relate to shorter shelf life, prolonged cooking time, and chewy texture. These could be overcome by proper storage practices and improved culinary skills.

Policy Advocacy and Consumer Awareness

The rice-eating countries should mutually take an effective step to overcome the hurdles to brown rice consumption. These steps should involve departments of agriculture, health, education, science and technology, trade and industry, central and local government, social welfare organizations, and private religious organizations.

One such novel positive action was initiated in the Philippines as a campaign, **"Be RICEponsible"** (RICEponsible (2014) **www.bericeponsible.com**). It is an advocacy campaign initiated by the Department of Agriculture-Philippine Rice Research Institute (DA-PhilRice) since 2014, which aims to promote the responsible rice consumption through three major objectives: (1) reduce rice wastage, (2) increase consumption of brown rice, and (3) motivate farmers to grow rice that will contribute toward the attainment of rice sufficiency and food security in the country. With the help of a few local executives and legislators, half cup rice program (i.e., reducing the serving size to half cup in restaurants and public places to prevent rice wastage) has been implemented in about 14 city/municipal ordinances. The campaign messages on awareness/health benefits of brown rice were shown as video clips, songs, prints, expert talk, and advertisements in TV, radio, Internet, and social networking sites like Facebook, Twitter, YouTube, Instagram, etc. The campaign also conducted cooking competitions, sale of brown rice, and other activities during festivities. The second Friday of November every year is celebrated as Brown Rice Day in the Philippines, and during this day, brown rice is served nationwide in restaurants and gatherings. Thus, the collective efforts of individuals, policy makers, and government of Philippines are considered as a role model for all the rice-eating countries to spread awareness and enhance the availability and consumption of authentic brown rice.

In India, brown rice can serve as a valuable tool to reduce the prevalence of NCDs. Both local and central governments should take necessary steps to encourage brown rice production, distribution, and marketing. Rice millers can be provided incentives or avail low taxes for the production of brown rice. The government should increase the availability of brown rice by introducing it in the public distribution system (PDS), and brown rice can be easily available to government distributors. Healthy delicious dishes could be prepared out of brown rice and can be served at home at individual level and in public eating joints like restaurants.

Conclusion

Despite scientific evidence showing innumerable health benefits, brown rice suffers from low market demand and consumption owing to its high cost, short shelf life, unappealing texture, longer cooking time, and perceived presence of anti-nutrition components. Further research is also needed to improve the shelf life and duration of the brown rice without altering its functional properties like glycemic property. Lack of awareness about brown rice and cultural factors further hinder its promotion. Hence, collaborative efforts by governments, health organizations, nutrition societies, the food processing industry, policy makers, and health seekers are needed to overcome these hurdles and encourage brown rice consumption.

References

Achaya KT (2009) The illustrated food of India A–Z. Oxford University Press, New Delhi
Anjana RM, Sudha V, Nair DH et al (2015) Diabetes in Asian Indians—how much is preventable? Ten-year follow-up of the Chennai Urban Rural Epidemiology Study (CURES-142). Diabetes Res Clin Pract 109(2):253–261
Bhattacharya KR (2011) Rice quality – a guide to rice properties and analysis. Woodhead Publishing Ltd, Oxford, pp 8–18
Burgess-Champoux T, Marquart L, Vickers Z et al (2006) Perceptions of children, parents and teachers regarding whole grain foods and implications for a school based intervention. J Nutr Edu 38:230–237
Callegaro MD, Tirapegui J (1995) Comparison of the nutritional value between brown rice and white rice. Arq Gastroenterol 33(4):225–231
Champagne ET, Wood DF, Juliano BO, Bechtel DB (2004) The rice grain and its gross composition. Rice Chem Technol 3:77–100
Chang TT (2003) Origin, domestication and diversification. In: Smith CW, Dilday RH (eds) Rice: origin, history, technology and production. John Wiley and Sons, Hoboken, pp 3–25
Chase K, Reicks M, Smith C et al (2003) Use of the think aloud method to identify factors influencing purchase of bread and cereals by low-income African American women and implications for whole grain education. J Am Diet Assoc 103:501–504

Chen Y, Jiang W, Jiang Z et al (2015) Changes in physicochemical, structural, and sensory properties of irradiated Brown japonica Rice during storage. J Agric Food Chem 63(17):4361–4369

Corpuz HM (2011) Ways to improve acceptability of brown rice to consumers. In: Policy workshop on mainstreaming brown rice to low and middle income families. Clark Field, Pampanga

Croy M, Marquart L (2005) Factors influencing whole grain intake by health club members. Top Clin Nutr 20:166–176

Dinesh Babu P, Subhasree RS, Bhakyaraj R, Vidhyalakshmi R (2009) Brown rice-beyond the color reviving a lost health food – a review. Am Eur J Agro 2:67–72

FAO (2016) Food based dietary guidelines. Source: http://www.fao.org/nutrition/education/food-based-dietary-guidelines/regions/countries/. Accessed 28 Sep 2016

Hadi A, Izani AH, Selamat J et al (2012) Consumers' demand and willingness to pay for rice attributes in Malaysia. Int Food Res J 19(1):363–369

Harris KA, Kris-Etherton PM (2011) Effects of whole grains on coronary heart disease risk. Curr Atheroscler Rep 12:368–376

Hesse D, Braun C, Dostal A et al (2009) Barriers and opportunities related to whole grain foods in Minnesota school foodservice. J Child Nutr Manag 33(1). Available at http://www.schoolnutrition.org/Content.aspx+id512511&terms5barriers1and1opportunities1relating1to1whole1grain1foods. Accessed 4 Nov 2016.

Horigane AK, Takahashi H, Maruyama S et al (2006) Water penetration into rice grains during soaking observed by gradient echo magnetic resonance imaging. J Cereal Sci 44:307–316

Hsu TF, Kise M, Wang MF et al (2008) Effects of pre-germinated brown rice on blood glucose and lipid levels in free-living patients with impaired fasting glucose or type 2 diabetes. J Nutr Sci Vitaminol 54(2):163–168

Imam MU, Ishaka A, Ooi DJ et al (2014) Germinated brown rice regulates hepatic cholesterol metabolism and cardiovascular disease risk in hypercholesterolaemic rats. J Funct Foods 8:193–203

Jonnalagadda SS, Harnack L, Lai RH et al (2011) Putting the whole grain puzzle together: health benefits associated with whole grains – summary of American Society for Nutrition 2010 satellite symposium. J Nutr 141(5):1011S–1022S

Juliano BO (1993) Rice in human nutrition. Food and Agriculture Organization, Rome

Kaur B, Ranawana V, Henry J (2016) The glycemic index of rice and rice products: a review, and table of GI values. Crit Rev Food Sci Nutr 56(2):215–236

Kumar S, Mohanraj R, Sudha V et al (2011) Perceptions about varieties of brown rice: a qualitative study from southern India. J Am Diet Assoc 111(10):1517–1522

Lang R, Thane CW, Bolton-Smith C et al (2003) Consumption of whole grain foods by British adults: findings from further analysis of two national dietary surveys. Public Health Nutr 6:479–484

Liong HLB (2011) Strategies to increase the availability of brown rice in the market. In: Seminar-workshop on mainstreaming brown rice to low-and middle-income families, p 79

Liu S, Willett WC, Manson JE et al (2003) Relation between changes in intakes of dietary fiber and grain products and changes in weight and development of obesity among middle-aged women. Am J Clin Nutr 78(5):920–927

Luh BS, Robert RM (1991) Parboiled rice. In: Rice. Springer US, New York, pp 470–507

Marshall WE, Wadsworth JI (1994) Chapter 1: introduction. Rice Sci Technol 1:1–15

Miyoshi H, Okuda T, Okuda K, Koishi H (1987) Effects of brown rice on apparent digestibility and balance of nutrients in young men on low protein diets. J Nutri Sci Vitaminol 33(3):207–218

Mohan V, Radhika G, Sathya RM et al (2009) Dietary carbohydrates, glycaemic load, food groups and newly detected type 2 diabetes among urban Asian Indian population in Chennai, India (Chennai Urban Rural Epidemiology Study 59). Br J Nutr 102(10):1498–1506

Mohan V, Spiegelman D, Sudha V et al (2014) Effect of brown rice, white rice, and brown rice with legumes on blood glucose and insulin responses in overweight Asian Indians: a randomized controlled trial. Diabetes Technol Ther 16(5):317–325

Mohan V, Anjana RM, Gayathri R et al (2016) Glycemic index of a novel high-fiber white rice variety developed in India—a randomized control trial study. Diabetes Technol Ther 18(3):164–170

Monge-Rojas R, Mattei J, Fuster T et al (2014) Influence of sensory and cultural perceptions of white rice, brown rice and beans by Costa Rican adults in their dietary choices. Appetite 81:200–208

Muhihi A, Gimbi D, Njelekela M et al (2013) Consumption and acceptability of whole grain staples for lowering markers of diabetes risk among overweight and obese Tanzanian adults. Glob Health 9(1):26

Rehman ZU, Shah WH (2005) Thermal heat processing effects on antinutrients, protein and starch digestibility of food legumes. Food Chem 91(2):327–331

Ricepedia. Source: http://ricepedia.org/rice-as-food/the-global-staple-rice-consumers. Accessed 24 June 2016

RICEponsible (2014) www.bericeponsible.com/. Accessed 28 June 2016

Richardson DP (2003) Whole grain health claims in Europe. Proc Nutr Soc 62:161–169

Romero MV (2011) Nutritional and health aspects of brown rice. In: Seminar-workshop on mainstreaming brown rice to low-and middle-income families, pp 7–12.

Seal CJ, Brownlee IA (2015) Whole-grain foods and chronic disease: evidence from epidemiological and intervention studies. Proc Nutr Soc 11:1–7

Seal CJ, Jones AR (2007) Barriers to the consumption of whole grain foods. In: Marquart L, Jacobs DR Jr, McIntosh GH, Poutanen K, Reicks M (eds) Whole grains and health. Blackwell Publishing, Oxford, pp 243–254

Shobana S, Malleshi NG, Sudha V et al (2011) Nutritional and sensory profile of two Indian rice varieties with different degrees of polishing. Int J Food Sci Nutr 62(8):800–810

Shobana S, Ramyabai M, Sudha V, Anjana RM, Unnikrishnan R, Krishnaswamy K, Mohan V (2015) Rice: nutritional facts and delusions. In Touch Heinz Foundation of India 17(2&3):12–18

Song SJ, Lee JE et al (2014) Carbohydrate intake and refined-grain consumption are associated with metabolic syndrome in the Korean adult population. J Acad Nutr Diet 114(1):54–62

Sudha V, Spiegelman D, Hong B et al (2013) Consumer Acceptance and Preference Study (CAPS) on brown and undermilled Indian rice varieties in Chennai, India. J Am Coll Nutr 32(1):50–57

Sun Q, Spiegelman D, van Dam RM et al (2010) White rice, brown rice, and risk of type 2 diabetes in US men and women. Arch Intern Med 170(11):961–969

Thane CW, Jones AR, Stephen AM et al (2005) Whole-grain intake of British young people aged 4–18 years. Br J Nutr 94:825–831

Thane CW, Jones AM, Stephen AM et al (2007) Comparative whole-grain intake of British adults in 1986–7 and 2000–1. Br J Nutr 97:987–992

USDA (2016) 2015–2020 dietary guidelines for Americans. Source: www.cnpp.usda.gov/2015-2020-dietary-guidelines-americans. Accessed 8 Jan 2016

WHO (2014) Global status report on noncommunicable diseases 2014. Source: www.who.int/nmh/publications/ncd-status-report-2014/en/. Accessed 6 Jan 2016

Wu F, Yang N, Touré A et al (2013) Germinated brown rice and its role in human health. Crit Rev Food Sci Nutr 53(5):451–463

Zhang G, Malik VS, Pan A et al (2010) Substituting brown rice for white rice to lower diabetes risk: a focus-group study in Chinese adults. J Am Diet Assoc 110(8):1216–1221

Chapter 16
Opportunities and Challenges in Marketing of Brown Rice

S. Selvam, P. Masilamani, P.T. Umashankar, and V. Alex Albert

In some parts of the world, the word "to eat" literally means "to eat rice." All varieties of rice are available throughout the year, supplying as much as half of the daily calories for half of the world's population. Rice is the basic grain consumed as a food in India which is found in almost every Indian kitchen. It is the most common grain and the most common food in India; however, India is not only a big consumer of rice, but also it is the second largest producer of rice in the world after China.

As defined by the Asia Rice Foundation (2017), brown rice is unpolished whole grain rice that is produced by removing only the hull or husk using a mortar and pestle or rubber rolls. The process that produces brown rice removes only the outer most layer, the hull, of the rice kernel and is the least damaging to its nutritional value. The complete milling and polishing that converts brown rice into white rice destroys 67% of the vitamin B3, 80% of the vitamin B1, 90% of the vitamin B6, half of the manganese, half of the phosphorus, 60% of the iron, and all of the dietary fiber and essential fatty acids.

Despite of several health benefits, brown rice currently comprises a very small share of the household rice basket. One reason is the apparent lack of information about brown rice. Cuyno (2003) noted several drawbacks in the brown rice business.

S. Selvam (✉)
Anbil Dharmalingam Agricultural College and Research Institute, Tamil Nadu Agricultural University, N. Kuttapattu, 620 009 Tiruchirappalli, Tamil Nadu, India
e-mail: malarselvamtnau@gmail.com

P. Masilamani • V. Alex Albert
Agricultural Engineering College and Research Institute, Tamil Nadu Agricultural University, Kumulur, 621 712 Tiruchirappalli, Tamil Nadu, India
e-mail: masil_mahesh@yahoo.com; alex.tnau@gmail.com

P.T. Umashankar
Consultant, Asian Development Bank, No. 1, 3rd cross street, Adayar, Chennai, Tamil Nadu, India
e-mail: ptumashankar@gmail.com

© Springer International Publishing AG 2017
A. Manickavasagan et al. (eds.), *Brown Rice*,
DOI 10.1007/978-3-319-59011-0_16

Brown rice is more expensive than white rice, is not readily or conveniently available, has a shorter shelf life and gets rancid if stored beyond 4–6 weeks due to the high fat content in the bran, has a coarse texture and a "dirty" look, is susceptible to storage weevil, and entails longer cooking time. In the study of consumers in Los Baños, Laguna, Reforma (2007) reported that there is limited knowledge on the exact nutritional benefits from brown rice and where the product could be sourced. On the whole, the market chain of brown rice is not well understood.

Market Drivers of Brown Rice

Brown rice market is increasing with respect to the growth of rice trade between the countries. Brown rice market is projected to expand at a higher pace compared to rice market, due to shifting preference toward healthy eating habits, and nutrients provided by brown rice, such as vitamins (B), phosphorus, and magnesium, make it a preferred choice for many health-conscious customers. Due to increasing demand, the manufacturers are focusing on launching new brown rice brands and making renewed efforts to attract health-conscious people. Demand for brown rice is also emerging from various Middle East and African countries, driving the growth of brown rice market. Recently innovation in packaging and changing food preference such as emergence of ready-to-eat food product will result in introduction of ready-to-cook brown rice. A majority of countries, where brown rice was earlier preferred by only diabetic patients, is now getting traction from a growing number of fitness-conscious consumers. Increasing per capita consumption of rice in countries where it is not a staple food is also expected to support the growth of brown rice market.

Market Segmentation of Brown Rice

Brown rice market can be segmented on the basis of length, type, and regions that constitute the key markets (USAID 2009). Brown rice on the basis of length can be segmented into long-grain brown rice, medium-grain brown rice, and short-grain brown rice. Long-grain brown rice when cooked becomes firm and offers fluffy texture that remains separate, whereas medium-grain offers more sweet and nutty flavor and remains more tender when cooked. Short-grain offers sweeter taste than medium- and long-grain brown rice and offers chewier and denser texture.

According to Trade Development Authority of Pakistan (2016), the world rice trend matrix (Fig. 16.1) classifies the export products according to their world market share and annual average growth rate. The size of the bubble represents the unit value for each of the rice products. It is evident from the graph that the semi-/wholly milled category of rice has the highest world market share which is growing above the average of the cereal sector. The growth rate of brown rice is below the average of the sector, which means the demand for brown rice is growing but at a lower rate than that of the sector; therefore, it is considered as a static product.

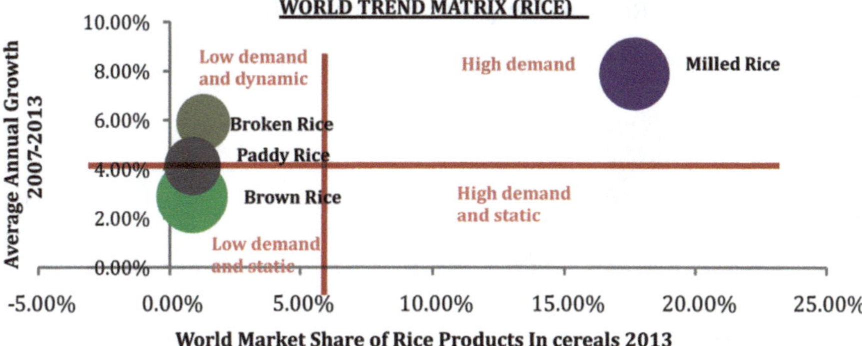

Fig. 16.1 World trend matrix

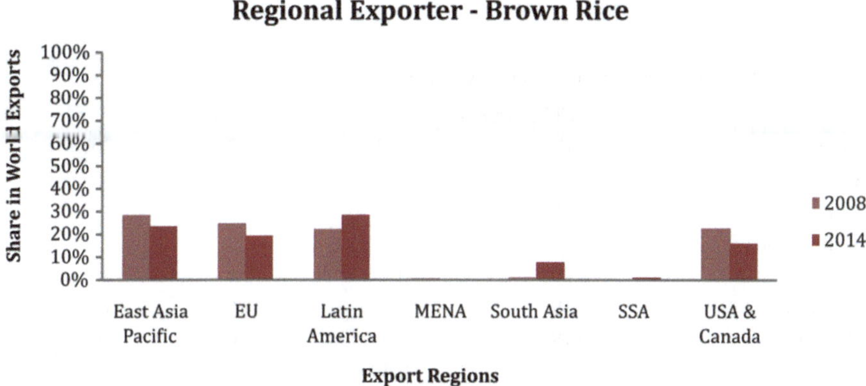

Fig. 16.2 Major exporting regions of brown rice

Figure 16.2 shows each region's share in the world exports of brown rice. East Asia Pacific, European Union (EU), Latin America, the USA, and Canada are the main exporters of brown rice. In 2013, East Asia Pacific, EU, and the USA decreased the exports of brown rice, whereas Latin America and South Asia increased it.

Figure 16.3 shows the major markets for brown rice. European Union, East Asia Pacific, the USA, and Canada are the major importing regions of husked or brown rice. It is evident that EU is the largest importing region of brown rice; however, EU's imports as a share of total world imports of brown rice decreased from 64 to 49% between 2000 and 2014. Between 2008 and 2014, the imports of husked rice increased in East Asia, Latin America, and sub-Saharan Africa.

Table 16.1 shows the export competitiveness rankings of the countries exporting brown rice. Guyana is the most competitive country in the world to export brown rice. Developed countries such as the USA, Italy, Australia, and Spain export brown rice.

With more people having been used to consuming white rice, the reintroduction of brown rice could be a challenge. People habitually buy and consume white rice.

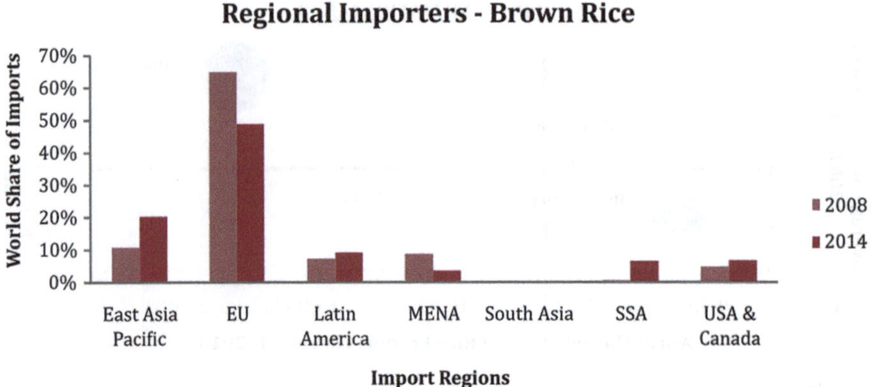

Fig. 16.3 Major importing regions of brown rice

Table 16.1 Export competitive index of brown rice

	ECI brown rice		Rank	
Countries	2014	2008	2014	2008
Guyana	0.92	0.62	1	1
China	0.50	0.17	2	6
United States	0.50	0.51	3	2
Spain	0.20	0.13	4	7
Pakistan	0.19	0.01	5	18
Paraguay	0.19	0.01	6	20
Thailand	0.18	0.46	7	3
Italy	0.17	0.25	8	5
Uruguay	0.15	0.42	9	4
Australia	0.13	0.01	10	19

A reason could be affluence associated with eating white rice. Even in the USA, white rice is preferred over brown rice, and part of the reason according to Pollan (2008) is prestige as for so long a time, only the wealthy could afford refined rice. Although there were initial efforts of positioning brown rice as a health food which is consumed by richer people who are exhibiting healthy lifestyle, acceptance of the product is still a slow process.

Marketing Channels of Brown Rice

Several key market players were involved in the production and marketing of brown rice, and these include (1) farmers, (2) market intermediaries, (3) processors (rice millers), and (4) consumers.

According to AMIBPC (2016), the primary stage of the brown rice supply chain is held by the paddy farmers who supply paddy to the rice-processing companies, which is supplied both directly and through intermediaries, depending on the type of farmers and their productivity.

There are three groups of participants in rice marketing, namely, (1) merchants, (2) commission agents/brokers, and (3) facilitators. Merchants, such as rice assemblers, wholesalers, millers/polisher, retailers, and middleman/brokers, are the main actors in the market. Normally, there are two types of assemblers such as local assemblers and assembler-wholesalers. Local assemblers are usually local people, who have an in-depth knowledge of cropping patterns, quality of different rice varieties, and the production traditions of local farmers. The assemblers may be farmers located in the village. They purchase paddy from farmers and transport it to millers or sell it at local trading places. Assemblers-wholesalers are middlemen, which are usually located near the good market places. They buy the paddy in large amounts from the rice millers. Sometimes, they ask millers for milling their paddy which they later sell to wholesalers or private agencies, usually in larger volumes as compared to the local assemblers.

Private wholesalers perform the wholesale functions. They usually have a long experience with paddy/rice trading, strong relationships with other traders, and good knowledge on rice quality and market prices. They buy a large amount of brown rice or milled rice from private rice millers and polishers and directly sell to retailers. Direct procurement from small farmers is not profitable for the rice-processing companies, as their cost of logistics and supply chain will be very higher comparative to the cost of sourcing through "mundies" and "agents."

Private rice millers are middlemen who are usually located near the town or market centers,purchasing paddy from farmers, assemblers, or wholesalers, and are engaged in milling/polishing activities and delivering milled rice to other polishers, wholesalers, and retailers in the market centers. Sometimes they do not purchase the product but just mill paddy or polish the brown rice, and they receive the payment as processing costs from rice traders and farmers.

Private retailers are sellers at the retail market. Private retailers are usually local people,who have an in-depth knowledge of consumer demand. Aside from producers, traders, and consumers, other market participants include state-owned enterprises/companies, cooperatives who are the main sources of raw material, milling service providers, and sellers of packaging and labeling materials used in brown rice.

The three types of marketing channels were present in the marketing of brown rice. In marketing channel I, the farmer grows his own paddy which is then milled with the help of milling service provider and sold to the consumers directly. In this market channel involving direct sale of brown rice by the farmers to the consumer, the producer's share is the entire product value (100%). Consumer also gets lowest produce price. This market channel is more prevalent in small towns.

In the case of marketing channel II, the farmer grows his own paddy which then has it milled by a milling service provider and sold to the retailers and then reaches consumers. The chains with the next biggest producer shares were those with only one intermediary involving the retailer. In this channel producer share was 79.17%. The remaining shares of the product value were for the retailers.

In the case of marketing channel III, the brown rice wholesaler (organic shop groups/rice millers) buys traditional variety of paddy from various sources, processes it into brown rice through a milling service provider/using his/her own milling facilities, and then sells it primarily on a wholesale basis and retail. The producer's share was smallest in the market channel involving two intermediaries (wholesaler-distributor and retailer) as part of the chain. However, the producer's share remained the biggest among the three market participants at 59.32%, followed by the distributor's share at 30.53%, and then by the retailer's share at 10.15%. Pabuayon and Quilloy (2011) reported a similar trend also in the Philippines.

Despite brown rice supplier's effort in encouraging the use of brown rice, there are still a lot of challenges that need to be addressed. With the help of product contact point audit analysis and product contact priority grid analysis, we understand challenges in brown rice marketing. This analysis was based on consumer and trader's survey.

Product Contact Point Audit Analysis of Brown Rice

According to Henry Lim Bon Liong (2013), product contact points are the experiences consumers have when they, directly or indirectly, come in contact with a product. The major product contact points consumers have for brown rice are word of mouth from family, friends, and doctors and in grocery stores or supermarkets. Recommendation is the number one channel in helping persuade consumers to consider brown rice to their diet. Word of mouth is the most powerful and inexpensive way of marketing a product because it is based on a person's firsthand experience. And most of the time, consumers listen to the advice of people who are close to them or doctors when it comes to health. The next important contact point is the retail store. The retail store is an important outlet because this is where consumers get to see, touch, and experience the product.

Product Contact Priority Grid Analysis of Brown Rice

A product contact priority grid contains four quadrants that reveals what consumers think about a certain product, whether it is good or bad. This grid explains that there are more negative aspects of brown rice than positive for consumers to consider brown rice. One of the main reasons consumers do not want to eat brown rice is because of the taste and, secondly, the appearance, followed by other elements such as cooking time, availability, and price. The characteristics of a cooked brown rice are hard to chew and dry to swallow, and it tastes like cardboard. The color of brown rice also connotes a "dirty" look, which does not appeal to people. The cooking process is also another aspect that annoys consumers because it is longer to cook compared to white rice. Others even soak it for an hour to overnight just so they can eat it. Brown rice eaters also have a limited choice of brands to choose from. They buy whatever is available in the supermarket even if the taste isn't to their liking. The only delighter found was that it is good for the health. Consumers also have this mindset that if it is brown, it is automatically organic (Henry Lim Bon Liong (2013).

Quick store check analyses were conducted in the selected cities of Tamil Nadu, India, to determine the availability and extent of distribution of brown rice in the town (AMIBPC 2016). The store checks were done in supermarkets and specialty shops. All stores were selling not only brown rice but also white rice and other types of rice. All the supermarkets carry branded brown rice like Patanjali brand/own brand names/without brand names of brown rice. It is available in 1 kg transparent plastic packaging with a small sticker bearing the supplier's name, rice variety, and price. The average price of brown rice in the supermarkets is Rs. 90–120 per kg (approximately 1.5–2 US$). This average price is 1.4 times higher than the price of selected varieties of white rice. These supermarkets usually receive consumer feedback mostly from new users, about the product, such as chewiness, coarse texture, and unattractive color.

According to trade sources, the average sale of brown rice in these supermarkets is not known, but the supply is usually in excess and most of its buyers are foreigners and health-conscious people. Excess brown rice is returned to the supplier but sometimes sold as surplus to its employees. The average sale of brown rice is 5–10 kg per week, and the supply is usually sufficient but sometimes in excess for the whole month (AMIBPC 2016).

The demand for brown rice in selected cities is generally not stable because more than 60% of the respondents are rarely buying the product. Only after doctors advise consumer interested purchase brown rice after 1 or 2 weeks of doctors advise consumer purchase again same white rice due to obvious reasons. Considering that the most common packaging of brown rice is one kilogram packs, the total purchase volume could be relatively low. The mentioned reasons for the occasional purchasing were high price and unavailability.

The positive feedback was that brown rice has more filling, has good aroma and mild taste, and is good. The negative feedback was it takes longer time to cook, is grainy, smells like rice hull, and is not liked by children (Mojica and Reforma (2010). Brown rice prices vary from one supermarket to the next, depending on the location, the size of the supermarket, the supplier, the clientele, and the time of the year. Prices also vary depending on the varieties. The price of local brown rice ranged from Rs. 90 to 120 per kg (approximately 1.5–2 US$). These prices are higher than the price of white rice by at least 80%. Brown rice is relatively more expensive than white rice for a number of reasons, namely, (1) the supply is limited and producers do not supply much because of limited demand; (2) it is marketed with all value-adding activities such as packaging, labeling, storage, quality control, and in some cases advertising, all involving higher costs compared to similar activities done for white rice sold in regular variety and retail stores and public markets;

and (3) there is perceived price premium associated with health benefits for which current users, generally the high income, express willingness to pay.

There is opportunity to lower the price of brown rice if supply could be increased, and this could be sold with less value and cost-adding activities and shorter chain but still maintaining the product attributes that consumers are looking for. The case in point is sesame oil. Although there are many branded cooking sesame oil products available in the supermarkets, there remains a big market for unbranded cooking oil available in retail stores as lose – it is affordable to the majority of consumers, is readily available, and has similar cooking qualities as the more expensive branded oil.

Brown rice can also be classified as a low-input product and considered as an eco-food because it requires less energy to process, thus consuming less power. There is less energy required because brown rice is minimally processed in that it passes the mill only once. According to Javier (2004), the power saved due to polishing and whitening is around 65%. In addition, the milling recovery in brown rice is 10% higher as compared to white rice. The economic benefits are less milling cost and higher returns due to higher recovery.

The trader's sources indicated though that while brown rice demand is picking up, the major barrier is still the acceptance of brown rice in the market. Many who do not like the taste commented that it tastes like old rice. There were also complaints regarding its texture which is harder than white rice. The current positioning of brown rice as health food presents both an opportunity and a barrier. The increasing health consciousness of the market will positively impact on the demand for brown rice. On the other hand, the current positioning is associated with higher prices. This may present a barrier for penetrating the lower segment of the market.

The present brown rice market is characterized as having low demand and low supply, and, therefore, the price is high and there is limited quantity in the market. We need to develop strategies that will shift both demand and supply to the right. A right shift means (1) higher demand, that is, at a given price, consumers are willing and able to buy more, and (2) higher supply, that is, at a given price, producers are willing to supply more; the latter also implies lower marginal cost of supplying the product in the market. The first equilibrium shows a higher price and a lower quantity; the second shows a lower price and a higher quantity, meaning a bigger market share for brown rice and more consumers availing healthy and nutritious rice. The latter scenario is consistent with food security and nutrition objectives, even poverty reduction if small farmers become a significant part of the brown rice market.

Factors Affecting the Availability of Brown Rice

After discussing the market, there are two major factors affecting the availability of brown rice in the market, which we can easily say as the two As: (i) awareness and (ii) acceptability.

Awareness

Health is one of the biggest topics being given importance today. Not only have people enrolled in fitness centers but they are becoming more involved in what they eat. People are now preferred switching to whole grains and wheat. It is because people are now realizing that today's modern and urban lifestyle leads to increased medical cases such as obesity, diabetes, and heart disease. And one way of battling it is to switch to a healthy diet and standard of living.

People must first be aware of the benefits that brown rice provides. Although a lot of people know that brown rice is healthy, how much do they really know? Education is the first step in letting people appreciate the product and let them realize what they can gain from it at a deeper level. An example will be explaining to consumers how fiber in brown rice binds to cancer-causing chemicals, keeping them away from the cell lining of the colon, thus preventing colon cancer, or how the vitamin B present can help prevent beriberi, or how brown rice helps increase mothers' milk supply for breastfeeding.

Aside from the nutritional benefits, it is also important to make people understand how brown rice can help make the country self-sufficient in rice and address the issue of rice shortage. Research has shown that the milling recovery of brown rice is 10% higher than that of white rice. And since the process does not require whitening or polishing, it can save at least 65% of energy. After being knowledgeable about the product, the next step is to accept it.

Acceptability

The popularity of white rice increased because of its appearance, texture, taste, cooking time, and shelf life. On the other hand, brown rice is quite the opposite, with taste being the main problem. If people do not accept the fact that brown rice has its own distinct taste and texture, it will be very difficult to convince them to switch or even buy it. Acceptability is one of the key factors in driving up demand, which will eventually influence supply. Because of the negative aspects mentioned by consumers, new entrepreneur or supplier will introduce one of the best-tasting brown rice in the market based on the feedback of loyal customers. In Tamil Nadu five important brown rice varieties in the market are (i) Kudavazhai, (ii) Mappilai Samba, (iii) Karthiyanam, (iv) Salem Sannam, and (v) Sixty karkuruvai. Among these brown rice varieties of Tamil Nadu, India, Mappilai Samba is the leading traditional variety because of its taste, texture, and other medicinal properties (AMIBPC 2016). By addressing the need for a better-tasting brown rice, new entrepreneur will convince white rice eaters to try brown rice and make them believe that eating healthy does not mean that one has to sacrifice one's taste buds.

Marketing Strategy for the Promotion of Brown Rice

After understanding the factors that hinder the demand for brown rice, which limits its availability in the market, focus will now be on how we can increase it. The strategy will be called the four Ps (product, price, place, and promotion).

Product

A product, no matter how good it is positioned, will always have flaws. Brown rice is more prone to weevil infestation and molding due to the high nutrition content of the bran, thus limiting it to small packaging. It would be very helpful if research institutions should come up with how to increase the shelf life of brown rice. Solving this predicament can help persuade millers, traders, and distributors to carry and offer the product. The shorter shelf life mainly due to its fat content makes it become rancid faster. Vacuum packing was found to be the solution to this problem but would also mean a high price due to high production cost. To avoid the cost of vacuum packing, brown rice suppliers in Tamil Nadu resort to drying after hauling process. Brown rice could also be kept in refrigerators to maintain its eating quality.

Entrepreneurs also think alternative product forms for brown rice, that is, develop snack foods and breakfast cereals using brown rice as basic raw material. Since these food products are already in the mainstream capturing both young and old and different socioeconomic groups, food companies could broaden their product lines by introducing brown rice-based options.

Price

The price of brown rice ranges from Rs. 90 to 120 per kg, almost double the amount of the most popular consumer-preferred varieties like Sona or Samba Masuri. The reason why brown rice is priced higher than polished rice is because of the low demand. Even though milling would give millers and businessmen higher savings and recovery compared to white rice, the demand cannot compensate for the cost in producing and storing it. For example, head rice recovery was around 65% for Mappilai Samba variety. And for demand to take place, an understanding of the product and the market must be taken into context, as what was pointed out in the contact point and grid, which will then lead to a lower price.

Place

In urban and rural areas of India, buying habits of rice have changed. Today, more than 40% of the consumers purchase their groceries from supermarkets. Hence, focus should be based on outlets like supermarkets and grocery stores nationwide. We have to make the small chains start to carry brown rice as part of their product line. Focus should not solely be based on outlets as there are three key potential areas that should be tapped in order to address the issue of availability. The first key potential areas are the hotels, restaurants, and catering. Nowadays Indians prefer eating out rather than cook at home due to practicality and lifestyle changes.

Restaurant and fast-food chains should take the initial steps in giving their customers a wide range of food choices. Today, only about less than 1 % of the whole industry offers brown rice as an alternative side dish. As an example, high-end restaurants and five-star hotels also offer brown rice but are mostly overlooked in the menu because customers would rather settle for mashed potato than rice.

Another key potential area is the fitness centers. Gyms and fitness centers are popping in every corner of the metro like mushrooms. Not only should it provide facilities for people to work out, it should also provide people with knowledge on the different food groups. Instructors and nutritionists should be well equipped with not only the proper ways of getting a good figure but on how to start and maintain a healthy lifestyle.

For the last key potential area, hospitals and clinics should spearhead in promoting awareness on brown rice. It is ironic that hospitals, a place where people go to seek medical advice, do not serve brown rice to their patients.

Promotion

Lastly, what is needed to address the awareness and acceptability is promotion. No campaign will be successful if it is not promoted properly and continuously. For retail, aside from just displaying the product on racks and pallets, stores should have a conscious effort in promoting healthy eating among its consumers – a healthy corner – or even making in-store displays to highlight brown rice like Shopwise.

Restaurants should put up pamphlets or tent cards on their tables and menus explaining the benefits of brown rice to one's diet. Chefs can even be creative in coming up with new and exciting recipes using brown rice to entice more people to try it and adapt to its taste.

Other medical organizations specializing in diabetes and obesity and even pediatricians should promote the benefits of brown rice to their patients and fellow practitioners.

Conclusion

With the four Ps' strategy, we can expect that more and more rice consumers in the world will have a change in mindset toward brown rice and may increase its share ofthe pie to 50%. Despite the efforts to improve the cooking and eating quality of brownrice and extending its shelf life, there are still no concrete or practical answersto these problems. These concerns must be addressed thoroughly through extensive research in order to break the barriers that prevent the popularization, commercialization, and substantial consumption of brown rice.

References

Agromarketing Intelligence and Business Promotion Centre (AMIBPC) wet season rice-traders survey report 2016. 25p

Asia Rice Foundation (2017) Brown rice: beyond the colour. Brown Rice Campaign Committee, The Asia Rice Foundation, Los Baños, Laguna. http://www.asiarice.org/sections/whatsnew/brbulletin.html. Accessed 02 Feb 2017

Cuyno RV (2003) The national campaign to combat hidden hunger through brown rice. Asia Rice Foundation, Los Baños, Laguna. http://www.asiarice. Accessed 25 July 2009

Javier EQ (2004). Let's promote brown rice to combat hidden hunger. Rice Today. International Rice Research Institute

Liong HLB (2013) Strategies to increase the availability of brown rice in the market. In: Proceedings of the policy seminar-workshop on mainstreaming brown rice to low and middle income families by DA-Philippine Rice Research Institute, pp 80–85

Mojica LE, Reforma MG (2010) An exploratory market study of brown rice as a health food in the Philippines. J Agric Sci, Tokyo Univ Agric 55(1):19–30

Pabuayon IM, Quilloy AJA (2011) Brown rice market chain and marketing practices, Luzon, Philippine. The Philippine Agricultural Scientist 94(1):54–65

Pollan M (2008) In defense of food. The Penguin Press, New York, 106

Reforma ML (2007) Market potential of brown rice in Los Baños, Laguna. Unpublished under graduate thesis, University of the Philippines Los Baños, p 133

Trade Development Authority of Pakistan (2016) Sectoral competitiveness and value chain analysis rice value chain analysis in Pakistan. pp 46

USAID (2009) Global food security response: West Africa rice value chain analysis, microReport# 161. USAID, Washington, DC

Index

© Springer International Publishing AG 2017
A. Manickavasagan et al. (eds.), *Brown Rice*,
DOI 10.1007/978-3-319-59011-0